William Graves Hoyt

and **MARS**

W9-AFJ-195

THE UNIVERSITY OF ARIZONA PRESS
Tucson, Arizona

About the Author...

WILLIAM GRAVES HOYT, for twenty-five years reporter, writer, and editor, holds a Bachelor of Science degree in anthropology and a Master of Arts degree in history. In 1966 he became a member of the professional staff at Northern Arizona University, at Flagstaff, as a writer specializing in the areas of science and education. Simultaneously he has been conducting research on the history of astronomy as a research associate at the Lowell Observatory in Flagstaff.

THE UNIVERSITY OF ARIZONA PRESS

Copyright © 1976
The Arizona Board of Regents
All Rights Reserved
Manufactured in the U.S.A.

I.S.B.N. 0-8165-0435-0 cloth
I.S.B.N. 0-8165-0514-4 paper
L.C. No. 75-9144

For Midge

Contents

Reference Material

List of Illustrations

Acknowledgments

The research upon which this study is based would not have been possible without the full cooperation of John S. Hall, a distinguished astronomer who since 1958 has been the director of the Lowell Observatory in Flagstaff, Arizona. Hall, with characteristic grace, granted access to what herein will be referred to as the Lowell Observatory Archives and to the extensive documentary materials they contain relating to Lowell, his work, and the work of his observatory. His knowledge, wisdom, and friendship also have been an invaluable resource.

I owe a debt to the late Roger Lowell Putnam, for forty years the sole trustee of Lowell Observatory, and to his son and successor as trustee, Michael C. J. Putnam, Brown University, for their active interest in my research. In addition, I thank the various members of the observatory's staff who helped me find materials, answered my questions, and generally tolerated my intrusions into observatory routine, often on a daily basis. These include, importantly, Henry L. Giclas, Otto G. Franz, William A. Baum, Charles Capen, Norman Thomas, Stuart E. Jones, Jean Scheele, and Lois Niero.

Alice Babbitt of Flagstaff, whose labors largely were responsible for compiling the microfilm edition of *The Early Correspondence of the Lowell Observatory 1894–1916*, (W. G. Hoyt and A. Babbitt, eds.; Flagstaff: Lowell Observatory, 1973), was particularly helpful in tracking down elusive documents in the observatory archives.

I am grateful to Joseph Ashbrook, the editor of *Sky and Telescope*, and A. M. J. ("Tom") Gehrels of the University of Arizona's Lunar and Planetary Laboratory, for their critical reading of my manuscript and for their valuable suggestions, although of course I remain solely responsible for the final content of the work. Others who gave counsel and encouragement include four members of the Northern Arizona University faculty. They are Dwight E. Mayo, historian of science; Lee M. Nash, whose field is intellectual history; William H. Lyon, American historian; and Arthur Adel, astronomer, astrophysicist, mathematician, and former member of the Lowell Observatory staff.

I gratefully acknowledge the cooperation of the following publishers and/or organizations for granting me permission to use material on which they hold, or control publication rights:

The American Association for the Advancement of Science permitted quotations from articles by Eliot Blackwelder, T. J. J. See, Joseph Barrell, F. R. Moulton and Percival Lowell in *Science* in 1909, and by Charles Lane Poor in 1910. The Lick Observatory, University of California, granted permission to use quotations from its *Bulletin 437* by E. C. Bower. The Astronomical Society of the Pacific gave permission to quote R. J. Trumpler's report of his 1924 martian observations in its *Publications,* and from its *Leaflet No. 209,* "The Discovery of Pluto," by C. W. Tombaugh. Quotations from articles by Andrew T. Young, Robert B. Leighton, John S. Hall and William K. Hartmann in *Sky and Telescope* are by permission of the Sky Publishing Co. *The New York Times* editorials quoted in Chapter 15 and first published October 18 and October 29, 1916 are © 1916 by The New York Times Company, and reprinted by permission. Also, excerpts from the *Times* editorial "Seeing the Unseen" of March 16, 1930, are © 1930 by The New York Times Company, and reprinted by permission.

The Macmillan Publishing Co., Inc., granted permission to use quotations from *Astronomy of the 20th Century* by Otto Struve and Velta Zebergs, © The Macmillan Company, 1962; from *Worlds Without End* by N. J. Berrill, © by N. J. Berrill, 1964; and from *Biography of Percival Lowell,* © The Macmillan Company, 1935. Houghton Mifflin Company provided permission to use the quotations from *The Lowells and Their Seven Worlds* by Ferris Greenslet. Cambridge University Press permitted the quotation from James H. Jeans' *The Universe Around Us.* Harper & Row, Publishers, Inc., gave permission to quote from Jacques Barzun's *Science: The Glorious Entertainment.* The quotation from *And Then There Was Light* by Rudolf Thiel, © Alfred A. Knopf, Inc., 1957, is used by permission of Random House, Inc.

Acknowledgment must also be made to the D. Reidel Publishing Co., of Dortrecht-Holland, The Netherlands, for giving me permission to quote from S. F. Dermott's and A. P. Lenham's paper on "Stability of the Solar System: Evidence From the Asteroids," which appeared in *The Moon;* to George Allen & Unwin Ltd., of London, England, for permission to quote from Antonie Pannekoek's *History of Astronomy,* specifically the translation from Flammarion's *La Planète Mars et ses conditions d'habitabilité;* to Faber and Faber Ltd., of London for permission to quote from *Physics of the Planet Mars* by

Gerard de Vaucouleurs; and to Gerald Duckworth & Co., Ltd., of London for permission to quote from R. L. Waterfield's *A Hundred Years of Astronomy*. The quotations from F. R. Moulton's *Consider the Heavens,* © 1935 by F. R. Moulton and published by the Doubleday, Doran Co. by arrangement with the University of Chicago Press, are reproduced by permission of the University of Chicago Press.

Typing and other secretarial chores were provided by Mrs. Bartley Kirst of Flagstaff, and by my daughter, Mrs. Charles Andrews Glass. Countless friends and colleagues in Flagstaff and at Northern Arizona University have also helped in many ways. I cannot thank all of them by name here, but they know who they are, and how much I am indebted to them.

Marshall Townsend and Elaine Nantkes, director and associate editor respectively of the University of Arizona Press, along with other members of its staff, provided invaluable assistance in the final preparation of this book for publication. Regina Wajda helped materially in compiling the index.

Most importantly, there was Midge, my wife, who contributed to this work more than I can say, but who did not live to see it brought to its conclusion.

W. G. H.

In Perspective

Gods and gargoyles, minotaurs and martians, perhaps, can all be considered particular manifestations of one of the seminal ideas underlying the whole of intellectual history—the idea that intelligence exists elsewhere in the universe.

In its most fundamental form, the idea is very old, and indeed through recorded time men have peopled their cosmos, however conceived, with all manner of thinking beings, just as they have peopled terrestrial lands beyond their direct knowledge and experience with demons, monsters, and other fabulous creatures of their imagination.

The idea clearly has an intrinsic fascination for the human mind, holding as it does profound psychological, philosophical, and religious implications for man and his relative importance in the cosmic scheme of things. Its oft-proven ability to provoke not merely curiosity, but such deep and contrastive emotions as hope and fear, can certainly be attributed in part to its inherent element of mystery, for it postulates the unknown and, probably, the unknowable. That its power through history has been immense is plainly evidenced in the vast and complex systems of thought and belief that men, from time to time, have built upon it. It is, furthermore, an idea which can be accepted, without violence to reason, entirely on faith. No amount of negative evidence can lessen the possibilities of its being true, which for all practical purposes are infinite, while the discovery of a living intelligence at any point in space and time beyond the here and now of earth will bring it instantly and permanently into the realm of human reality.

In the decades bracketing the turn of the present century, it was widely believed that just such a discovery had been made, that positive evidence had been found for the existence of intelligent life on earth's neighboring planet Mars. This belief stirred an extraordinary controversy that for a time involved much of the literate world. Eminent scientists, philosophers, and moralists joined in the unprecedented debate, while ordinary people everywhere followed its course in books, at lectures, and through lengthy and detailed accounts in newspapers and magazines of the day.

Housewives and hard-headed businessmen, store clerks and visionary crackpots, the young, the old, the halt and the blind were touched by the swirling excitement. Poets and popular song writers penned rhymes and rhapsodies about Mars and the martians. The planet at once became the subject of sermons, science fiction, and satire.

The central, dominant figure in this tantalizing but transient interlude in scientific and intellectual history was Percival Lowell, a wealthy, well-educated, much-traveled Boston businessman with a long-standing interest in astronomy. No one before him or since has presented a case for intelligent life on Mars so logically, so lucidly, and thus so compellingly, to a skeptical science and a wondering laity. Nor, it may be added, has such a case ever been presented on such respected authority. "Fantastic though the theory ... appear," the leading historian of astronomy of the day wrote at the turn of the century, "it is held by serious astronomers. Its vogue is largely due to Mr. Lowell's ingenious advocacy."[1]

The phenomenal popularity of Lowell's provocative martian theory, first promulgated in 1894, lasted hardly more than a dozen years, and by 1910 public interest in it clearly was on the wane. This may have been partly because his ideas had by then lost much of their novelty, their ability to startle, for Lowell was not only persistent but consistent in advancing them, reiterating the main points of his argument in much the same terms, whatever his forum, with single-minded intensity. It is, of course, entirely possible that the public appetite for such out-of-this-world sensations simply became jaded. The "life on Mars" theme had been heavily worked by the more imaginative writers for a number of years, particularly after England's H. G. Wells published his science fiction classic, *The War of the Worlds,* in 1898. In the spring of 1910, Halley's Comet, spectacular enough for some and briefly outshining Mars, had come and gone without fulfilling any of the dire prophecies made in its name. It may also reflect a general decline in activity within astronomy itself that began about this time and has been attributed to an increasing concern over more mundane matters in a world on the verge of its first global war.[2]

Lowell himself by this time considered his martian theory completely and repeatedly confirmed by his own and others' observations of the planet over fifteen years at the observatory he had founded in 1894 at Flagstaff, Arizona Territory. Time, he was sure, would eventually erode the ignorance and prejudice that stood in the way of full recognition of his Mars work and bring it general acceptance. Meanwhile there were other problems to solve. Increasingly in the final years of his life he let his restless mind range farther into space—to Jupiter, Saturn, Uranus, and to a then unknown planet beyond distant Neptune that he had already been referring to for several years as "Planet X."

Short-lived as was the furor over Lowell's insistence on the existence of intelligent life on Mars, this singular episode has significance for the history of science and for intellectual history. It may surprise not a few scientists today to find that Lowell's work and thought, much of it now forgotten or scornfully rejected as merely fanciful speculation, was given serious consideration by leading astronomers and others prominent in the intellectual life of his day. Some of it, indeed, was sound. But none of it can be fairly judged out of context with the times in which he lived, or in the far brighter light of subsequent scientific knowledge. Lowell's martian theory and his other provocative ideas have been collected from his various public and private writings and summarized herein in considerable detail. This is done not to defend them but to better define them and to place them in perspective in an attempt to explain their peculiar fascination and their undeniable impact on his own and later generations.

It should be noted that this book is not a biography in the ordinary sense of the term. Rather it is a study of a man's mind, a man's ideas. That these ideas were sometimes superficial and often erroneous is beside the point. Right or wrong, they were influential, a stimulant or perhaps a goad to other minds, and thus they belong to intellectual history.

If a further rationale is needed for this book, it is provided by V. M. Slipher, a long-time Lowell associate and a highly-respected astronomer who knew Lowell far better than most. "One needed to know Lowell well to appreciate him," Slipher wrote shortly after Lowell's death in 1916, "and I have often thought that scientific men would have held him in even higher regard had they known him personally. He had unbounded energy, enthusiasm and perseverance, somewhat the pioneer spirit and courage. He entered science by an unusual course, but he made and filled a large place in astronomy and in the world as well."[3]

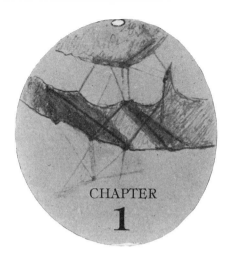

CHAPTER

1

Mars

MARS, the fourth planet from the sun in the solar system, has stirred the curiosity, wonder, and awe of man through all history. In antiquity the red planet, a brilliant fiery wanderer in the heavens near its periodic oppositions,* was propitiated by vaguely known peoples as the fearsome god of war; its brighter apparitions were taken as portents of cataclysmic events bringing death and devastation in their wake. Although no longer a god to most men, Mars in its anthropomorphic Greco-Roman deification is known even today to millions of the world's people as the graphic symbol of war.

Through the ages, too, the planet's apparently erratic movements and relative positions in the firmament have been held to exert sinister influences over men's destinies, to prophesy individual tragedy and collective calamity, by those for whom the future is somehow limned in the stars. For well over two thousand years, Urania, astronomy's Muse, has been handmaiden to occult astrology.

Only with the advent of the first scientific revolution in the early sixteenth century and the quickening of the spirit of reasoned inquiry that motivated and sustained it, has man with any degree of dispassion been able to view Mars for what it is—a neighboring planet in a vast and seemingly limitless universe. Descriptive astronomy, indeed, began in 1609 when Galileo Galilei first turned the

---

*Mars is said to be "at opposition" when it is opposite the earth from the sun. Opposition occurs at approximately 26-month intervals, and because the planet's orbit is an ellipse and not a circle, its distance from the earth at such times ranges from 34.6 to more than 60 million miles. Its closest approaches, termed "most favorable" oppositions, occur at intervals of just over fifteen years.

newly-invented telescope toward the night sky and discovered there not the star-spangled, crystalline spheres of the ancients but other, strangely earthlike, globes.

Since Galileo's time countless curious and often gifted observers have studied the planet with varying degrees of skill and zeal, seeking thereby to learn what secrets it may hold about our place among the infinities. Many of the major figures in the history of astronomy—Huygens, the Cassinis, Maraldi, the Herschels, Schroeter, Schiaparelli, and others—have turned their telescopes toward Mars at one time or another and have speculated on its mysteries.

Yet such are the practical difficulties of observing Mars through the earth's turbid atmosphere and across tens of millions of miles of space that little certain knowledge about the planet has been accumulated over more than three centuries of telescopic observation. This paucity of undisputed fact, moreover, has prevailed despite the progressive sophistication of instrumentation and technique that earth-bound astronomers have been able to command.

Even with the dawning of the space age and after four successful Mariner spacecraft missions to Mars, fundamental questions about the planet remain. Certainly the most intriguing question—whether life exists or has existed there—has not been answered to anyone's full satisfaction. Few astronomers would categorically deny that "it is not impossible that living organisms, specially adapted to conditions on the planet, can exist, or could have existed in the past." None would dispute the statement that "the discovery of life, past or present, on Mars would constitute one of the most exciting scientific events of all time."[1]

In the waning decades of the twentieth century, however, it is clear that this and other enigmas concerning Mars will not be resolved by the telescopes of terrestrial watchers of the stars. Rather their eventual solution will come through man's new-found technological capabilities to transcend his own sphere, as it were, by hurling himself, or electronic extensions of himself, into space.

Still, through the centuries, some things were learned about Mars.[2] That the planet's visible disk bears certain darkish markings was realized in Galileo's lifetime by Francesco Fontana in 1636, although these markings appeared as little more than blurred blotches in the small, crude telescopes of the time. The most prominent of these markings—a large, triangular area known to modern areographers as the Syrtis Major—was recognizably sketched in 1659 by the Dutch physicist and astronomer Christiaan Huygens.

Believing the dark markings to be permanent surface features, Huygens used them to estimate the length of the martian "day" at about 24 hours. In 1666, Giovanni Domenico Cassini confirmed and refined this, arriving at the surprisingly accurate figure of 24 hours and 40 minutes, less than three minutes above the currently accepted value.

Cassini also first clearly discerned the glistening white polar caps of Mars, the most spectacular single feature of the planet's telescopic aspect and one which more than any other lent itself to close comparisons between Mars and the earth. It was quickly assumed that the martian caps were analogous to the great masses of ice and snow that cover the earth's Arctic and Antarctic regions. Observers in the late seventeenth and early eighteenth centuries soon found that the martian caps periodically advanced and retreated in apparent response to the martian seasons, which, because of the tilt of the planet's axis to its orbital plane and its greater distance from the sun, are quite similar though nearly twice as long as those of the earth.*

It was also noted that the dark markings, thought to be like the dark *maria* areas of the moon and thus "seas," were subject to fluctuations in color and configuration, and indeed very early it was commonly believed that Mars was the most terrestrial of all the planets. In the 1780s, William Herschel, the great English astronomer and discoverer of the planet Uranus, stirred no sensation when he suggested to the august Royal Society that Mars experienced seasons like those of the earth and, from his observations of what he took to be "clouds and vapours," that it possessed "a considerable but moderate atmosphere." Its inhabitants, he concluded, "probably enjoy a situation similar in many respects to our own."[3] It does not detract from Herschel's monumental achievements in astronomy to note that he, like others in his time, thought that all the planets might be inhabited, and not only opted for a populated moon at an early point in his career but later declared the sun to be "richly stored with inhabitants."[4]

Herschel effectively summarized what was known and believed about Mars in his time. The picture of the planet he had sketched for the Royal Society in 1783 prevailed for many years, at least in its broad outlines. Even his younger contemporary Johann Heironymous Schroeter, the "Herschel of Germany" and the leading lunar

---

*The martian axial tilt is approximately 25 degrees compared to 23.5 degrees for the earth. The martian year is 687 earth days and 669 martian days long.

and planetary astronomer of the late eighteenth and early nineteenth centuries, accepted an inhabited Mars although in some respects he saw the planet quite differently than other observers had seen it. He thought, for instance, that Mars was encased in a shell of clouds and thus that the dark markings were not surface features at all but were atmospheric in origin—clouds, in short, deriving a certain permanence of position from the aerodynamic effects of their movement over an unseen martian topography.

The years of the late 1700s and early 1800s, however, although highly productive for many other areas of astronomy, are not particularly notable for contributions to martian lore. Indeed, Schroeter stands almost alone during this period as a systematic observer of the planet. But by 1830 interest in Mars was again on the ascendant and, with the favorable opposition of that year, investigation of the planet took a somewhat different turn.

Through the middle decades of the nineteenth century, a number of conscientious observers began to produce a series of increasingly detailed maps of Mars—maps which not only fixed many of the planet's visible features cartographically but which tended to enhance the terrestrial analogy to which Sir William had lent his very prestigious authority. For on these areographic charts, the dark, bluish-green markings, through now long-established usage, were labeled "seas," "oceans," and "bays," while the more extensive reddish-ochre areas, which cover five-eighths of the planet's surface and give it the distinctive ruddy hue it displays in the heavens, became "continents" and "islands."

As a result of the 1830 opposition, the German observers Wilhelm Beer and J. H. von Maedler made the first map of Mars that was more than a mere sketch of the planet. They followed it in 1840 with an improved and expanded chart based on their careful observations of Mars in its various aspects at oppositions during the preceding decade. In the 1860s, with Mars again in a favorable position, there was a flurry of martian map-making. Norman Lockyer, the Reverend William R. Dawes, and Richard A. Proctor in England; Father Pierre Angelo Secchi in Rome; and F. Kaiser in Holland, turned out notable charts. Father Secchi's 1863 map was the first to depict areas of the planet in color.

Gradually, through the efforts of these and other pioneers, the major features of Mars became more or less set in their areographic latitudes and longitudes. These early martian maps were in general agreement as far as the grosser martian topography was concerned, but they differed quite radically in their style of presentation and their nomenclature according to the areographer. Some followed the practice already established for the moon, of naming martian features after well-known scientists, particularly astronomers. On Nathaniel Green's 1877 map, for example, Syrtis Major was "Kaiser Sea," Sinus Sabaeus was the "Herschel II Strait," and the Sinus Meridiani, which defined the zero meridian on Mars, was "Dawes' Forked Bay." This system of names did not prevail, however.[5]

It remained for the respected Italian astronomer Giovanni Virginio Schiaparelli, on his 1877 map, to apply names from ancient geography and mythology to the martian features and thus set the nomenclature that—with only a few variations but many additions—is in general use. Percival Lowell, it may be noted, helped to gain acceptance of Schiaparelli's nomenclature by following the Italian's precedent in the many maps and globes he made of the planet in subsequent years. Lowell considered Schiaparelli's system of names an "at once appropriate and beautiful scheme, in which Clio [history's Muse] does ancillary duty to Uranis [sic]."[6]

The year 1877, which marked a "most favorable" opposition of the planet, is certainly a most auspicious one in the observational history of Mars. Two notable events occurred.

In August, American astronomer Asaph Hall, working with the fine 26-inch refracting telescope at the U.S. Naval Observatory in Washington, D.C., discovered two tiny moons orbiting Mars. Fact thus caught up with fiction; for Jonathan Swift, with strangely accurate prescience, had imaginatively assigned a pair of moons to Mars nearly two centuries before, and astronomer-mystic Johannes Kepler had suggested two satellites for Mars a century before Swift. These whirling bits of rock, each only a few miles in diameter, were appropriately named Phobos (Fear) and Deimos (Panic) after the attendants of the Olympian war god Ares—the Roman Mars.

The discovery of the moons not only permitted the determination of Mars mass with greater precision but posed a new problem for cosmologists. For Phobos, the larger and innermost of the two satellites, circles Mars in just 7 hours and 39 minutes, making three revolutions while the planet rotates but once on its axis. It is thus the only one of the more than thirty known natural satellites in the solar system to revolve faster than its primary rotates.

But a second discovery during the 1877 opposition was much more far-reaching in its implications and has proved to have a deep and long-lasting fascination for the lay as well as the scientific mind. As Mars swung close to the earth late that summer, the keen-eyed Schiaparelli, observing from Milan with a telescope of only 8¾ inches aperture, caught glimpses of a vast and intricate network of fine lines criss-crossing the reddish-ochre "continental" areas of the planet—lines that have since become famous as the so-called "canals" of Mars.

Schiaparelli, in reporting his discovery, called the lines *canali,* which in Italian more properly refers to "channels," and he carefully refrained from drawing any conclusions as to their nature and possible purpose. The term *canali,* indeed, had been used by Father Secchi in his 1863 map of the planet to label certain vague, streak-like features he had observed—features also depicted by several other early areographers, including Dawes and the pioneers Beer and von Maedler. But Schiaparelli's lines were finer, sharper and more systematically arranged on the surface of the planet. In a world long accustomed to easy comparisons between the earth and Mars, it was perhaps inevitable that *canali* would be translated into the more connotative "canals," and Schiaparelli's admirable caution notwithstanding, so it was.

In describing these lines later, the Italian emphasized at the outset they "may be designated as canals although we do not yet know what they are." Sometimes, he wrote, they appeared "vague and shadowy; other times clear and concise." Their "aspect and degree of visibility are not always the same," but their "arrangement appears to be invariable and permanent." In general, the canals were "like the lines of great circles," and he estimated their width at about two martian degrees, or about 120 kilometers (75 miles). Several of them, he noted, extended for 80 degrees over the planet's surface, a distance of some 4,800 kilometers (3,000 miles).[7]

At the opposition of 1879, just after the advent of the martian spring, he observed that one of his *canali* appeared as a double line; at the next opposition, in the period between January 19 and February 19, 1882, he recorded twenty examples of this curious paralleling, or "gemination," as he called it. The phenomenon, he declared, was "not an optical effect. . . . I am absolutely sure of what I have observed."[8]

Schiaparelli's "canals" and the strange "doubling" of some of them marked the beginning of what was to become one of the most intense and widespread scientific controversies of all time, an unprecedented debate which, as word of Schiaparelli's discoveries filtered out beyond the bounds of disciplined astronomy, would involve the public at large in many areas of the world. Although Schiaparelli's reputation as an astronomical observer was not assailed, and indeed is unassailable even today, the existence of the delicate, reticular lines which he had so carefully described was strongly questioned. His announcement of their doubling only served to increase the skepticism. Indeed, as the planet swung to the farthest reaches of its orbit at successive oppositions early in the 1880s, other observers failed to see anything like the Italian's *canali*. It was not until 1886 that another astronomer—J. Perrotin at Nice, France—discerned them with a 15-inch refractor. Within the next few years, as martian oppositions became more favorable again, others saw, or said they saw, the lines. Since then, many competent observers have described them in varying degrees of detail and complexity. Yet—and this seems puzzling—many others, equally competent, have never seen them.

In the final years of the decade of the 1880s, the discovery of the so-called "canals" on Mars, which had already stirred new astronomical interest in the planet, began to have a dramatic impact on the minds and emotions of people throughout the world. The question of whether the lines were actually on Mars or merely a psychological or optical effect incident to observation was a scientific one and presumably could be settled eventually by specialists. But this question soon paled before their significance if, in fact, they were real. For the existence on Mars of such a complicated and geometrical network as Schiaparelli had described, let alone "canals," seemed to imply the work of intelligence.

There were those even within science who pointed this out. Most notable and perhaps the most authoritative of these at first was the flamboyant French astronomer Camille Flammarion who declared in 1892 that "the present inhabitation of Mars by a race superior to ours is very probable."

The considerable variations observed in the network of waterways testify that this planet is the seat of an energetic vitality. These movements seem to us to take place silently because of the great distances separating us; but while we quietly observe these continents and seas slowly carried across our vision by the planet's axial rotation and wonder on which of these shores life would be most pleasant to live, there might be at the same time thunderstorms, volcanoes, tempests, social upheavals and all kinds of struggle for life. . . . Yet we may hope that, because the world of Mars is older than ours, mankind there will be more advanced and wiser. No doubt it is the work and noise of peace that for centuries has animated this neighboring home.[9]

This single paragraph, in essence, summarizes almost all of the major premises of the subsequent theories and philosophies of martian life that were soon to explode into a brief but intense worldwide uproar and touch the thought and feelings of millions of people in virtually every land. The "canals," an "energetic vitality," "mankind," the earth-like aspect of the planet with its "continents" and "seas," the presumptions of a hostile nature and human frailty, even the idea that Mars was older than the earth and thus martians were wiser and more peaceful than men, would become so familiar over the next two decades as to serve as subjects for both moralists and humorists.

Although this picture of martian life has since been dimmed, and perhaps the moral concepts embodied in Flammarion's florid prose have eroded, vestiges certainly remain. "Men from Mars" and the "little green men" of the "flying saucer" reports of the late 1940s and early 1950s are not the responsibility of science fiction writers alone, and peace is still at least one of man's dreams. Indeed, more than thirty years after the Mars furor reached its peak, the extent to which the idea of intelligent martian life had penetrated the popular mind was dramatically illustrated in a widely-publicized, well-documented incident that occurred in the southern New Jersey area on the eve before All Hallow's Eve in 1938. This was the spectacle of thousands of supposedly sensible Americans fleeing their homes in panic, instantly crediting a dramatized radio broadcast of a fictional invasion from Mars as not only possible, but all too true.[10]

The new surge of interest in Mars that resulted from the discovery of the "canals" began what has been described as a "Mars mania" that burst into the public purview in the late 1880s and gained accelerating momentum with the approach of the favorable opposition of Mars in 1892:

Now a veritable tumult began. Novels about Mars poured from the pens of imaginative writers. Telescopes were built for the specific purpose of observing the next opposition. Mars observatories sprang up by the dozens. Amateurs detected light signals. Millionaires offered prizes for communication with the Martians. It was proposed that a drawing of the Pythagorean theorem half the size of Europe be traced on the Sahara Desert that Martian astronomers might see it and realize that the creatures of Earth were also intelligent. Never had there been such a wild enthusiasm for astronomy.[11]

Jules Verne, of course, had already written his *Dr. Ox's Experiment* and other fantasies. New telescopes were in fact being built, although not solely to observe Mars. Nice had installed its 30.3-inch refractor in 1886, the Lick Observatory on California's Mt.

Hamilton began using its 36-inch refractor in 1888, and Meudon Observatory near Paris had its 32.5-inch refractor in 1891. These are still among the largest and finest refracting telescopes in the world. Reflecting telescopes had not yet regained their earlier popularity, but silvered glass was already beginning to replace speculum metal in their optical systems. England's A. A. Commons had turned out a 60-inch glass mirror in 1891, and G. W. Ritchey in America would produce another of this size within ten years. There were several new observatories, too, most notably the Harvard College Observatory's Boyden Station high in the Andean foothills at Arequipa, Peru, far more favorably situated for observing the 1892 opposition of Mars than the older established observatories in the northern hemisphere.

But the business of the supposed "signals" from Mars and the grandiose, often absurd, schemes advanced to respond to them are what give the Mars furor its early manic taint. These short-lived but intense flurries of interest in the planet, beginning about 1888, stemmed from reported observations of what astronomers took to be bright "clouds" projecting over the planet's terminator* and into the dark side of its partially sunlighted disk. In apparently one instance, at least, the agent was the flashing reflection of the sun on the so-called Mountains of Mitchel, a feature periodically observed on the tawny martian plain as a bright white patch isolated by the annual spring retreat poleward of the south polar cap.

These projections and flashes, grossly misconstrued and sensationalized by the press, were seen as prima facie evidence that intelligent martians were attempting to communicate with their neighbors on earth. Some of the proposals that were advanced for answering the "signals," such as the Pythagorean theorem one, were preposterous even in that technologically over-confident era. Popular interest in interplanetary communications also was considerably enhanced when such inventive geniuses of the day as Thomas Edison, Sir Francis Galton, Nicola Tesla, and Guglielmo Marconi publicly took a hand in such goings on, lending credence to the speculations.[12]

In vain did some astronomers attempt to convince the more credulous members of the press and public through carefully written explanations that the supposed "signals" were purely imaginary. Reports of "messages" from martians continued to appear for many years, some of them quite fantastic. In 1895 for example, the

---

*The "terminator" is the line separating the bright, sunlit portion of the visible disk of a planet or satellite from its dark, or night, side.

New York *Tribune* published the claim that certain dark markings spelled out the words "The Almighty" in Hebrew on the surface of the planet, drawing the indignant comment from the journal *Popular Astronomy* that "it is a burning shame that such nonsense finds place in our best and greatest daily papers."[13]

It was on this background of "canals" and "signals" that astronomers and a fascinated public alike looked forward with heightened anticipation to the martian opposition in 1892 and the new and dramatic discoveries it was expected to reveal. But, though closer to the earth than it had been for fifteen years, the planet was poorly situated in the sky for observers in the northern hemisphere where nearly all existing observatories and all the largest telescopes were located. As a consequence, few new facts and fewer sensations were immediately forthcoming from the observations in 1892. "Public expectation, which had been raised to the highest pitch by the announcements of sensation-mongers," Agnes M. Clerke, the most authoritative historian of astronomy of that time would later write, "was somewhat disappointed at the 'meagerness' of the news authentically received from Mars."[14]

This was not entirely true, however, for Harvard astronomer William H. Pickering and his young assistant, Andrew E. Douglass, in Peru, as well as observers at the Lick Observatory and at Nice, reported more "projections" on the planet's terminator, setting off a new round of speculation about martian "signals." Pickering also discerned a number of small, round, dark spots at the intersections of two or more "canals" and promptly labeled them "lakes." And he and Douglass, as well as observers at the Lick Observatory, noticed faint darker markings within the bluish-green martian "seas" which suggested that these dark areas were not bodies of water at all and which, when seen more clearly and identified by Douglass in 1894 as "canals," effectively delivered a coup de grace to this particular aspect of the terrestrial analogy. All in all, however, 1892 was not a banner year for Mars, and both science and sensation had to await the 1894 apparition when the planet would still be quite close to earth, and far better positioned in the northern sky.

Astronomical interest as that year of opposition dawned was intense, for, as the Lick Observatory's W. W. Campbell noted, "nearly all the problems concerning Mars are still awaiting solution. It is difficult to mention any, indeed, except some of those relating to the polar caps, that have been solved satisfactorily."

And to this, he added the sardonic but prophetic comment:

Many astronomers are preparing to expend a great deal of energy on the planet, and are expecting more valuable results than have been obtained at any previous opposition. . . . We shall undoubtedly have interesting accounts of the progress of the work, both in the scientific journals and in the public press, which everyone, astronomers included, will welcome. It is, unfortunately, perfectly safe to predict that we shall also hear from the sensationalists, astronomers included; and that fact is a source of sincere regret to all healthy minds.[15]

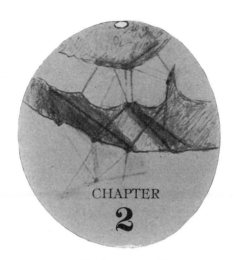

CHAPTER

2

Lowell

O F ALL THE MEN through history who have posed questions and proposed answers about Mars, the most influential and by all odds the most controversial was Percival Lowell, born in Boston on March 13, 1855, into a singularly illustrious generation of two of America's great and wealthy families—the Lowells and the Lawrences of New England.

Perhaps more than any other individual, Lowell shaped the broad conceptions and misconceptions regarding Mars in particular and extraterrestrial life in general that have been part of the intellectual milieu of Americans and Europeans since before the turn of the twentieth century. For Lowell was both the most intense student of the planet up to his time and the most prolific and persistent popularizer of martian lore before or since.

From 1894, when he founded an astronomical observatory at Flagstaff in the Arizona Territory primarily to study the planet, until his death on November 12, 1916, Lowell and a succession of assistants carried out what even in his lifetime was publicly acknowledged to be "one of the most remarkable investigations of modern astronomy."[1] Over more than two decades, he and others at Flagstaff accumulated an unprecedented body of systematic data regarding the red planet. "Whether or not astronomers agree with his conclusions," the *Scientific American* conceded at the peak of the Mars furor he had generated, "it cannot be denied that he has been by far the most indefatigable observer of our planetary neighbor."[2]

Almost single-handedly, Lowell escalated what had been a relatively polite and formal debate largely within astronomy itself over the so-called "canals" of Mars into a raging public controversy over their possible significance which at its height overshadowed in popular interest the severe economic depression of 1907.

"Think back on '07," the staid *Wall Street Journal* urged its readers as that year of astronomical sensations drew to a close. "What has been in your opinion the most extraordinary event of the twelve months?" For the *Journal* at least, it was "not the financial panic which is occupying our minds to the exclusion of most other thoughts" but "the proof afforded by astronomical observations ... that conscious, intelligent human life exists upon the planet Mars."

The proof is indeed circumstantial ... of the same kind and of the same strength as might convict a criminal accused of murder in cases where there had been no actual witnesses to the deed. ... There could be no more wonderful achievement than this, to establish the fact of life upon another planet. ... That a more complete knowledge of the planet Mars may possibly have a profound effect upon life on our own globe goes without question.[3]

There were some then, as there are many now, who could scoff incredulously at a statement that the existence of intelligent martian life had been proven. But in 1907, there were many thousands of more or less literate people around the world who, like the anonymous *Journal* writer, were convinced, and many thousands more who were intrigued enough by the possibility to be unwilling to reject it out of hand.

Lowell himself, who provided the "circumstantial" evidence in the form of photographs purporting to show the long-disputed "canals," would object only to the use of the word "human" in reference to martian life. For although he repeatedly insisted on the presence of local intelligence on Mars, indeed of beings politically, socially, and technologically advanced over those on earth, he consistently stopped short of the analogy, so easily drawn by so many others before and since, between martians and men.

The year 1907 was clearly the high point of the Mars furor, a curious and fascinating thirty-year phenomenon that is unique in the annals of science and in intellectual history. Never before had an astronomical issue so stirred the active interest and the imagination of so many people in all walks of life throughout the literate world. Not even the great sixteenth-century clash over the Ptolemaic and Copernican cosmologies equals it in scope and intensity, although historically, of course, Copernicanism has had vastly greater significance. A reason for this, perhaps, is that while both controversies involved fundamental concepts, beliefs, and emotions, and had religious and philosophic overtones, the Mars furor still remains unresolved and hasn't really changed anything. In the one instance, subsequent science sustained Copernican heliocentrism, and Christian man over the centuries has had to reorient himself gradually to a new and different universe from the geocentric one to which he had become accustomed over nearly two thousand years. Science has not as yet—in the early 1970s—demonstrated the existence of extraterrestrial life on Mars or anywhere else. But when and if this comes to pass, the impact on man's thought and beliefs will be immediate and profound. The Mars furor itself is presumptive evidence for this.

It began in all innocence in 1877 when Schiaparelli discovered the "canals." But, it was not until 1894, with Mars again at favorable opposition and with the sudden appearance of Lowell on the astronomical scene, that the Mars furor began to take on the dimensions of an international cause célèbre.

The "canals" of Mars, one historian of the planet has stated bluntly, "would have been left to the scientists had it not been for the activities of Percival Lowell."[4] Nearly every other commentator on Mars and on Lowell's life and work has made the same point, although usually somewhat more effusively. "This might have been the end of them [the canals] had it not been for one of the Boston Brahmins, Dr. Percival Lowell, a man of many interests and many moods, financially independent, well trained in mathematics and physics, but not a professional astronomer," another recent writer on the subject has pointed out:

The canals caught his imagination and he built an observatory at Flagstaff, Arizona, to study them. He was sure he could see canals, and drew elaborate maps of their courses; he was convinced that they had been constructed by intelligent people to conserve a dwindling water supply. His ideas evoked headlines in newspapers and worldwide discussion, so much so that the discovery of anything less on Mars may seem an anticlimax.[5]

Lowell's abrupt plunge into astronomy early in 1894 was surprising enough to most people at the time, for there was little in his publicly-known background that appeared to qualify him as an astronomical authority. Then and in retrospect, his apparently impulsive move seemed to be a "hideously unconventional proceeding."[6] But in reality, astronomy had long been one of his "many

interests and many moods," and later in his life he would recall
that his first conscious memory was of an astronomical event:

Consciously, I came into this world with a comet, Donati's Comet
of 1858 being my earliest recollection and I can see yet a small boy
half way up a turning staircase gazing with all his soul into the
evening sky where the stranger stood.[7]

Donati's Comet was indeed a dramatic sight, one of the great
astronomical shows of the nineteenth century. Visible to the un-
aided eye for 112 days, its "scimitar-like curve" stretched at one
time across a full third of the night sky, "the most majestic celestial
object of which living memories retain the impress," Miss Clerke
later noted.[8] Such a spectacle cannot have failed to leave a vivid
image on the mind of a bright three-year-old boy. Nevertheless
Lowell's interest in astronomy crops out only occasionally during
the first thirty-eight years of his life, and these occasions hardly
hint at the curious dominance it would have in later years.

Lowell was the oldest of the five children resulting from the
1854 marriage of Augustus Lowell, a leading figure in business,
education, science, and the arts in mid-nineteenth century Boston, and Katharine Bigelow Lawrence, daughter of Abbot Lawrence, the U.S. minister to Great Britain in 1851.[9] The cities of
Lowell and Lawrence in Massachusetts are named for these two
prominent New England families. Augustus was a cousin of James
Russell Lowell, the distinguished poet, essayist, and editor who
also served as America's envoy to the Court of St. James.

Eleven generations before, while Galileo still lived, the first
Percival Lowle (1571–1664) had emigrated from England during
the first wave of Puritan migration to the New World, and had
settled in what later was to become Newbury, Massachusetts, in
1639. The Lowells prospered and by the early decades of the
nineteenth century had already founded one of America's great
fortunes on textile manufacturing. After 1870, it has been said,
"the popular impression that the words, 'Lowell,' and 'millionaire,'
were synonymous was not wide of the mark."[10] The Lowell
motto—*Occasionem Cognosce*, that is, "seize your opportunity"—
was added to the family coat of arms by the Reverend John Lowell
(1704–1767) and certainly appears to have been taken to heart by
at least some of his descendants.

The second Percival's generation represents an unusual intellectual flowering of the Lowell clan. His brother, Abbot Lawrence
Lowell, became a successful and prominent New England lawyer

and, as president of Harvard University from 1909 to 1933, one of the nation's leading educators. He also eventually became Percival's biographer. Amy, the youngest of his three sisters and his junior by some twenty years, won renown as a cigar-smoking avant-garde poetess, a leader in the free verse movement of the early twentieth century, and as literary biographer of the nineteenth century English poet, John Keats. Only sisters Elizabeth and Katherine shunned the public's eye.

Born to wealth and high social prestige, Percival was reared as a proper Bostonian and exposed to the manners and modes of thought of the plush secure Victorian world that flourished, particularly in eastern America, during the decades after the Civil War. He throughly absorbed the pragmatic conventional wisdom of New England's affluent industrial aristocracy with its strong strains of Puritan ethics and, latterly, social Darwinism.

As a child, he attended a "dame school" in Boston run by a "Miss Fette." In 1864 Augustus took his growing family to Europe. Percival, at boarding schools in France, began to learn the rigorous discipline of the scholar and to acquire a working knowledge of other languages. He is said to have been fluent in French at age ten and in Latin at eleven. Returning to Boston with his family in the fall of 1866, he continued an essentially classical college preparatory education for the next five years. It is during this period that his interest in astronomy first shows up in the available record. Of these adolescent years, his brother has written:

He took to astronomy, read many books thereon, had a telescope of his own, of about two and a quarter inches in diameter, with which he observed the stars from the flat roof of our house; and later in his life he recalled that with it he had seen the white snow cap on the pole of Mars crowning a globe spread with blue-green patches on an orange background.[11]

Percival himself later fixed the date of his interest more specifically in correcting a well-known British popularizer of science on the point. "So I may say that my interest in astronomy did not date, as you suppose, from 1894 but 1870 when I used to look at Mars with as keen interest as now," he wrote.[12]

Lowell entered Harvard in 1872, studying under the distinguished mathematician Benjamin Peirce and taking second year honors in the subject. He won the Bowdoin Prize for his essay on "The Rank of England as a European Power from the Death of Elizabeth to the Death of Anne" and performed well enough in

Percival Lowell and mother in 1855

physics, history, and the classics to be elected to the academic honorary Phi Beta Kappa. At his graduation in 1876, interestingly enough, his commencement part was a short talk on "The Nebular Hypothesis" by which Immanuel Kant and later Pierre Simon Laplace had sought to explain the origin of the solar system.

Following graduation, he made the "Grand Tour" of Europe, traveling as far as Syria before returning to Boston and the office of his grandfather John Amory Lowell to spend the next six years, somewhat restlessly one presumes, managing family interests and business affairs.

By 1883, when he had "by shrewd investments acquired a portly pocketbook of his own,"[13] the urge to travel apparently seized him again and he set sail for the Far East on the first of a series of extended trips that over the next ten years would take him thrice to the Orient, once to Spain, and twice around the world coming and going. In between these excursions he returned to Boston, to business, to writing about his Far Eastern adventures, and to the sport

Lowell's first telescope, a 2¼-inch refractor

of polo which he played at the nearby Dedham Polo Club which he and a group of friends founded in 1887 and which, for a period, he came to regard as his home.

In 1883 the Orient was a region of the world that had been accessible to Occidental travelers for only a few years, and about which little was known in western countries. The Far East, as Lowell saw it then and in his subsequent visits, excited and delighted him. He quickly learned the Japanese language and began quite intensive studies of the ethos and customs of the Oriental peoples. The Koreans and Japanese he met and lived with there apparently held him in high regard. At any rate, his reputation was such that he was asked to serve as foreign secretary and counsellor for a special Korean diplomatic mission to the United States in August of 1883. The position "amounts to his having complete charge and control of the most important legation from a new country that has visited the United States since the opening of Japan," and was one he filled to the apparent satisfaction of everyone concerned.[14]

Lowell's experiences on his trips to the Orient resulted in several articles describing Korean and Japanese events in the *Atlantic Monthly* and the publication of four books dealing with that exotic part of the world. *Choson–The Land of Morning Calm* (1886) and *The Soul of the Far East* (1888) were the result of his first visit; *Noto* (1891) of his second; and *Occult Japan* (1894) of his third and final trip.

Of these, *The Soul of the Far East* became the most popular and was quite widely read not only in America but in Europe as well. The book was hailed as an excellent summary and analysis of the Oriental mind and ways of life and brought Lowell a considerable reputation not only as an Orientalist but as a clear and colorful writer. It also, incidentally, has been credited with helping to make an Orientalist of the gifted New Orleans journalist Lafcadio Hearn, whose enthusiasm for the Far East, at least partially inspired by Lowell's writings, took him in 1891 to Japan to live out the remainder of his life writing about that country and its people. [15]

The Soul of the Far East contains, interestingly enough, one of Lowell's early excursions into the synthesis of ideas, a somewhat vague theory that human progress is a function of the qualities of individuality and imagination. Civilization, he contended, becomes progressively advanced as one proceeds geographically around the world from east to west, from Asia to the New World. As evidence of this, he pointed to the backwardness of the politely impersonal and imitative Orient vis-à-vis the obvious superiority of the intensely personal, highly individualistic cultures of the west and more especially that of the United States. It may be noted here that Hearn was intrigued by this idea when he first read Lowell's book. But after living for a while in Japan, he felt that Lowell had missed "the most essential and astonishing quality of the race: its genius for eclecticism." [16]

Lowell's assessment, read today, is quite ethnocentric and condescending. But at the time it was, like his other Far Eastern writings, highly readable as travel lore. And as in his later books about Mars and the solar system, it was spiced with pungent comments on morals and mores ("To the west belongs the credit for making articles to kill people instead of killing time," he remarks on the Chinese invention and use of fireworks), and sometimes atrocious puns (the Japanese family with its deep reverence for ancestors is to Lowell "a relative affair," and Oriental politeness reminds him unchivalrously of the "knight-times of the Middle Ages"). [17]

Lowell's interest in science and things scientific surfaces occasionally in his Far Eastern writings. His discussion of "imagination" in *The Soul of the Far East*, for instance, seems to provide some insights into the subsequent controversies over his own "imaginative" ideas about Mars:

Imagination is the single source of the new ... reason, like a balance wheel, only keeping the action regular. For reason ... compares what we imagine with what we know, and gives us the answer in terms of the here and now, which we call the actual. But the actual ... does not mark the limit of the possible.[18]

We readily impute imagination to art, he points out, "but we are by no means ready to appreciate its connection with science." Yet:

The science of mathematics ... might be called the most imaginative product of human thought; for it is simply one vast imagination based upon a few so-called axioms, which are nothing more or less than the results of experience. It is nonetheless imaginative because its discoveries always accord subsequently with fact, since man was not aware of them beforehand. Nor are its inevitable conclusions inevitable to any save those possessed of the mathematician's prophetic insight.[19]

A passage in a letter apparently written about this same time gives additional glimpses into Lowell's thinking along these lines:

Somebody wrote me the other day apropos of what I may or may not write, that facts not reflections were the thing. Facts not reflections indeed! Why that is what most pleases mankind from philosophers to the fair, one's own reflections on or from things. Are we to forego the splendor of the French salon which returns us beauty from a score of different points of view from its mirrors more brilliant than golden settings. The fact gives us but a flat image. It is our reflections upon it that make it a solid truth. For every truth is many sided. It has many aspects. We know now what was long unknown, that true seeing is done with the mind from the comparatively meagre material supplied us by the eye.[20]

How Lowell's later critics would have chortled over that final statement had it been available to them, for that Lowell saw the "canals" of Mars with his mind rather than with his eye was one of the major thrusts of the attacks on his theory of intelligent martian life.

It is necessary to point out here that while the word "imagination" has a number of definitions, Lowell seems to have used it consistently in a rather restricted sense, that is, "the mental ability to create original and striking images and concepts by recombining the products of past experience."[21] This perhaps explains the fact that while his subsequent theories about Mars and the solar system were frequently ridiculed as "imaginative," the circumstance doesn't appear to have bothered him. The adjective, in fact, may have pleased him, for he did not use it in the ordinary pejorative sense, implying "fanciful" speculation, as his critics intended, although he certainly understood their intent. Rather, he firmly believed that his theories were "imaginative" in the intellectually creative sense implying a reasoned synthesis; this, indeed, was one source of his pride in them. Hear, for instance, what he has to say on the same subject some thirty years later, within a few weeks of his death, while lecturing to West Coast college students on the basic prerequisites for a successful career in science:

A good education is indispensable, one as broad as it is long; without it he runs the risk of becoming a crank. Then enters the important quality of imagination. This word to the routine rabble of science is a red rag to a bull; partly because it is beyond their conception, partly because they do not comprehend how it is used. To their thinking to call a man imaginative is to damn him; when, did they but know it, it is admitting the very genius they would fain deny. For all great work imagination is vital; just as necessary in science and business as it is in novels and art. ... The difference between the everyday and scientific use of it is that in science every imagining must be tested to see whether it explains the facts. Imagination harnessed to reason is the force that pulls an idea through. Reason, too, of the most complete, uncompromising kind. Imagination supplies the motive power, reason the guiding rein.[22]

For Lowell, then, imagination as he defined it was essential to greatness in science or in anything else. The idea is fundamental to his thinking and runs through all his work. It even crops up, for example, in a "Commemoration Day" speech he delivered in Flagstaff in 1903:

The bundle of ideas, past and present which man has had, constitute his brain. The greatness or littleness of the man depends first upon the quality of his ideas, whether simple or abstruse, and secondly, upon the cooperation of those ideas with one another. The genius is the man who not only has the requisite ideas but who contrives to combine them. The history of science, pure or applied, abstract or inventive, teems with illustrations of this, and what is specially instructive, shows that it is not so much in the quality of the idea as in the combination of them that men differ, the duffer from the deft. Two thoughts will have been known for centuries until someday a genius comes along to whom it occurs to put them together, and behold we have a nitroglycerine in the world of thought which revolutionizes humanity. [23]

Imagination, although the sine qua non of greatness in science, is not its sole ingredient. Lowell detailed this point in the same 1916 lecture mentioned above, and it will be instructive to quote from it here at greater length, for what he had to say throws considerable light not only on his own thinking about science but on his concept of himself as a scientist:

Now in science there exist two classes of workers. There are men who spend their days in amassing material, in gathering facts. They are the collectors of specimens in natural history, the industrious takers of routine measurements in physics and astronomy or the mechanical accumulators of photographic plates. Very valuable such collections are. They may not require much brains to get, but they enable other brains to get a great deal out of them later. . . . The rawer they are the better. For the less mind enters into them the more they are worth. When destitute altogether of informing intelligence, they become priceless, as they then convey nature's meaning unmeddled of man. [24]

Those who do this "important work," Lowell calls the "day-laborers in the temple of science" without whom "the edifice would largely lack bricks." But:

The second class of scientists are the architects of the profession. They are the men to whom the building up of science is due. In their hands, the acquired facts are put together to that synthesizing of knowledge from which new conceptions spring. . . . Though the gathering of material is good, without the informing mind to combine the facts they had forever remained barren of fruit. [25]

Lowell illustrates the point with the classic example of Kepler's use of Tycho Brahe's observational data on Mars in the discovery of the laws of planetary motion. But, he adds, "at times, a man who reasons from his facts is the first to get them too," and here his historical reference is to the Greek astronomer Hipparchus. His own work in astronomy, of course, would have served well enough, and the implication probably was not lost on his young listeners.

It is wholly typical of Lowell that in writing or speaking on things scientific, Oriental, or whatever, he frequently went outside the subject at hand for supporting arguments and thus, not entirely incidentally, revealed his thinking on the social, political, and economic issues of his time. For instance, in telling his 1916 audiences that he who would become a great scientist must not only pursue quality but pursue it alone, Lowell also contrives to air his views on such diverse topics as the preoccupation of Americans with sheer size, the increasing tendency toward mechanism in human affairs, and unionism. Again it will be informative to quote at some length:

Be not a collector only, be a creator too. Especially is this to be remembered now in these times and in this country where in popular estimation mere size bulks so large. To put quantity above quality is our national besetting sin. Bigness is not greatness. No amount of mediocrity can equal a little of the best. ...

Now of all times is the moral timely. For science is subject to the universal wave-motion of the universe. It grows by fits and starts; fits of easy reflection and fits of hard. At times men riot in reflection with but scant regard for facts; and then thought falls into disrepute and nothing but undigested facts will do. We live in one of the latter periods when the pendulum has swung as near to mere mechanism as man can make himself go. ... [26]

Lowell's remarks about unionism also carry some implied criticism of the then rapidly expanding professional scientific societies, where much of the skepticism about his own scientific ideas was centered:

Everywhere cooperation is the order of the day. Unionism is in the air. We unionize everything except the United States, and we are even thinking of unionizing that. It is therefore quite in the fashion that scientific unions are now in vogue. Now unions are excellent in their way for routine work. But though they push the lower men up, they pull the upper down. To be willing to cooperate is to admit that one can do better in conjunction with others than he can do alone. No great man ever cooperated with another in the idea that made him great; the thing is unthinkable. ... That banding together should be held advisable is a sad comment on the paucity of genius of the age. [27]

Lowell's basic plea here reflects his own intense individualism which was clearly of the "rugged" variety and in the Social Darwinist tradition of Herbert Spencer and William Graham Sumner. Individualism is his major propagandistic pitch in *The Soul of the Far East,* and in making it he reveals an equally profound commitment to Darwinian evolution in its nonsocial aspects. As usual, he injects various and sundry other ideas into his argument, some of them astronomical.

In the Orient, Lowell found what he considered to be "the

survival of the unfittest." The Far East, he felt, was "only half civilized in an absolute rather than a relative sense; in the sense of what might have been, not what is. It is so," he added, "compared not with us, but with the eventual possibilities of humanity" because "as yet neither of the systems, Western nor Eastern, is perfect enough to serve in all things as standard for the other."

The evolution of these people seems suddenly to have come to an end in mid-career ... and shows all signs of having fully run its course ... the same spectacle we see cosmically in the case of the moon, a world that died of old age.[28]

The reason for this is that the peoples of the Far East are impersonal, and lack individuality, which is saying much the same thing. "The soul of the Far East may be said to be impersonality," he declared. This trait, which he found in Oriental customs, language, art, and religion, logically precluded the development of individualism and insured an imitative, rather than an imaginative bent of mind:

Take away the stimulus of individuality, and action is paralysed at once. ... They [Orientals] remind us forcefully of that happy-go-lucky class in the community which prefers to live on questionable loans rather than work itself for a living. For with most men, the promptings of personal advantage *only* afford sufficient incentive to effort. ... Destroy this force, then any consideration due it lapses, and socialism is not only justified, it is raised instantly into an axiom of life. ... Socialism, then communism, then nihilism follow in inevitable sequence.[29]

For Lowell, individuality is the prerequisite of imagination; imagination, in turn, is the prerequisite for the evolutionary progress of civilization. His first chapter, indeed, is entitled, "Individuality," and his book moves progressively through to the last, entitled, "Imagination," in which he expresses his conclusions in Darwinian terms:

Dissimilarity of Western and Eastern attitude of mind shows that individuality bears the same relation to the development of mind that the differentiation of species does to the evolution of organic life: *that the degree of individualization of a people is the self-recorded measure of its place in the great march of mind.* All life ... whether organic or inorganic, consists in a change from a state of simple homogeneity to one of complex heterogeneity. ... The process is apparently the same in a nebula or a brachiopod. The immediate force is ... a subtle something which we call spontaneous variation. ... What spontaneous variation is to the material organism, imagination, apparently, is to the mental one.[30]

Thus Lowell's thought was thoroughly if uncritically Darwinian. It also will be seen to have been pragmatic in the tradition that was soon to receive its fullest formulation at the hands of his friend and fellow New Englander William James. Perhaps, more importantly, it was broadly Baconian in a simple straightforward way. He apparently never concerned himself with the highly advanced scientific thought introduced during his lifetime by such complex thinkers as Planck and Einstein. He believed quite simply that scientific truths could best be reached by reasoning logically from facts—positive facts—and this uncomplicated approach he found to be entirely adequate for his work throughout his life. He stated this position early and often in his astronomical writings:

All deduction rests ultimately upon the data derived from experience. This is the tortoise that supports our conception of the cosmos. For us, therefore, the point at issue in any theory is not whether there is a possibility of its being false, but whether there is a probability of its being true. This ... is too often lost sight of in discussing theories on their way to recognition. Negative evidence is no evidence at all, and the possibility that a thing might be otherwise, no proof whatever that it is not so. The test of a theory is, first, that it shall not be directly contradicted by any facts, and secondly, that the probabilities in its favor shall be sufficiently great. [31]

The matter of sufficiency, he adds, involves the geometric, rather than the arithmetic, increase in the "odds that a thing is true," and that if this is remembered, "it will be seen how rapidly the probability of the truth of a theory mounts up from the amount of detail it explains."

For whatever the cogency of each detail of the argument in itself, the concurrence of all renders them not simply additionally but multiplicitly effective. That different lines of induction all converge to one point proves that point to be the radiant point of the result. [32]

Scientific truth, or proof, for Lowell is always a matter of probability, and indeed he understood clearly, as many people then and since have not, that science does not deal in absolutes:

Between the truths we take for granted because of their age, and those we question because of their youth, we are apt to forget that in both proof is nothing more than a preponderance of probability. The law of gravitation, for example, depends eventually, as recognized by us, upon a question of probability; and so do the thousand and one problems of daily life upon which so many of us act unhesitatingly and should be philosophic fools if we did not. [33]

Lowell was consistent in his thinking, and once he had adopted an idea he not only held to it tenaciously and without significant modification but expressed it henceforth in essentially the same terms and frames of reference. The point is pertinent, as will appear, to his astronomical theories, and it can be illustrated here with his definition of scientific proof more than two decades later:

Outside of mathematics, which is formulated logic, proof consists in an overwhelming preponderance of probability. Take, for instance, the law of gravitation, as being inversely as the square of the distance of the bodies apart. We call it proved and rightly because almost everything stands explained by it and the few things that do not, like the motion of the apsides of Mercury, we are confident will eventually fall into line. This accordance is the test of truth. We apply it every day in our lives and act upon it in cases far less demonstrated than science requires and we should be fools if we did not.[34]

The major wellsprings of Lowell's thought, then, were Darwinian evolution, which he applied broadly to both science and society; an energetic, common-sense pragmatism, the product of his family background and business experience; a romanticism derived from his classical education and wide-ranging travels; and finally the Baconian "scientific method" which, perhaps, he applied somewhat more imaginatively than Sir Francis had intended.

These intellectual strains were already deeply entrenched in Lowell's thought by 1892 when he sailed on what would be his final trip to Japan, carrying with him a fine six-inch telescope which he used there to observe the planet Saturn and presumably other celestial objects. His friend and fellow Boston scientist George Russell Agassiz accompanied him and helped him investigate the mysteries of Shinto occultism. It was during this trip, according to Agassiz, that Lowell learned that Schiaparelli had been forced to abandon his planetary studies because of failing eyesight, and Lowell decided to take up the Italian's work. Lowell himself, however, apparently never documented this decision.[35]

Returning to Boston in the autumn of 1893, Lowell set to work to implement his resolve with typical energy, enthusiasm, and impatience. Time was short, for he intended nothing less than to build, equip, and staff a major new astronomical observatory in the best possible location he could find, and Mars was swinging inexorably toward its favorable opposition of October 13, 1894, less than a year hence.

Lowell was ready for the red planet, with time to spare.

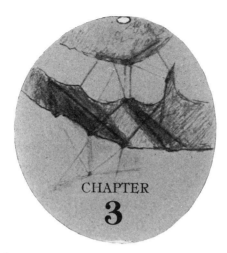

CHAPTER
3

The Observatory

FROM THE WINTER OF 1893–94, and for the remaining twenty-two years of Percival Lowell's life, the Lowell Observatory and its work were to be the focal point of his interest. Once his final book on the Orient, *Occult Japan,* was finished, he turned with fateful finality from the Far East to the far out—to make a Lowellian pun. Henceforth astronomy would dominate his thoughts, yielding to other interests only briefly and then usually only when those interests were somehow related to his observatory or his astronomical purposes.

In the years of more favorable martian oppositions, Lowell was to spend as many as eight months at a time at his observatory at Flagstaff, carrying out a rigorous, self-imposed, observational routine. Mars was the principal object of this work, for the continuation of Schiaparelli's investigation of the planet was his first reason for establishing the observatory. But the rest of the solar system also came in for a goodly share of attention. To Lowell, the problems of the one were part of and inseparable from the problems of the other.

In addition to these observations, there was also continuing experimentation to improve telescopic, spectrographic and photographic techniques, and some of this pioneering work was quite productive. A part of Lowell's work day at the observatory was, of course, spent in writing about the results of these various researches for both public and private consumption.

His relaxation during these sojourns in Flagstaff was equally vigorous. He was, for one thing, an enthusiastic gardener, a hobby he pursued in a properly scientific manner by keeping daily records of the progress of his flowers and vegetables. In particular, the pumpkins and squashes grown on Mars Hill were something of a source of wonder both locally and in Boston, where a portion of each year's crop usually wound up.[1]

He was also a voracious reader, devouring everything from detective stories and romantic novels to technical scientific treatises. In 1903, for example, along with a full set of James Fenimore Cooper's works, he ordered books on subjects ranging from after-dinner speaking through gardening to astronomy; in 1907, his reading included the *The Maulevener Murders* and Savante's *Theories of Chemistry;* and in 1909, his choices ran from *Mr. Justice Raffles* to Sir Ernest Shackleton's *Antarctic Voyage,* and *Conquest of the Air* by his meteorologist cousin, A. Lawrence Rotch.[2]

There were frequent picnics, camping trips, and scientific fact-finding expeditions into the wildly beautiful mountain-canyon-plateau country of northern Arizona. On these excursions, which were comfortably outfitted and handsomely provisioned, he explored roadless wildernesses of forest and desert, botanized,[3] and gathered geologic and geographic data for subsequent use in his theories on Mars and on planetary evolution, or "planetology," as he liked to call it.[4] A favorite companion on these trips was the convivial Edward M. Doe—attorney, judge, fellow Republican and his closest confidant in Flagstaff.[5]

In addition, and increasingly as his fame and that of his observatory spread, Lowell served as a gracious host to those who were professionally interested in his work there, as well as those who were simply curious. Not alone his friends, who sometimes stayed for weeks or even months,[6] but strangers were cordially received and, if at all convenient and the weather amenable, were given a glimpse of Mars or some other celestial object through the telescope. In 1901, for instance, paleobotanist-sociologist Lester Frank Ward, while studying the area's Triassic era fossil flora, visited Mars Hill and spent an hour observing the planet and sketching what he saw. Ward, who was already becoming famous as one of the pioneers of American sociology, also became an avid supporter of Lowell's martian ideas.[7]

Even those who vehemently disagreed with Lowell's thinking astronomically or otherwise were warmly welcomed. Among the leading astronomers who visited the observatory at one time or another was his arch-critic W. W. Campbell of California's Lick Observatory.[8] In 1907 newspaper magnate William Randolph

The Baronial Mansion, viewed from the east.

Hearst, then one of the nation's most prominent Democrats, was an appreciative guest with Henry Fountain Ashurst, a Democratic territorial legislator who was soon to become one of Arizona's first U.S. senators. "When it is remembered that your view of politics and economics is the reverse of Mr. Hearst's," Ashurst wrote in thanking Lowell for his courtesies, "your hospitality becomes magnanimous."[9]

Only in the years between martian oppositions, and from 1897 through 1900 and again in 1912–13 when nervous exhaustion severely sapped his mental and physical vigor, did Lowell fail to take up at least a brief residence at what came to be known as "The Baronial Mansion" on Mars Hill.[10] Yet whether in his State Street office in Boston or on his periodic visits to Europe, the observatory, its work, and even the most minute nonastronomical details of its life remained uppermost in his mind.

"By the bye," he wrote in 1904 to his then chief assistant in Flagstaff, Vesto Melvin Slipher, "I trust you have seen to the cow's calving regularly since I went away? As I have heard nothing about it I think it well to ask. If not, see to her at once, please, *and let me know*."[11] And again, only two days later, he wrote:

By the bye, I received a capital picture of Harry [Hussey] and his young wife on the piazza and apropos of that may I just hint to you that dining room chairs are only to be used on the piazza for photographic purposes. I have no doubt that it is all right, but I thought it just as well to call your attention to it.[12]

The Baronial Mansion, viewed from the west, with the San Francisco Peaks in the background.

"Venus," the observatory's cow, and "Satellite."

This, it must be noted here, is not evidence of parsimony or picayunishness, but of Lowell's personal fastidiousness, his strong sense of propriety, and his pride in his possessions. For in concluding this same letter, he added:

Is everything else in running order at the observatory. For instance: Have you a servant, etc., etc.? Is Mr. Lampland happily provided for; also Harry and his wife? I want everything to be kept up and if there is anything you need I want it got as a matter of course.[13]

What was to become the Lowell Observatory had its beginnings in the winter of 1893–94; from the fragmentary record available, it appears to have gotten off to something of a false start. At first, Lowell apparently proposed the establishment of an Arizona observatory as a joint venture with Harvard College Observatory. Early in February 1894 the *Boston Herald* announced that the Harvard observatory was planning an expedition to Arizona to observe the opposition of Mars scheduled in October of that year. The expedition, the paper reported, was to be directed by William H. Pickering, the Harvard astronomer who discovered the martian "lakes" from Peru in 1892, and to be financed by Lowell who, it added, was "going along."[14]

But within two weeks, Edward C. Pickering, the director of the Harvard observatory and William's brother, wrote Lowell that "it is desirable that a short statement should be made by the [Harvard] observatory to show that its connection with the Arizona expedition has ceased." Pickering suggested a correction in the press stating that "the proposed expedition, to be undertaken jointly by Mr. Percival Lowell and Harvard College Observatory, has been abandoned by mutual consent."[15]

There is, in the Lowell Observatory Archives, a fragment of an undated, unsigned, penciled draft of such a statement in which the words, "no such plan as was there described was even contemplated," have been crossed out. A final documented word on this episode came some ten months later when E. C. Pickering, after reading a newspaper account of Lowell's Arizona activities, wrote him asking for a summary of his observations for inclusion in the Harvard observatory's reports. Lowell replied:

As "observations in Arizona" are the work of this observatory and have, as the paper itself states, no relation to Harvard College, I fail to see their place in the report. The expedition to which the paper refers was not planned exactly as stated and was proposed by me to Harvard College. This for certain reasons was subsequently given up. With it, furthermore, my present observatory has nothing to do.[16]

Whatever these "certain reasons" may have been, it would appear they were not serious or lasting ones, for Harvard wound up loaning Lowell a 12-inch refracting telescope for his expedition and cooperating in a number of other ways. Lowell's assistants in his project, too, were both Harvard astronomers—W. H. Pickering and Pickering's young colleague in 1892 in Peru, Andrew E. Douglass. Although Lowell would repeatedly disavow any connection between his observatory and Harvard, it would in fact be many years before the idea of a formal association between the two would finally vanish from newspaper reports and begin to fade from the popular mind.[17] It was perhaps the first of many subsequent frustrations in his relationships with the press, and one of the minor ones at that.

Although his initial proposal for a joint venture had fallen through, Lowell, drawing on the younger Pickering's knowledge and experience, moved swiftly to finalize plans for an expedition of his own. By the first week of March, Douglass was aboard a westbound train, carrying with him the six-inch telescope Lowell had used in Japan, and taking brief notes on the geology of the countryside to relieve the tedium of the journey.[18] His assignment was to make observational tests of the atmospheric conditions at various points in the southern part of the Arizona Territory, using a "scale of seeing" from zero to ten devised by Pickering in Peru,[19] and to report his results back to Lowell in Boston.

Lowell and Pickering, in the meantime, were busy making arrangements to borrow not only the Harvard 12-inch but a fine 18-inch refractor that had just been completed by instrument-maker John A. Brashear of Pittsburgh's Allegheny Observatory. Lowell contracted with the renowned telescope-making firm of Alvan Clark and Sons, which has produced his six-inch glass as well as some of the world's finest refractors, to provide a dual mounting for the two instruments. Work was also begun on a temporary wooden dome to house them. The dome was to be shipped west in sections and erected on a foundation to be built under Douglass' supervision at whatever site was finally selected for the observatory.[20] Lowell also borrowed a spectroscope from Brashear, and various other pieces of observational equipment from Harvard and other local sources.[21]

Douglass made his first atmospheric test in Arizona at Tombstone on March 8, 1894, as the Tombstone *Prospector* duly reported two days later, and found the "seeing" to be "seven and six" on a scale of ten.[22] During the next two weeks, he tested Tombstone several more times, and then moved to Tucson where he ran into stormy weather, and to Tempe and Phoenix, finding the seeing at all three places no better than seven. Turning northward, he stopped at Prescott and Ash Fork, where the best seeing was six, and then went on to Flagstaff at 7,000 feet on the Coconino Plateau and a few miles south of the 12,600-foot San Francisco Peaks, the highest point in Arizona. There, on April 5 and 6, the seeing ranged up to eight at "site eleven," an "opening in the woods" on a low mesa just to the west of the town.[23]

Flagstaff was one of Lowell's additions to Douglass' itinerary which originally had been restricted to southern Arizona points. When, in mid-March, Douglass had suggested Prescott as a likely place for better results, Lowell had replied on March 23: "After Prescott try Flagstaff locating in forest. Then Santa Fe. Height important." On April 6, he telegraphed Douglass to stay in Flagstaff and test other possible sites in the vicinity, suggesting the "summit north of spring station." The next day he asked Douglass to "try six miles northwest," and on April 9 he advised: "Try highest available spot in neighborhood." Finally, on April 10, he telegraphed: "Site probably Flagstaff. Prospect for best seeing. Report climate compared with Tucson."[24] And in a letter that same day, he wrote:

I have just sent you a telegram that Flagstaff would probably be the best site for the observatory, but the final decision will depend on this week's observations. Other things being equal, the higher we can get the better.[25]

During the next few days, Douglass again tested the atmospheric conditions at various sites in and around Flagstaff. He also reported, both to Lowell and the local newspaper, that a comet was visible in the southeastern, predawn sky "as large as a star of the third magnitude."[26] There is apparently no further reference to this particular comet, if indeed that is what it was. But some two weeks later, on April 26, Douglass became the first observer in the northern hemisphere to sight Gale's Comet, discovered by an Australian astronomer April 2 in southern skies. Gale's Comet, Lowell noted in congratulating Douglass on the observation, was not sighted from Europe until April 28.[27] Apparently it was not observed again in the United States until May 5 when it was picked up by the 26-inch Clark refractor at the U.S. Naval Observatory in Washington, D.C.[28] A few nights later, Lowell himself was able to get "a good view" of the comet through his two-and-a-quarter-inch telescope from Boston.[29]

Lowell came to a final decision on the observatory's location on April 16:

Your telegram just received decides me; Flagstaff it is, as I shall now telegraph you. If No. 11 seems to you the best available station push on the work there as fast as possible. If however you find after observations a better station nearby take that. What does the town propose to do in the matter of a road or other conveniences?[30]

The town, a ranching and lumbering community of some 800 persons on the Atlantic and Pacific Railroad's main line to California,[31] was prepared to make a generous offer in the event it won the observatory site competition. But it had held back from making the offer immediately in order that Douglass would be "entirely free from any influences that would effect [sic] the choice ... or cause the selection to be made for any reason other than a scientific one." On the same day, however, that Lowell's decision was announced, Douglass received a pledge signed by more than eighty of the community's leading citizens to provide land "to the extent of ten or fifteen acres" and to build and maintain "a good wagon road" from the railroad depot to the site.[32] On April 21 Lowell telegraphed Douglass selecting "site eleven"—the present location of Lowell Observatory—as the best of the Flagstaff sites tested.[33] Two days later Douglass wired back: "Ground broken. Town gives land and builds road."[34] Lowell immediately wrote Flagstaff's Board of Trade thanking the community for its courtesies and cooperation.[35]

By May 3 the new road up the east side of the low mesa west of the town soon to be known as "Mars Hill" was far enough along that it was "passable for a light conveyance"; local workmen also had finished the foundation for the dome. The dome itself, along with the telescopes and other apparatus, had also arrived in Flagstaff, having been originally shipped to Tucson in anticipation that the eventual site would be somewhere in the southern Arizona area.[36]

Douglass, at Lowell's direction, kept a photographic record of progress on the observatory and continued to send regular reports back to Boston both by letter and by wire. Along with the observational facility, a residence of four rooms was rising atop the mesa.[37] Lowell himself, with Pickering, arrived in Flagstaff on May 28 during a brief period of cloudy weather which delayed the start of martian observations with the 12-inch until May 31, and with the 18-inch Brashear refractor until June 1.[38]

Douglass' expedition to survey atmospheric conditions in Arizona was followed with intense interest by the territorial press and, in fact, set off a flurry of frontier community boomerism of a

Tube of 18-inch refractor atop Mars Hill, May 1894, awaiting installation in the new Lowell Observatory. Note that the telescope originally was consigned to Tucson.

type that can still be discerned in some southwestern cities and towns today. Arizona then and since has marketed its climate as one of its major resources, and town boosters of the mid-1890s were no less enthusiastic, although perhaps less sophisticated, in proclaiming its virtues than various and sundry modern chambers of commerce. Certainly, these pioneer promoters were quick to realize the publicity and prestige that might accrue to a community from the presence of a major scientific facility. They had a point, as the Lowell Observatory soon proved, and as Lowell himself acknowledged some years later after the peak of the Mars furor had passed:

In a direct sense the observatory is thus the child of the land whose name it has since spread over the face of the Earth. For you will find Flagstaff is now known wherever Man reads his daily paper and even in corners out of the stress and press where he does not. [39]

The citizens of Flagstaff were first appraised of Douglass' expedition to Arizona and its purpose while he was still making his first atmospheric tests in the southern part of the territory. On March 15, the *Coconino Sun* in Flagstaff took note of the Tombstone *Prospector's* initial report of these observations by editorializing with breezy pride:

The clearness of the atmosphere of Arizona is being tested by Prof. A. E. Douglass of Harvard University. His first test is at Tombstone. ... If Prof. Douglass will come to Coconino County he can obtain a perfect atmosphere during the entire year, and he will find also a perfect all the year round climate. [40]

Such blurbs, some mildly and some blatantly boastful, appeared throughout that spring in most of the newspapers of the territory. The prose in praise of the meteorological blessings of this or that community was often richly profuse. Because most of these papers kept up with Douglass' Arizona activities and travels through the weekly exchanges—got their information, in short, from each other's pages—inaccuracies and exaggerations were frequently repeated, or even compounded. [41] References to Harvard in connection with Douglass' site survey appeared repeatedly and seem to have particularly annoyed Lowell. "Simply call it the Lowell Observatory," he wired Douglass in mid-March soon after the first press reports from Arizona began to filter back to Boston. And a few days later he fired off another telegram:

Lowell Observatory has nothing to do with Harvard you having left the latter for the former. Prevent misconception. [42]

Douglass tried his best to straighten out the editors, but with indifferent success. [43]

The reports of Douglass' one-man expedition in the press resulted in a steady stream of telegrams and letters from officials, businessmen, and other boosters throughout the territory, which were usually addressed to Douglass rather than Lowell. In these epistles, with varying degrees of insistence and urgency, Douglass was invited to visit particular communities whose advantages, atmospheric and otherwise, were described in glowing terms. Most of the writers suggested some prominent local landmark, usually the area's loftiest topographical feature, as a potential site for the observatory.

As early as February 18, one G. W. Hull of the Arizona copper-mining town of Jerome wrote to push for the Black Range of the Mingus Mountains as a suitable location for astronomical observations.[44] Only two days later, Tempe banker W. L. Van Horne advocated Tempe Butte, adding that "it may not be amiss to say that the Territorial Normal School is located here."[45] C. G. Bell, the district attorney of New Mexico's Grant County, urged Bear Mountain near Silver City;[46] and W. A. Langhorne, a Southern Pacific railroader at Red Rock, Arizona Territory, proposed Picacho Peak between Tucson and Casa Grande and, incidentally, close to the SP's tracks.[47] Mayors F. D. Sugalls of Yuma[48] and Anton Proto of Nogales[49] simply offered "clear days, cloudless skies and generally clear atmosphere."

The poor weather that Douglass had encountered during his Tucson visit brought apologies, including one from the U.S. Weather Bureau's observer there, and pleas for his return to make new observations under more favorable conditions.[50] Even after Lowell's selection of Flagstaff was made known, Douglass received entreaties to consider other communities as possible observatory sites, including an all-expenses-paid offer from a group of citizens in Williams, Flagstaff's nearest neighbor some thirty miles to the west.[51]

Despite this vigorous intercommunity competition, Lowell's final decision was generally well received throughout the territory, with most of the losers conceding defeat gracefully. "The site is well chosen and will be of immense benefit to Arizona," the Tombstone *Prospector* commented on the Flagstaff choice, while the Tucson *Enterprise* declared that "the Skylight city is to be congratulated upon the acquisition, and it is to be hoped that the cause of science will be greatly advanced by the new venture."[52]

Most editors found consolation in the fact that the new observatory would at least be located in the territory. Williams, one of the late entrants in the site selection sweepstakes, even drew a lesson in boomerism for its citizens from the decision:

Again an example has been set for Williams by the enterprising people of Flagstaff. It requires more than mere talk in order to secure public institutions. Energy supported by a small amount of capital will accomplish wonders, and the people of Flagstaff have demonstrated that they are wide awake to the importance of concerted, harmonious action in all things pertaining to the upbuilding of their town.[53]

Only in Tucson was there apparently some negative reaction in the press, and this was quickly answered both humorously and indignantly by northern Arizona newspapers. "How the Tucson papers are chewing the dobies over northern Arizona's unexcelled weather," one Prescott editor chortled in print, "but let those sectional brigands roar just the same."

The Tucson papers make us feel sleepy when they mention the "poor weather" which prevailed when Prof. Douglas [sic] visited them with his telescope. The professor is going there again in June, and we would advise him to take his observations through a refrijerator [sic] if he doesn't want to look into the hereafter.[54]

The "refrijerator" reference here, of course, is to the 100-degree-plus summer daytime temperatures that prevail in southern Arizona desert areas.

The Tucson *Star* saw the selection of Flagstaff as something of an eastern capitalistic plot; noted darkly that "the observatory will be a Boston institution and the Atlantic and Pacific Railroad is controlled by Boston capitalists."

This may have had something to do with the selection of Flagstaff, for it is but natural that, other things being equal, the amount of travel to which the establishment of the university [sic] will give rise, would be directed toward any line of railroad so closely associated with Boston interests.[55]

To which the *Coconino Sun* appended a pious editorial note that "this discourteous and narrow-minded comment... shows a feeling which was hardly to be expected from the editor of that paper, who has been but a few months in Arizona."[56]

The proximity of a major transcontinental rail line such as the Atlantic and Pacific or the Southern Pacific was perhaps a minor consideration for Lowell in choosing his observatory site. Altitude was a far more important factor and "site eleven," some 250 feet higher than 7,000-foot-high Flagstaff, was two thousand feet above any site elsewhere in Arizona that Douglass had tested.

But over and above this, atmospheric quality was Lowell's overriding concern. He was convinced that in just such a place as Flagstaff, on the high forested Coconino Plateau where the air should be not only clear but unusually steady for extensive periods during the year, a whole new instrument could be brought to bear on astronomical observations—the air itself.

CHAPTER

4

The Atmosphere

THE FOUNDING OF THE LOWELL OBSERVATORY in Arizona in the spring of 1894 is in itself enough to give Percival Lowell some place in the history of astronomy. But it is in his reasons for locating his observatory where he did that he made his first contributions to astronomical research. As with so many of the ideas with which Lowell's name has been associated, these reasons were not original with him, nor did he ever claim them as his own. Yet he was among the first to fully recognize the importance of overall atmospheric quality to efficient telescopic observation and to act on this recognition. He was also among the first to study the complex phenomena involved and to attempt to systematize the problem for the benefit of astronomy in general. Certainly he was the most vigorous early advocate of the proposition that "good seeing," astronomically speaking, is not simply a matter of a cloudless sky:

A steady atmosphere is essential to the study of planetary detail; size of instrument being a very secondary matter. A large instrument in poor air will not begin to show what a smaller one in good air will. When this is recognized, as it eventually will be, it will become the fashion to put up observatories where they can see rather than be seen.[1]

This prediction, made in 1895, appears in one of the more frequently quoted passages in all of Lowell's writings; it has, as Otto Struve pointed out in 1962, "certainly come true."[2] The recognition that Lowell so confidently prophesied was, perhaps, slow in coming; but nevertheless for many years now the quality of "seeing" has been, as a matter of course, a dominant factor in the selection of sites for major new observatories. The relative excellence of astronomical seeing in Arizona, first established by the pioneering work of Lowell and his observatory, has made the state the center of one of the world's largest concentrations of astronomical research facilities.* The search for "good air," which Lowell so persistently urged upon the astronomers of his own day, has also been pursued at an accelerating pace, focusing primarily on the arid desert areas along the western coast of South America where, early in the twentieth century, Lowell prospected successfully.[3]

It is important to understand Lowell's thinking on the subject of atmosphere vis-à-vis astronomical observation, or more simply "seeing," for it bears an intimate relationship to the controversies that raged around his work on Mars and the other planets.

To begin with, Lowell's choice of Arizona as the general location for his proposed observatory was not a chance one. Given the short time available for the establishment of such a facility and an already astute grasp of the kind of atmospheric conditions required for the best planetary work, the arid American Southwest was his only practical alternative. His brother recalls:

Percival was convinced that the most favorable atmospheric situations would lie in one of the two great desert bands that encircle a great part of the earth, north and south of the equator, caused by the sucking up of moisture by the trade winds; and that a mountain, with the currents of air running up and down it, did not offer so steady an atmosphere as a high table-land. The height is important because the amount of atmosphere through which the light travels is much less than at sea level. He was aware that the best position of this kind might well be found in some foreign country; but again there was no time to search for it, or indeed to build an observatory far away, if it must be equipped by the early summer.[4]

Surprising as it may seem, the idea of a steady rather than simply a superficially clear atmosphere was quite new at the time. Lowell had come to it second hand and had immediately realized its import from his own experience observing with small telescopes, from his knowledge of the problems encountered by Schiaparelli and others in their early planetary studies, and particularly from his familiarity with the 1892 Peruvian observations of Pickering and Douglass, both of whom had joined him in 1894 in his Flagstaff

---

*It should be noted, however, that the center has shifted to the southern part of the state with the establishment of Kitt Peak National Observatory near Tucson in the 1960s.

venture. Always willing to give credit where he felt credit was due, Lowell often acknowledged these debts in discussing the subject:

Intrinsically important as was Pickering's work [in Peru], it was even more important extrinsically. Schiaparelli's discoveries were due solely to the genius of the man,—his insight, not his eyesight, for at the telescope eyes differ surprisingly little, brains much; Pickering's brought into cooperation a practically new instrument, the air itself. For at the same time with his specific advance came a general one,—the realization of the supreme importance of atmosphere in astronomical research. To the Harvard Observatory is due the first really far-reaching move in this direction, and to Professor W. H. Pickering of that observatory the first fruits of carrying it out.[5]

Nonetheless it was Lowell who gave this essentially simple idea its first full formulation, and who directed the systematic accumulation of evidence in its behalf. And it was he who subsequently promoted it far and wide, not only at scientific meetings and in astronomical journals but in his popular books, magazine articles, and public lectures. Taking what Douglass had learned about atmospheric conditions in Peru, in Arizona in 1894, and in Mexico in 1896, Lowell expanded and elaborated the data into a full-blown theory that he later used with great effect in defense of his embattled theories on Mars.

Like many other ideas that he accepted and exploited in the astronomical field, the realization that there was more to atmospheric quality than an obvious transparency of the air struck his mind with something of the force of revelation. To Lowell, this hitherto unsuspected circumstance represented nothing less than "a new departure in astronomy."

... A new departure due to the detection of a factor in telescopic research which proves to be as important as it has been unknown to wit: the part played in telescopy by the air itself. That the vast importance of the air from an astronomical point of view is quite unexpected is conclusively shown by the incredulity with which the results which study of it has accomplished have been received. Nor is it odd that they should have been met with such skepticism. Nothing is so sure to escape recognition at second hand as what has itself escaped recognition at first hand, the existence of a new quality in an old thing.

So virgin a field indeed is the matter at hand that I find myself obliged at the very outset to dispel an illusion. From Newton on down, the far-seeing have all preached the abstract importance of good air, good air being to them synonymous with a clear sky. That one must have a clear sky to observe in ... is a truism. But in the first place it is not always a truth, some of the best definition being got through a fog. ... What I refer to is not an inability to see which is itself patent but one which has eluded the most practiced observers: not to obstacles which shut out a clear sky but to conditions in the sky itself which render it telescopically fit or unfit. In short it is the intrinsic quality of the air itself that I am dealing with. [6]

With this introductory statement, written in 1897 after his martian theory and some of his observations of Venus had already provoked strong controversy, Lowell went on to outline in detail the role of the atmosphere in astronomy, drawing much of his data from what he considered to be the key papers on the subject published in *Popular Astronomy* by Pickering in May 1896 and by Douglass in June 1897: [7]

The matter proved eminently deserving of the pains bestowed upon it. Not only did the practical results surpass the writer's anticipation, to say nothing of distancing contemporary belief, but through the ability of Mr. Douglass there was disclosed the theoretic side of the subject. From an attempt one evening at Arequipa [Peru] to discover the reason of a nightly visitation of poor seeing came the detection of a phenomenon which, followed up first at Flagstaff and then in Mexico has led him to the foundations of the true theory of the subject: what causes good or bad air telescopically; how it may be tested; and how avoided or applied. [8]

It perhaps will be well to follow Lowell's exposition of "the theoretic side" of the matter here at some length, and to a large degree in his own words, for he expressed these ideas in much the same form on numerous occasions throughout the rest of his life. It also will illustrate, better than any abstracted description could, his wide-ranging style and his typical weaving of various ideas and theories together, massing them, as it were, to repel attacks on the citadel of his thought:

Between what lies outside the Earth and man's knowledge of it interpose three things: the air, the instrument and the human eye. Through these three media must pass the waves of light that give us all our extra-terrestrial information. Now while much skill has been spent upon the perfecting of instruments and some thought upon the training cf observers almost no heed, until recently, has been given to the third factor in the case, the air itself; and in view of the cheerful way in which the location of the largest recent glass [the Yerkes 40-inch] has disregarded this fact, it will surprise most people to learn that the factor which has thus been overlooked proves to be the most vital one of all: that compared with atmosphere the instrument used in an astronomical observation, so far as all but illumination is concerned, is a very secondary consideration.[9]

The reason for "this scientifically criminal neglect," he declared with characteristic gusto, "is to be attributed to the averseness or the inability to go away from home."

Most observatories are dependent upon universities and are consequently located near the latter, while private founders have followed the same bent. Both, for not knowing anything better, believe their air to be good enough. How hard it is to break away from this facile prejudice has just been shown by the location of the new Yerkes Observatory which has just buried its glass in Wisconsin and from which Dr. [E. E.] Barnard writes me that he has been unable to make out the markings on Venus discovered ... last autumn in Flagstaff and seen there and in Mexico through three glasses and by every member of the staff, six persons in all, to say nothing of several outsiders. The same innocence was shown by a critic at the Lick [Observatory] who with more polemic zeal than optical knowledge actually imputed impossible effects to the Flagstaff glass in his anxiety to explain his own inability to see the new discoveries, ignoring what he had himself stated some years ago, that the day-air at the Lick is wholly impossible.[10]*

Parenthetically, Lowell's pointed remarks in these passages regarding the location of the Yerkes Observatory seem to have been well-founded. At any rate within a few years, Yerkes director George Ellery Hale, the eminent solar physicist, found the atmospheric conditions in Wisconsin inadequate for his observational purposes and moved his base of operations, if not the 40-inch refractor, to southern California's Wilson's Peak where in 1904 he established the Mt. Wilson Solar Observatory, later to become a part of the astronomical research complex known as the Hale Observatories.[11]

Having fired this verbal volley at those who vehemently questioned his observations, Lowell turned to a somewhat more technical discussion of "seeing," starting from the puzzling phenomenon of "shadow-bands," parallel waves of shadows of varying amplitude

---

*Astronomer Edward Emerson Barnard, who had discovered Jupiter's fifth satellite in 1892 at Lick Observatory and had joined the Yerkes staff in 1896, remained Lowell's friend for many years, maintaining a cautious neutrality in Lowell's astronomical wars. Lick director Edward S. Holden was the "critic" who claimed the Venus markings were a defect in Lowell's 24-inch lens; see Chapter 8 herein. The final reference is the fact that Lowell had adopted Schiaparelli's practice of observing the inner planets, Venus and Mercury, in daylight.

that are seen traveling with great speed over the ground just prior to totality during a solar eclipse:

Hitherto these have baffled explanation. Now this phenomenon Mr. Douglass believes to be parallel in explanation as it is certainly parallel in appearance to the one he discovered in Peru and has since been investigating at Flagstaff and in Mexico. Upon searching for the cause of the periodic ruination of the seeing upon the advent of certain cold air he was minded to take out the eye-piece while still keeping the telescope pointed on a bright object and to place his eye at the focus. Whereupon he found the field of view, the illuminated object glass itself, traversed by sets of shadow-bands travelling rapidly across it. Having once detected these shadow-bands he was not long in finding out that they bore a most intimate connection to the quality of seeing at the moment. While the seeing was good certain sets alone were visible whereas as soon as the seeing went off he found on taking out the eye-piece that certain others had usurped their place. Now to make a long story short these shadow-bands are the visible effects of condensation and rarefaction of the air-current which is passing at the time. Some are small; some larger; some swift, some slow; some regular like long billows; some as choppy as a cross sea. I need not point out ... how the refraction due to these waves distorts the image. [12]

These shadow-bands or waves, Lowell noted, were found to vary in length from crest to crest from half an inch up to two feet, and to range in refractive power from half a second up to twenty-five seconds of arc, depending on their type, the direction of the currents producing them, their speed, and their height above the surface of the earth. From Douglass' studies, he added, several conclusions could be drawn:

One of these is that the length of the waves determines the aperture [of the telescope] which at the moment will show the best definition and therefore the most planetary detail. It will be evident ... that if the wave-length be greater than the aperture of the objective we get bodily motion of the image with little blurring of detail; if it be less we get steadiness of image with confusion of detail. If the aperture is big enough we may have what apparently is excellent definition and yet not be able to see any markings at all. The size of the glass to be used is thus a direct function of the wave-length. The revolution in our ideas about telescopic vision consequent upon this is patent at a glance. [13]

A second discovery incident to Douglass' atmospheric studies and one of particular pertinence to planetary work, he declared, was that the definition of limb* and of detail "do not by any means go together."

---

*The "limb" is the circumference of a planet, satellite, or other object visible as a disk.

In other words we may seem to have excellent views if we judge by the limb, without having really any view at all. As astronomers are in the habit of judging by the aspect of the limb in the case of a planet, the subversion of the ordinary estimate is evident.[14]

One of the first results of this investigation into "seeing," Lowell noted, was the development by Pickering in Peru of the standardized "scale of seeing" which Douglass had used in his survey of potential observatory sites in Arizona early in 1894. This scale was based on an interference phenomenon in which a spurious disk surrounded by a number of concentric rings is formed when light from a bright point is passed through a refractive lens. On Pickering's scale, the disk and rings appeared as a confused, violently moving, mass at zero, and perfectly defined and motionless at ten.[15]

Such a scale, Lowell contended, made it possible for the first time for astronomers to evaluate the seeing at widely-separated locations on a comparative basis:

The sanction to any statement about "seeing" consists in some absolute test to which its relativity may be put. Hitherto the only test has been the *ipse dixit* of the observer graded if you will by what he sees but based on what he thinks he ought to see. One observer, for instance, will mark the seeing perfect while he is unable to detect detail which at another station is visible in air tabulated as mediocre and the world decides which of the two is right by abstract reference to the size of their object glasses, a standard only preferable to that of the diameters of their respective domes.[16]

Pickering's scale, Lowell conceded, "is not the last word on the subject for its principle of precision is not exactly that for planetary detail but it has the important quality of being absolute and so of being applicable as a comparative test by everyone."[17]

In this same discussion, Lowell touched briefly on his conviction of the importance of locating observatories in arid regions:

It is evident that to secure definition the air currents must be at a minimum and what there are of them as steady as possible. Now water in the air is the great unsettler. As dry a place as may be is therefore the first *desideratum*, i.e., a desert. But as diurnal charge [sic] in deserts is great we must have a shelter from the charge or in other words an oasis in a desert. These conditions are fulfilled by the great pine oasis of Flagstaff in the midst of the Arizona desert. From the above we see incidentally why mountain tops will not do on the one hand nor anything situate above 40° of latitude on the other. For the roaring forties are quite as disastrous to telescopy on land as they are to shipping at sea.[18]

In 1897, Lowell was stricken with what his biographer calls "neurasthenia,"[19] and which he confided to Douglass was "what I never supposed would overtake me, a complete breaking down of the machine."[20]

It would be four years before he returned to full vigor and could resume his astronomical work. But once fully recovered, he again took up his one-man campaign to get astronomers to see the importance of "seeing."

In 1902, a non-opposition year for Mars, he seems to have devoted much of his energy to the subject, preparing papers and proposals for astronomers and astronomical groups in the United States, and making his campaign a major reason for an extended trip to Europe in June of that year. He sailed, he wrote at the time, to achieve two objects:

To get adopted by astronomers generally a standard scale of "seeing;"

To get the cooperation of astronomers in a scheme for sending observers armed with 6-inch glasses and printed directions to examine for telescopic purposes various parts of the world that we may know in the future where observatories had best be put.[21]

On his return, he asked the eminent American astronomer Simon Newcomb to present these proposals to the Astronomical and Astrophysical Society of America, noting that the second[22] "is the suggestion which I hope to see carried out of fitting out an expedition for locating the best places for observatories. We are much handicapped now by the atmosphere and shall be more so in the future and it behooves us to take the first step for preparing for it."[23]

Lowell failed to persuade the astronomers of his own day to accept these specific proposals, yet the broad ideas on which they were based, if not Lowell's early advocacy of them, have long since been recognized as having fundamental importance in observational astronomy. Extensive research has shown the subject to be far more complicated than Lowell had even suspected, yet apparently his basic premises still stand in the 1970s.[24]

From 1897 on, Lowell was to repeat his arguments many times to explain why others could not see all the detail that he could see on Mars, Venus, and other solar system objects. He would, by 1903, add another dimension to his exposition of the subject of seeing. Eyes, he would insist, do differ.

Lowell himself apparently had exceptionally keen eyesight,[25] which he declared to be "acute" rather than "sensitive." It was this

Lowell observing Venus by daylight with the 24-inch telescope.

quality of "acute" vision that he believed enabled him to see planetary details that other observers, possessing a more "sensitive" eye, failed to see. He cited this distinction, for instance, in 1903 in a letter to one of his former associates at Flagstaff:

An eye may be either *sensitive* or *acute*. The two qualities do not go together. The first is the quality that enables faint stars or satellites to be seen. In this, [E. E.] Barnard is well nigh unrivalled. For planetary detail, however, it is an acute eye that is needed. ... Most of the adverse criticism to the surprising fineness of the canals [of Mars] comes evidently from a failure in the observers to recognize these two distinct qualities of vision in themselves or in others.[26]

Two years later he enlarged on this point in what is perhaps the most comprehensive summary of his ideas on astronomical observation:

By a sensitive eye is designated one which is peculiarly susceptible to the action of light. This eye has the quality to perceive faint luminous effects or distinguish delicate differences of illumination. ... This kind of eye is, therefore, of primary value in star detection. It is the eye which makes the visual satellite discoverer or the finder of comets and nebulae. Just as it is paralleled in its capabilities by the large telescope so it is peculiarly at home with a big instrument. The greater the illumination the better it is pleased.
     On the other hand what is called an acute eye is one peculiarly perceptive to form. Definition plays with it the like part that illumination does with the other. The trait is what is commonly known as keenness of sight. ... This, then, is the eye necessary to planetary work. What to the sensitive eye appears as a confused and blurred mass is by the acute eye resolved into its constituent parts. That the detail thus deciphered should to the possessor of the sensitive eye seem illusory is a matter of most natural wonder for not only is he incapable of perceiving it himself but has been led to credit himself with an eye superior to most people, and therefore to be believed against others' assertions. [27]

In this extensive exposition, Lowell also declared that planetary studies, as a distinct and specialized area within astronomy, required distinct and specialized techniques involving "improvements and adaptations of old and neglected conditions resulting in refinements of procedure of which the conservative can hardly bring himself to believe that he does not know all there is to be known." He then proceeded to describe some of these relatively radical departures from established astronomical custom and practice:

Large telescopes are not always to be preferred for planetary work. On the contrary, small ones are not only sometimes but nine times out of ten more efficient, more powerful in planetary visual research, strange as the fact may seem, than large ones.
     So much light is grasped by any glass over 12 inches aperture that workers on planetary detail have to employ appliances to get rid of it ... the most effective perhaps is the interposition of a piece of colored glass between the eye-piece and the eye. Schiaparelli made use of a yellow screen of the sort; but the writer has found a neutral-tinted one the best in his case. [28]

This problem of excess illumination, coupled with that of turbulent air currents, demanded a second modification, Lowell pointed out: "The adapting of the telescope to the atmospheric air-waves of the moment by means of diaphragms, thus converting a large glass to a smaller one at will."

This observatory has had diaphragms constructed of 18, 12, 9 and 6-inch diameters respectively which are put off and on as the seeing changes. The importance of this device to counteract nature can not well be overestimated. Detail which would remain hopelessly hid with the full aperture of the 24-inch, starts forth to sight by the mere capping down and consequent curtailing of the means which permits its being seen at all.[29]

There were other tricks and techniques for the planetary observer, like the one pioneered by Schiaparelli:

Daylight is another agent which has been made to do duty in the search. To this natural diluter of illumination is directly due the inspiring knowledge acquired by Schiaparelli upon Mercury and Venus. Instead of viewing these planets by twilight ... he struck out a new departure by scanning them by day. In this position he was followed by the writer and to it, joined to selecting steady day-air for his search, are to be attributed the detections of the markings other than shades upon their surfaces and the consequent determination of their rotation periods.[30]

Nor was the time-honored habit of astronomers of observing the outer planets only at favorable oppositions as productive as was believed, as Lowell noted with specific reference to Mars:

Now it is a rather curious fact that the so-called favorable oppositions are the most unfavorable ones for the detection of the very detail critics are at labor to prove non-existent. For it is then that such detail chances to be very ill-placed for observation and to come the nearest to extinction as well.[31]

Finally, Lowell argued, the relative rarity of moments of excellent seeing, even in superior locations, demanded that the planetary astronomer be both highly experienced and rigidly systematic in his work. "Unless a man become a systematic worker in the field," he declared, "he has no means of realizing the deception that has been put upon him."

Many aspects of this value of persistence may be cited. In the first place the ability to correct the necessary data is thereby enormously increased. ... Familiarity with the phenomena to be seen makes detection of the phenomena easier.
Such is the case with the branch of astronomy of which I am speaking. Because its province is the planets and because every astronomer has at some time or other looked at them he is tempted to feel himself qualified to pass on what may or may not be seen, and we know with what difficulty we resist temptation. Hence arose the skepticism to Schiaparelli's work, and hence comes the incredulity

of today. . . . It is not only quite possible but highly probable that one scanning the planet [Venus] at most observatories casually hit upon a moment so unfavorable meteorologically that to an unpracticed eye nothing of the recently detected detail is to be seen. If after two or three such attempts the would-be observer be thereby convinced, as he usually is, that the object is in truth featurelesss he goes away with the feeling that the reported detail is illusory and forthwith begins to promulgate the idea.

Though the detail detected on Mars for example is wonderfully apparent to one who has acquired experience from prolonged study of it, it is only under suitable atmospheric conditions that its characteristics are unmistakably evident to the casual observer. Even the practiced observer has to have one fine view before he himself credits *all* he sees. For the non-expert more is needed. For should he chance upon one of these revealings he is quite likely to doubt it. It is only on iteration and re-iteration of the same thing that he grows to believe his own eyesight. [32]

Lowell never publicly expressed any reservations over his choice of Flagstaff as the site for his observatory, or about the quality of atmospheric conditions there for astronomical work. But both he and Douglass had some doubts privately within a year after the observatory was founded. Conditions for observations that first summer and fall of 1894 were, indeed, excellent, but by December, after both Lowell and Pickering had returned to Boston leaving Douglass alone as assistant in charge, the weather took a turn for the worse. Douglass' monotonous reports of overcast skies and storms stirred Lowell to brittle sarcasm:

You have had a long spell of cloudy weather apparently. You must hold the inhabitants up to their climate and not let them go off in the quality of their one and only product. [33]

And again, some three weeks later, he inquired sardonically: "What do the oldest living inhabitants prophesy the weather to be?"[34]

Douglass replied in kind:

The younger Sykes [Stanley] has devised a new scale of seeing which he thinks will better suit our present climate.
    10, he says, is when you can see the moon.
    5, is when you can still see the telescope.
    1, is when you can only feel the telescope but not see it.
    He thinks it advisable to always state which scale we are referring to when we put down the seeing. [35]

By March, in view of the continuing poor weather and the fact that the 1896 opposition of Mars would come in December, Lowell decided—temporarily as it turned out—to abandon the Flagstaff observatory site:

The seeing seems to be so perpetually poor now that I see little use in keeping up the observatory longer. I wish however that you would get title to the land as soon as possible. I have no desire to make anything out of the town, but I should like the deed of land if only to deed it back. . . .

My plan at present is to find another site for the next opposition and to ship the dome as it is to whatever that may be—ship it in pieces, of course. The house I should be willing to sell—the above is between ourselves so also is the fact that I have agreed with Alvan G. Clark for a 24-inch to be finished June 1st, 1896. As to the best time of closing the present observatory I am not particular.[36]

Within two weeks, Lowell wrote suggesting the Mexico City area as a likely observatory site,[37] and Douglass agreed. "Such places should be investigated in the winter, it seems," he replied. "Thus, I think the tableland of Mexico is perhaps the most promising place."[38]

A few days later, Douglass drafted for the Flagstaff newspapers an announcement of the closing of the observatory. In this draft, he declared that the observatory "for the present . . . is to be discontinued," and he cited the unfavorable weather from the end of November through March in which "there was not a single perfect night . . . and scarcely one or two which could be called good."

Nevertheless I see no reason for changing our original opinion that this is the best place in the territory for astronomical work, but when it is once understood that we are not seeking for clear skies so much as for skies in which the air has no wind . . . it will be immediately perceived that we are not dealing with a simple matter like weather. . . . I consider that the real reason for the poor winter to be that we are too far north and that for improvement we must go farther south, by 10° or perhaps more.

Mr. Lowell hints that he will continue observation of Mars in the winter of '96–97.[39]

Some years later, Douglass would revise this estimate of Flagstaff, for in 1901 he confided to Pickering:

By the way, by studies of the atmosphere with my mirror, I have come to the conclusion that surprises me not a little. It is that of all the places I tried in Arizona in '94, Flagstaff is probably next to worst. The trouble seems to be that these mountains [the San Francisco Peaks] project up into the great stream of air moving overhead and cause eddies of various kinds, so that in their vicinity I get much worse effects than say thirty miles from them.[40]

The gloomy weather of March, however, was broken by at least one bright spot—a lunar eclipse on the night of March 10 which Douglass invited some of the townspeople to watch from Mars Hill. Douglass described their reactions to Lowell:

The different explanations I have heard for the recent eclipse are marvelous. Several have thought it was caused by Jupiter; one by a star, and many have asked me to what planet it was due. The "Flagstaff Democrat" reported that it was an important event and would not again be witnessed until the year 2193 (!!).[41]

Douglass shut down the observatory as a functioning unit in April, formally ending observations on April 3,[42] and by mid-May he was in Mexico for a brief preliminary survey of potential observatory sites in the vicinity of Mexico City. Early in 1896, Lowell took the opportunity of a trip to Europe to visit Algiers in North Africa to check out the Sahara Desert for astronomical purposes. But in March of that year he cabled Douglass from Cannes on the French Riviera: "Better go Mexico now Sahara mediocre."[43]

Although Lowell tried Mexico in 1896–97, and then again returned to Flagstaff, the search for new and better observatory sites continued. In May 1900 he would test the Sahara again and would still find it wanting.[44] In 1903 Carl Otto Lampland, who became an aide at Flagstaff late in 1902, tested the air at Glorieta and at Santa Fe, New Mexico.[45] In 1907 Lowell sent an expedition to the Andean highlands of South America to observe Mars, noting that "great heights have never yet been well explored for seeing and it seems a good opportunity to do so."[46] In 1908, he even considered establishing a branch of his observatory at the Grand Canyon, eighty miles north of Flagstaff, and authorized Lampland to make atmospheric tests there with the six-inch.[47]

In the final analysis, however, Lowell considered Flagstaff superior for observational purposes to any other place in the world that he knew anything about, with the exception of the Andean deserts of South America which, in any case, were too remote, too inconvenient to allow the close communication and continuous supervision that he wanted to maintain. He summed up his conclusions about the atmospheric conditions at Flagstaff quite candidly in 1903:

Of course the "seeing" in Flagstaff is not perfect; at times I am tempted to call it not even good. But what I am finding more and more certain every day is: that it is better than elsewhere—where any large observatory is, bar possibly Arequipa. I have recently obtained indirect evidence that it is better than the Lick. This is shown by the fact that the Lick not infrequently has the stellar image a blur. We had this but once during this last winter, spring and summer. One thinks more of the "seeing" at Flagstaff after experiments in other places.

The "seeing" averages better than any place I know. ... It permits systematic observations extending over a long period and thus enabling the growth or the development of the canals [of Mars] to be studied. [48]

One final aspect of Lowell's selection of Flagstaff as the site for his observatory and the reasons that led to his choice is of some importance and can be mentioned here for it has a peculiarily modern ring to it. Again in 1903 and again privately, he noted:

Every time I come east I am struck with the necessity of being out west for astronomic purposes. Every year man with his machinery pollutes the sky more and more and renders an outlook into space more difficult. [49]

He expanded this idea considerably in his later public writings:

To get conditions proper for his work the explorer must forego the haunts of men and even those terrestrial spots found by them most habitable. ... As little air as may be and that only of the best if obligatory, and securing it makes him perforce a hermit from his kind. [50]

Lowell perhaps over-romanticized this requirement for isolation in his popular writings. For Flagstaff was not all that remote, being on a transcontinental rail line only hours away from the populous areas of the west coast and only a few days' ride from Boston. But he had a point in arguing against the location of astronomical observatories in or near centers of population, and one which has been brought forcefully to the attention of astronomers in later years, even in Flagstaff. He had, indeed, some up-to-date things to say about what later became the "environmental problem."

In addition to what nature has done in the matter, humanity has further differentiated ... by processes of its own contrivance. Not only is civilized man actively engaged in defacing such part of the earth's surface as he comes in contact with, he is equally busy blotting out the sky. In the latter uncommendable pursuit he has in the last quarter of a century made surprising progress. ...

With a success only too undesirable his habitat has gradually become canopied with a welkin of his own fashioning that has rendered it largely unfit for the more delicate kinds of astronomical work. Smoke from multiplying factories ... has joined with electric lighting to help put out the stars. These concomitants of an advancing civilization have succeeded above the dreams of the most earth-centered in shutting off sight of the beyond so that today few city-bred children have any conception of the glories of the heavens which made of the Chaldean shepherds astronomers in spite of themselves.[51]

Both Europe and North America, he noted grimly, "vie with each other as to which sky shall be most obliterate" and "America is not behind in this race for sky extinction. ... Not until we pass the Missouri do the stars shine out as they shone before the white man came."[52]

Subsequently Lowell cited a series of controversial "tests" made in 1905, for which he claimed confirmation by photography, to argue not only the greater credibility of observations at Flagstaff but to prove the deleterious effects of this man-made "welkin" on astronomical work in general. The tests, he contended, showed that while 172 stars were visible in a given field of the sky with the Lowell 24-inch refractor, only 161 could be seen with the Lick 36-inch on Mt. Hamilton near San Francisco, and only 61 with the U.S. Navy's 26-inch telescope at Washington, D.C. The seeing at Washington, he declared, "may be taken as a fair representation of observatories generally in the United States and Europe."[53]

The Lowell Observatory, Lowell said even later and more eloquently, was located where it was because "for its work the country must be fresh, not to spoil the heavens."[54]

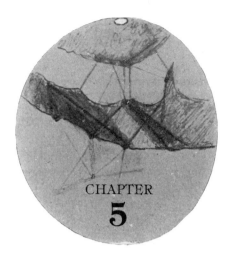

CHAPTER

5

Mars Again

T HE FUNDAMENTAL MOTIVATION underlying Percival Lowell's seemingly impulsive resolve early in 1894 to turn his energies, talents, and resources to planetary astronomy and specifically to the problems of Mars appears to have been an intellectual fascination with the possibility of the existence of extra-terrestrial life.

As a practical matter, this possibility loomed larger in the waning years of the nineteenth century than at any time before in history, as Lowell clearly understood. The broad revolutionary principles of Darwinian evolution, which had come to dominate the thought of the late Victorian era, appreciably enhanced the theoretical odds for life elsewhere in the universe through their logical implication that life, and indeed intelligence was an inevitable consequence of evolutionary processes. Lowell was convinced by the work of Schiaparelli and others, and particularly Schiaparelli's 1877 discovery of the so-called canals of Mars, that the problem was now susceptible to investigation by the rigorous methods of physical science and that in fact its solution might be found as close to hand, astronomically speaking, as the red planet itself.

At this time, however, Darwinian implications notwithstanding, the question of the existence of life or intelligence beyond the earth was widely considered to be also beyond the realm of science, and more properly a concern of metaphysics or theology. Few astronomers—France's ebullient Flammarion being a notable exception—cared to speculate publicly on the subject. Even Schiaparelli, when asked if intelligent beings might be inferred from the martian "canals," remained circumspect, saying only that "I should carefully refrain from combatting this supposition, which involves no impossibility."[1]

A reason for this reluctance, perhaps, can be found in the uproar that followed the 1859 publication of Darwin's *Origin of Species*. In view of the bitter theological opposition that Darwinism stirred, it is hardly surprising to find scientists in subsequent years shying away from public controversy, particularly when religious dogma was involved. It may also be noted that up through the late nineteenth century the most serious challenge to the idea of extraterrestrial life had come not from science but from religion and was embodied in a corollary to the doctrine of special creation, the doctrine to which Darwinian evolution directed its main thrust. This corollary held that God had created life only on earth and nowhere else in the universe. Man, so the logic ran, was the only reasoning creature in the whole sub-angelic scale of being and was thus uniquely favored among all of God's creations, and the center of His attention.

The argument, set in the matrix of mid-nineteenth century science, received a classic statement in the controversial book *Of the Plurality of Worlds,* first published in England in 1854 and written by the Reverend William Whewell, doctor of divinity, master of Trinity College at Cambridge, and the best known historian of science of his day.[2] The lively intellectual debate that Whewell's work touched off was soon inundated by the high tides of religious emotionalism set running by Darwin's book. But, as will be seen, the Whewellian thesis was destined for revival early in the twentieth century to repel the threat to man's divine cosmic uniqueness posed by Lowell's postulated martians.

That astronomers of the 1890s were at least aware, however vaguely, of the religious aspects of the question of extra-terrestrial life is perhaps evident in an article that appeared in a professional astronomical journal early in 1894 as Mars began to swing toward its favorable opposition due in October of that year. Its obscure author, after summarizing contemporary planetary lore, confidently concluded:

We cannot help admiring the inductive acumen of the theologians who considered the Earth the most important of the planets, and the center of creation. Although their opinions were not based on scientific facts, they arrived at the truth, nevertheless.[3]

Whatever other astronomers may have thought of this, the irrepressible Flammarion was not one to let such a statement pass unchallenged. His satirical response, in the form of "A letter from a citizen of Mars," undoubtedly provoked some private chuckles within the astronomical community when it was published six months later in the same journal. His parody concluded:

We cannot help admiring the inductive acumen of the theologians who considered our Mars to be the most important of the planets and the center of creation. Although their opinions were not based on scientific facts, they arrived at the truth, nevertheless. [4]

How unconventional Lowell's thinking was at this time is revealed in a paper he read before the Boston Scientific Society that spring while his new astronomical observatory was still a-building in Arizona. The paper subsequently was published in the Boston *Commonwealth* on May 26, 1894—two days before he arrived in Flagstaff to begin his observations. This paper, a wide-ranging, free-wheeling anticipation of the controversial theories he would soon announce regarding intelligent martians and planetary evolution, drew immediate criticism from no less an authority than Edward S. Holden, then president of the University of California and, as director of its Lick Observatory, one of the leading American astronomers of the day. Holden quoted Lowell at length in scoring, and indeed underscoring, his "overstatements" and "conjectures" as "misleading and unfortunate." [5]

Darwinian and Schiaparellian themes are clearly dominant and Whewellian conclusions conspiciously absent in this early outline of Lowell's astronomical ideas as he publicly explained why the main object of his new observatory was to be the study of the solar system: *

This can be put popularly as an investigation into the condition of life on other worlds, including last but not least, their habitability by beings like [or] *unlike man. This is not the chimerical search some may suppose. On the contrary, there is strong reason to believe we are on the eve of pretty definite discovery in the matter.* In the first place, analogy warrants us in conceiving this little ball on which we dwell in the sea of space no more the sole vehicle of intelligent life than it was once thought to be a pivot on which the whole cosmic system turned. Just as it is now known to be but one of many bodies revolving around the sun, so, doubtless is it but one of many worlds evolving in due course the phenomena of intelligent life. *If the nebular hypothesis is correct, and* there is good reason at present for believing in its general truth, *then*

---

*The italicized statements in the quotations that follow are those to which Holden particularly objected.

to develop life more or less distinctly resembling our own must be the destiny of every member of the solar family which is not prevented by purely physical conditions, size and so forth, from doing so. [6]

This much, Lowell reasoned somewhat speciously, could be deduced from Laplace's nebular hypothesis, published in 1796, and his celestial mechanics. But since Laplace's time "we have got collateral evidence of another kind, from the astrophysical contributions of modern astronomy."

The first bit of this *new* evidence may be said to have been the observation of the great red spot on *Jupiter*. Investigation into this led to the discovery that the giant globe still glows by its own inherent heat. Photometric observations of *Saturn* suggested a like state of things there. ... In the inner planets we see a very different state of things; a state more nearly that which our own Earth has reached. While in the moon we gaze upon the last sad age of decreptitude, a world almost *sans* air, *sans* sea, *sans* life, *sans* everything. [7]

A second body of new evidence, he contended, had resulted from "Schiaparelli's great discovery of the so-called canals of Mars."

Scouted at first by less penetrating observers, the existence of the canals has since been amply confirmed. Speculation has been singularly fruitful as to what these markings on our next nearest neighbor in space may mean. Each astronomer holds a different pet theory on the subject, and pooh-poohs those of all others. Nevertheless, *the most self-evident explanation from the markings themselves is probably the true one; namely, that in them we are looking upon the result of the work of some sort of intelligent beings.* In short, just as the great red spot of *Jupiter* implies that the planets may become habitable, the amazing blue network on *Mars* hints that one planet besides our own *is actually inhabited now.* What further heightens interest in the matter is that *Mars* has, from other considerations, undoubtedly reached a much later stage in planetary development than our Earth has yet attained, and that in gazing on him we are in a sense peering into futurity. [8]

"Mars," he declared, "is the most likely object for results to such planetary investigation," but:

There is evidence that something in the same general evolutionary line may be learnt from our other relatives in space—*Venus*, for example. From certain observations it would appear that, belying her name, she is the most modest of all the orbs, keeping herself constantly cloaked in clouds—a state of things corresponding to our carboniferous age. ... *The increasing apertures of modern objectives,* taken in connection with what has been recently done and with the revelations of the spectroscope, show that we stand upon the threshhold of a knowledge of our closest of kin in the world of space of the most important character.[9]

It was with such provocative ideas already firmly in mind that Lowell in June 1894 launched the most prolonged, systematic study of Mars ever before undertaken, an investigation which has been pursued at Flagstaff at every subsequent opposition and is still, in the early 1970s, a major part of the work of the observatory he founded there.[10]

Lowell's plan of attack was eminently pragmatic. He intended, in effect, to lay telescopic siege to the planet by making methodical observations on a virtually continuous basis over extended periods of time, not only through a single opposition but over a succession of oppositions. He was convinced that only through such a rigorous observational routine, faithfully and persistently carried out, could knowledge of Mars and perhaps the other planets as well be significantly advanced. Mars was a dynamic planet, and for Lowell this was the key that would unlock its secrets:

Underlying all the particular phenomena presented to us by the planet's disk is the fundamental one of change. The appearance of the planet does not remain the same day after day, or even hour after hour. ... Fundamental to the production of the phenomena, the fact of change is no less fundamental to an understanding of what these phenomena mean.[11]

That Mars displays broad changes in its telescopic appearances, of course, had long been recognized by astronomers in a general way. Lowell, however, planned to study these changes with the same careful attention to detail that the biologist gave to the observation of the life cycle of a protozoan under his microscope. It was not merely the "fact of change" but its progress and process that concerned him:

Were the object observed an inert mass the advantage of systematic observation would be chiefly that of quality. We should have more data upon which to draw and the increase of data, if it do nothing else, tends to minimize error. But Mars is not a dead body; on the contrary its features are in a continual state of change. To the detection of laws governing the change systematic, continuous study is absolutely vital; just as occasional observation is of absolutely no account. Indeed that latter is often worse than none at all since by missing of a link the cycle of change may be wholly misconstrued.[12]

Success, he noted, depended upon persistence, but "certain as persistence is to be rewarded, the difficulty inherent in the observations is ordinarily great."

Not everybody can see these delicate features at first sight, even when pointed out to them; and to perceive their more minute details takes a trained as well as an acute eye, observing under the best conditions.[13]

"Experience," he pointed out at another time, "makes expert, and perception eventually stands secure where it tiptoed at the start."[14] This is a well-turned truism, perhaps, but one which Lowell frequently turned to his purpose in charging those who criticized his martian work with inexperience in planetary observation. "The detail detected on Mars for example, is wonderfully apparent to one who has acquired experience from prolonged study of it," he insisted. [15] And again to the skeptics who questioned the very reality of the martian canal network, let alone his interpretation of it, he declared that the canals "stand out at times with startling abruptness" to those few observers "practiced in their detection," adding with heavy sarcasm:

I say this after having had twelve years' experience in the subject— almost entitling one to an opinion equal to that of critics who have had none at all.[16]

The attitude, Lowell once remarked to a newspaper reporter, was analogous to "that of claiming that because a man may see stars without looking at the sky, ergo, those in the sky do not exist, or that because a small boy could not distinguish Oolong and Souchong tea by taste, it is not yet safe to assume that the professional tea taster can so distinguish."[17]

Nor was it simply a matter of being a professional astronomer, he pointed out. For the study of planets was a highly specialized field, differing sharply in its requirements from other branches of astronomy, "and a man whose occupation is celestial mechanics is not on that account an authority on planetary observation."[18]

Lowell's systemization of his martian studies involved primarily the keeping of exhaustive graphic records of all observations in the form of detailed drawings of the planet by each observer during prolonged sessions at the eyepiece of the telescope. Specific areas and particular features were sketched over and over again, night after night, whenever seeing conditions and the planet's aspect permitted. Because these drawings were the product of the same mind, hand and eye, and were made at the same telescope at similar times and under similar atmospheric conditions, Lowell considered that all the major observational variables were controlled and that a given sequence of such drawings were internally consistent and scientifically comparable:

If Mars is to be many, his draftsman must be one. . . . Only the dates of the drawings differ. As therefore the same personality enters into all of them, it stands as between them eliminated from all; thus allowing any change which may have taken place on the planet's surface during the interval to disclose itself.[19]

Thus the drawings constituted "a kinematical as opposed to a statistical study" of the planet.[20]

Lowell repeatedly stressed the necessity of having a given series of observations made by the same person, and he took his own turns at the telescope for the long grueling routine this demanded. When, during the unfavorable opposition of 1898–99 he was forced by illness to leave all martian observations to Douglass, he prefaced Douglass' report of them with this warning:

The change in observer must be constantly borne in mind if one would interpret them aright inasmuch as the personal equation enters as distractingly into observations of planets as into all others which man may make.[21]

Over the years, Lowell and a succession of assistants at Flagstaff turned out thousands of such drawings and sketches, some of them quite artistic in their way. These became the basis of the intricate areography which he constructed and described in his books and astronomical publications and displayed on the numerous globes and maps of Mars he produced. He also frequently found them useful as enclosures in correspondence with astronomers and others on both sides of the great furor over Mars to illustrate key points in his arguments on behalf of his controversial theory.

Through that first summer and fall in Flagstaff, the weather cooperated fully to Lowell's purpose so that Mars was observed "almost every night . . . June through mid-December."[22] Lowell himself was not at his new observatory all this time, however, observing only through the month of June, again briefly late in August, and then extensively through October and November, returning to Boston in the intervals. During his absences observa-

Page from Lowell's logbook, recording his first observations at Flagstaff. Note that he observed both Mars and Saturn.

tions were carried out by his assistants, Pickering and Douglass, both experienced planetary observers.

The statistics alone of that first season's work at Flagstaff, while they do not tell the full story, nevertheless are impressive. In all, 917 drawings of the planet's disk were made. Pickering and Douglass drew fifty-seven sketches of the south polar cap—the one presented to the earth at the 1894 opposition. A total of 464 micrometric measures of the planet's equatorial and polar diameters were taken, all but sixty-one of them by Douglass. Lowell made 124 determinations of the latitude and longitude of various martian surface markings. Some 736 "irregularities" on the planet's terminator were recorded, again mostly by Douglass. Pickering also made thirteen observations with a polariscope in an attempt to determine whether the light reflected from specific areas on the planet's surface was polarized in the particular way it should be if these areas were actually bodies of open water.[23]

From these polariscope observations, Pickering concluded that "the permanent water area upon Mars, if it exists at all, is extremely limited in its dimensions."[24] But his polariscope did seem to indicate some temporary water surfaces in rifts that developed in the south summer cap, for example,[25] and transiently in an area just south of the prominent triangular marking, Syrtis Major, long believed to be a martian "sea."[26] He also found some indication of water in the polarization of light reflected from a dark blue band that Lowell discovered encircling the shrinking perimeter of the southern cap. Such a band had been noticed around the northern cap as early as 1830 by the pioneer areographers Beer and von Maedler, but Lowell was the first to claim it as a south cap feature.[27]

Douglass, from his measures of martian diameters, calculated the oblateness of the planet at about 1/190,* a figure which, as Lowell later noted, was derived independently within a year by astronomer Herman Struve in Europe from measurements of the orbital motion of Mars' two satellites.[28]

These two tiny moons, incidentally, also came in for their share of telescopic attention at Flagstaff. In late September and in October, Douglass made thirty-one observations of the relative brightness of Deimos and Phobos; from these data Lowell subsequently

---

*This is the ratio of the difference between the polar and equatorial radius of a planet to its equatorial radius and thus the index to the degree of flattening at the pole. More recent data from satellite observations, confirmed by the 1971 Mariner 9 mission, indicate that the oblateness of Mars is 0.0057 or about 1/175, compared to an oblateness of about 1/296 for the earth.

calculated their diameters at ten and thirty-six miles respectively.[29] These estimates, as it now turns out, were quite close to the mark, particularly for the smaller moon Deimos. From the historic photographs of the satellites taken during the 1971 Mariner 9 mission to Mars, the major dimensions of these irregular crater-pocked bits of rock have been found to be about 13.5 by 12 kilometers (8.4 by 7.4 miles) for Deimos and about 25 by 21 kilometers (15.5 by 13 miles) for Phobos.[30]

Douglass, on twenty-three nights during this same period, searched systematically for undiscovered martian moons and, in a negative result that is still valid in 1974, found none within a radius of 180 seconds of arc from the center of the planet.[31]

In general, the sustained observations over more than five months tended to confirm the suggestions advanced by a number of astronomers over the years that the long-observed changes in the color and configuration of the surface markings on the planet were seasonal in character.[32]

From the outset, the much-disputed "canals" were seen in relative profusion, at least by Lowell and Douglass. Pickering, apparently not possessing the required "acute" eye, never seems to have seen these features as the complex geometrical network of fine lines that Schiaparelli, Lowell, and some others described.[33]

During the 1894 opposition, a total of 183 canals were seen at Flagstaff, some only once but others as many as 127 times. Of these, 116 were not on Schiaparelli's maps, the most detailed of the planet up until that time. More importantly, sixty-seven of them were among the seventy-nine charted by the sharp-eyed Italian.[34] The fact, as Lowell was to point out later, "shows that the canals are for the most part stable detail as the larger markings."[35] Over the years, the number of canals reported by Lowell would rise to over 700.[36]

The "double" canals that Schiaparelli had first reported in 1879 were also an object of prime interest for Lowell in 1894; during October and November as and after the planet came to opposition, he made them his special concern.[37] Initially, he turned up only eight instances of this phenomenon which Schiaparelli had called "gemination," but by the time the favorable opposition of Mars in 1907 approached, the number had grown to fifty-one.[38] Interestingly enough, even Douglass, although he had no difficulty seeing single canals, was dubious about the "doubles," and eventually expressed his doubts in the observatory's *Annals:*

Pickering has given a strong argument against the objective reality of duplication based on the existence of a definite relation between the width of the double canal and the aperture of the telescope. Mr. Lowell writes that he is inclined to agree with this result. M. [E. M.] Antoniadi and others before him have attributed the doubling of a canal to bad focussing of the object on the retina of the eye.

My opinion is, that they are right who attribute duplication of the canals chiefly to a subjective effect; I am inclined to think that the cause is not poor focussing, but a misinterpretation of crowded canals. I have drawn double canals a number of times, but nearly always have expressed in the notes at the same time some reservation about them. Sometimes this reservation is a statement that I would not have suspected the duplication unless told of it. Another time it was that I would have suspected the duplication even if it had been unheard of. Again, and only once, I said "Positively see this double, but not all the time."[39]

The reference to Lowell in this passage is indeed curious, for later he would have much to say in defense of the objective reality of the double canals and in derogation of those who considered the phenomenon to be in any way optical or psychical in origin.

Among the canals observed in 1894, forty-four were seen by Douglass within some of the larger dark markings on the planet's surface. These seemed to confirm and clarify observations of streak-like features which he and Pickering, as well as observers at the Lick Observatory in California, had glimpsed in the so-called martian "seas" in 1892 from Arequipa, Peru, and which Pickering had considered to be possible "river systems" in line with long-held terrestrial analogies. In Flagstaff, Douglass searched the more prominent bluish-green markings for these streaks. Early in July he made some "suggestive" sketches, and finally, on August 14, he drew a chart of the Syrtis Major region "showing within it a clearly defined system of canals."[40]

The small, roundish, dark spots that Pickering had first sighted in the bright martian "deserts" in 1892 and had called "lakes" also turned up again in 1894 at Flagstaff. A total of fifty-three were seen and recorded, including a number in the larger dark areas.[41]

And finally, Lowell reported the discovery of yet another martian feature—a series of small, triangular dark spots along the inner edges of the "seas" where the canals of the bluish-green areas seemed to join with those of the brighter areas to form a single, continuous network. In all, thirty of these were seen, and because they looked like typographical check marks to Lowell, he called them "carets."[42]

There were also some more dramatic, if isolated, occurrences observed on the planet. On October 13 Douglass reported the disappearance of the south polar cap for the first time in the observational history of Mars[43]; this temporary vanishing was confirmed the next night by Lick astronomer E. E. Barnard. But while Lowell claimed the cap had melted to extinction, Barnard felt its disappearance was caused by some kind of atmospheric obscuration.[44]

On two occasions, Douglass observed unusually large "projections" that were "suggestive of cloud." The first of these was sighted on September 25–26, the second on November 25–26. The November "cloud," which appeared as a detached prominence on the planet's limb, he estimated to be about thirteen miles above the martian surface which compares to an average altitude of five or six miles for terrestrial cirrus clouds.[45] Douglass considered the November prominence in particular a valid observation and a significant one in view of the relative rarity of such "genuine" phenomena on Mars.[46] Lowell promptly released his report to the press in Boston, along with the news of Douglass' discovery of "canals" in the dark regions, and just as promptly, it would seem, regretted the action, fuming to Douglass:

If you want a thing done, do it yourself! I unfortunately did not, thinking I could entrust it and the result is this enclosed paragraph as announcement of your discoveries which is not what I intended. I asked to have your detached prominence and your canals in the dark regions noticed in the *Transcript* as facts not as gossip as to what anybody else can or cannot see. To take any notice of the thing now however, would be simply to mess it. So I have left it alone.[47]

Douglass was not only "eagle-eyed," as Pickering called him,[48] but apparently indefatigable as well. During the first year at Flagstaff he attempted to photograph Mars and to make spectroscopic observations of the planet.[49] While these early efforts were not at all productive, they presage the direction Lowell would give to his martian researches in the future.

In addition to their observations of Mars, his two assistants did some work on Jupiter and its satellites.[50] Douglass also spent some time observing such celestial phenomena as zodiacal light (the post-twilight or pre-dawn glow extending above the horizon along the ecliptic) and *gegenschein* (the counterglow sometimes visible in the night sky opposite the sun).[51] He also continued his investigation of the effects of atmospheric conditions on astronomical observations.[52] Some of his work, however, was of a more practical, mundane character. For instance, he found time to design a number of devices, including a "permanent bicycle lock," and a new kind of tennis racquet, inventions which apparently came to naught but for which both Lowell and Pickering attested to his originality.[53]

Given Douglass' many activities, it was with tongue in cheek perhaps that Lowell, noting the long spell of bad weather that winter in Flagstaff, wrote him that "you have my condolences. I will send you out some books to beguile your leisure."[54]

Yet it was not all work. Not surprisingly, there was apparently some convivial socializing with members of the opposite sex, for young Douglass had not yet married, and Lowell, although now 39 years old, was still a bachelor and somewhat the bon vivant and would remain so for another fourteen years. Soon after Lowell's return to Boston after only one month of observing, Douglass wrote him that "I am ernestly [sic] requested by the young ladies to state that in our parties and social gatherings your absence is deeply felt."[55] Lowell replied: "Pray convey my continued regrets to the fairer, the picnic half of our world,"[56] and later urged: "Keep the girls till the end of the month if prayers can avail."[57] Again, in noting Douglass' terminator work, he declared: "My compliments on your 'irregularities' of the terminator. Do you connect them in any way with the (wild) oats which alone we observed planted by man in the neighborhood of Flagstaff?"[58]

What "(wild) oats" were sowed at Flagstaff in 1894, Lowell Observatory's Archives do not reveal.* What they do reveal, however, is that from that first season of observation at Flagstaff, Lowell reaped a highly provocative theory that would stir worldwide controversy and bring him worldwide fame.

---

*References in the Archives to Lowell's social life, as well as his attitudes toward and relationships with women, are rare and almost always impersonal. It would seem that the Archives can shed little light on these aspects of Lowell's life which, in any case, are incidental to this study.

CHAPTER
6

The Theory

PERCIVAL LOWELL evolved his formal theory of the probable
existence of intelligent life on Mars within two months after
observations were begun at his new Flagstaff observatory and after
he himself had observed the planet for barely a month through the
18-inch telescope there.

He based the theory in part on his extensive knowledge of the
observational history of Mars, and particularly on his familiarity
with the work of Schiaparelli, the discoverer of the martian "canals,"
whom he respected highly both as an astronomer and as a friend.
Also, he drew selectively both factual and theoretical material from
such other scientific disciplines as biology, geography, geology, and
paleontology, in which he was well read, if not always up to date.

But the major foundation of his theory was the observational
data, both new and confirmatory, that he, Pickering, and Douglass
gathered early in the summer of 1894 as Mars' disk gradually grew
in telescopic size with the approach of that year's favorable opposition.

Lowell had already arrived at some positive and quite sensational conclusions about life on Mars in particular and extraterrestrial life in general before he ever looked through a telescope
from Flagstaff. But on his own testimony at least, he did not for-
malize his thinking into what he considered to be a full-blown
scientific theory until late in July of 1894 when he returned briefly
to Boston after observing the planet during June in Arizona. Early in
August, Douglass discovered that the network of canals and as-

sociated features in the bright, reddish-ochre regions of Mars extended into the dark, bluish-green areas as well. This discovery, Lowell later noted, came just after his theory "had been thought out."[1]

The network of canals and oases in the dark regions comes in with as singular oppositeness as it comes in apropos. For other phenomena had brought me to the framing of the general theory before, and yet only just before, this system turned up to support it.[2]

For the remaining twenty-two years of his life, Lowell would direct his own energies and the work of his observatory to the accumulation of an incredibly detailed body of data to bolster his martian theory as, increasingly, it provoked worldwide controversy. It is ironic, surely, that this sustained and systematic fact-gathering effort, although eventually of some significance to astronomy, proved to be insufficient for Lowell's purpose.

It again illustrates the consistency of Lowell's thought that once his "general theory" was framed, neither subsequent observations nor mounting criticism during the ensuing Mars furor ever caused him to abandon or alter any of its essential points. He did, of course, embellish it as new supportive evidence and arguments became available. But within weeks of his death in November 1916 he was still postulating intelligent life on Mars in much the same logical form, and indeed in virtually the same language, as he had used in his early expositions of his thesis in 1894–95.[3]

Lowell's theory was and is magnificently simple. It is the amount of observational data and his highly effective use of material culled eclectically from other fields of knowledge that made it appear quite complex and, to the lay mind, gave it scientific authority.

Basically, Lowell dealt with only two questions: (1) Are the physical conditions that prevail on Mars such that life in some form could exist there? (2) Is there any evidence on the planet now that life exists or has existed there sometime in the past?[4] To both of these questions of course, Lowell's answer was an emphatic affirmative. But he went further. Not only was the planet habitable and inhabited but, he insisted, the peculiar nature of the evidence dictated the conclusion that martian life must possess a high degree of intelligence.

Although in his formal public expositions of this theory he went into exhaustive detail and reasoned therefrom at great length, he could on other occasions express his ideas simply and concisely. Early in 1895, while the fading planet was still under sporadic observation at Flagstaff, he summed up his theory briefly in a chatty letter to Douglass:

Roughly speaking the evidence seems to be that Mars has (1) some but not much atmosphere; (2) is an aged world with no water to speak of except what makes the polar caps; (3) is provided with an elaborate system of line markings which are best explained by artificial construction, cases of assisted nature . . .; (4) shows what seem to be artificially produced oases as the termini of the canals—what we see and call canals being merely strips of vegetation watered by the canals, the canals themselves being too narrow to be seen.[5]

On the artificiality of the canal network, Lowell added parenthetically that "no natural explanation yet advanced will, in another sense, hold water. This system is what would occur on an old waterless planet if inhabited. . . . *Voila,* quite unconvincing till you go into details when the argument proves very strong."[6]

During the summer of 1894, the world, astronomical and otherwise, was given only brief glimpses of what Lowell was up to at his new observatory through occasional notices in the press and in a short, largely descriptive, article in *Astronomy and Astrophysics* in August. In this, he referred to the canal network as having "the clear cut character of a steel engraving," and found its *"raison d'etre"* in the "vernal freshets" set flowing by the spring melting of the polar cap.[7] He first set forth his full martian theory in a detailed systematic way in a series of six articles in *Popular Astronomy,* then a relatively new journal which published reports by some of the leading astronomers of America. This series began in September before Mars reached opposition and ended in April the following year.[8] Douglass provided a seventh article about the observatory and its work in May 1895.[9] A more general public heard of his theory and its sensational conclusions through a similar four-article series starting in May 1895 and running through August in the *Atlantic Monthly.* There were also talks and lectures—most notably a series in Boston's Huntington Hall in February 1895 which drew capacity audiences—as well as a swelling number of reports and comments in the press.

But his earliest complete statement of his martian theory came in the first of the three popular books he wrote about the planet. Entitled simply *Mars,* and published late in 1895, this work added new and explosive fuel to the smoldering fires of the Mars furor that Schiaparelli had kindled innocently some eighteen years before and that others since—particularly France's Flammarion—had from time to time fanned into flame.

For astronomers and others in science, Lowell followed this with a more conservative version with tables, drawings, and maps, for the first volume of his observatory's *Annals*. Publication of this, however, was delayed by the nervous exhaustion that overtook him in the spring of 1897 and forced the virtual suspension of his astronomical activities over the next four years. The volume, assembled and edited largely by Douglass, finally appeared in 1898 in a limited edition distributed principally to astronomers. It added little to the wonder and excitement, and the skepticism and sarcasm, that *Mars* and the earlier expositions of his theory had already stirred.

In all of his major popular writings on astronomical subjects, Lowell at the outset sought to provoke the interest and pique the imagination of his readers, using his flare for the dramatic with considerable skill. In his *Mars and Its Canals* (1906), for example, he set the stage for his revelations by colorfully romanticizing the adventurous spirit of the fifteenth and sixteenth century explorers of the New World, likening it to that of modern astronomers journeying telescopically to new worlds in space. Indeed, he dedicated the volume to Schiaparelli as "the Columbus of a new planetary world."[10] In *Mars as the Abode of Life* (1908), and a year later in *The Evolution of Worlds* in which Mars plays only a walk-on role to planetary evolution or "planetology," he began with highly effective descriptions of the birth of the solar system in the near collision of the sun and a "dark star" as a stunning prologue to his evolutionary theory of "the progress of a planet from sun to cinder."[11] In *Mars,* his opening gambit was the ancient enigma of extraterrestrial life:

Are they worlds, or are they mere masses of matter? Are physical forces alone at work there or has evolution begotten something more complex, something not unakin to what we know on Earth as life? It is in this that lies the peculiar interest of Mars.[12]

Lowell's answers to these questions, of course, are embodied in his controversial theory. His presentation is always methodical and would be quite pedantic were it not for his easy breezy style and his liberal use of pointed allusions and analogies. Faced with setting out a massive array of factual material, he avoided textual tedium by keeping his readers constantly alert for pungent, often irreverent, comments on life in general and science in particular, and for small jokes and large puns, usually made at the expense of his critics in astronomy.

He had at the outset one warning for his readers, and one he would reiterate consistently over the course of the Mars furor. This point was that in postulating life elsewhere in the universe and specifically on Mars, he was not postulating the form such life might take. He never thought martians were men. "Extraterrestrial life," he wrote in *Mars,* "does not necessarily mean extraterrestrial *human* life. Under changed conditions, life itself must take on other forms."[13] A few months after his book appeared, he made the same point in a newspaper interview:

When Mr. Lowell is asked about possible life and vegetation on Mars he becomes eloquent. All astronomers do not agree with his conclusions, he admits, but he holds to them tenaciously.
    "I have no doubt," he says, "that there is life and intelligence on Mars. No creatures resembling us are there. Local conditions, such as the thinness of the atmosphere ... forbid it, but there are creatures of intelligence."[14]

Such early statements of the invalidity of the terrestrial analogy in relation to martian life, however, failed to dampen exuberant speculation in the more sensation-oriented newspapers of the day,[15] and Lowell was forced to restate his position repeatedly. In 1907, for instance, he concluded a popular review of his ideas by declaring:

That life inhabits Mars now is the only rational deduction from the observations in our possession; the only one which is warranted by the facts.
    As to what the life inhabitant there may be like I should not pretend to say. As yet we have insufficient data to infer. For just as it is unscientific to deny observations because we fear the seemingly startling conclusions to which they commit us, so it is not the province of science to speculate where observation is wanting, however interesting and even useful such speculation may be.[16]

Yet the general belief that he considered martians earth-like beings persisted, in the public mind at least. Only days before his death, after receiving news clippings of his 1916 West Coast lecture tour, he wrote:

I notice one from Berkeley [Calif.] which makes me talk nonsense, to wit: that there are human beings on Mars, a thing of course I never stated. Our evidence is simply of intelligence there. How bodied we do not know.[17]

Lowell even refused to speculate on the nature of the vegetation that might exist on the planet. In acknowledging a publisher's copy of the book, *Journey to the Planet Mars* (1905), he noted that in reading it he was "especially interested in the character of Martian vegetation as its form somewhat transcends telescopic perception."[18]

Lowell founded his theory on the physical condition of Mars. He set the stage for canals and canal-building martians by first methodically setting forth the evidence for the basic habitability of the planet, much of it less in dispute now than in Lowell's time. He began with the general characteristics of the planet—its diameter (4,200 miles), its orbital eccentricity (0.09), its axial tilt which he took to be about 23.5 degrees or almost exactly that of the earth, the existence of long-observed surface markings including the polar caps, and the fact that these markings changed their telescopic appearance in apparent response to the seasons of the long martian year.[19]

Because he thought in evolutionary terms, he noted the implications of polar flattening which "hints at a past molten state" and "is impressed by cooling on planets."[20] His point was that at some time in its history, Mars was warmed by internal heat, as data from the 1971 Mariner 9 Mars mission suggest that it was.[21]

Yet he also thought that the surface of Mars must be quite flat and far less rugged in its topography than the earth, drawing his conclusion from the appearance of the planet's terminator which was not only relatively smooth but seemed to him to give its sphere somewhat the look of an irregular polygon or, as he put it, a pared apple.[22] Lowell would have been as surprised as most other astronomers were at the 1971 Mariner 9's finding that the martian topography is even more extreme than the earth's.[23]

But a more pertinent condition for life was the presence of an atmosphere. Here Lowell cited Douglass' observations of a twilight arc on the terminator which, he declared, was "due to air pure and simple," and of "limblight," a brightening around the perimeter of the planet's disk, which he attributed to "ice particles in the martian atmosphere."[24]

More generally, however, and quite obviously, the fact that "changes occur on Mars on a vast enough scale to be visible from earth" argued strongly for an atmosphere, for these changes seemed to be of such a nature that they could not take place without it. The waxing and waning of the polar caps was particularly germane here, for "change in the polar cap implies the presence of an atmosphere, and does so neatly, as in the absence of an atmosphere, change would progress but in one direction to its own destruction."[25]

The martian atmosphere, Lowell noted, was exceptionally clear. Yet despite the fact that "the first and foremost of its characteristics is cloudlessness,"[26] it contained some clouds, as shown by occasional sightings of "projections" on the terminator or limb and of bright patches in the reddish-ochre regions which he felt might be either "cloud," "mist," or "hoarfrost" but which in any case implied an atmosphere. Dew and frost probably were the most common form of precipitation on Mars in view of its general aridity, and the polar caps were "doubtless formed by successive precipitations of dew."[27]

Mars' atmosphere too, viewed from more than forty million miles away, seemed quite placid to Lowell, and he consequently assumed that it was stirred by "no very palpable wind."[28] Here it should be noted that the occurrence of vast dust storms on Mars had not yet been recognized, although such storms were observed later telescopically from earth. The 1971 Mariner 9 mission's success was threatened by such a storm which raged over most of the planet in the final months of 1971. Mariner 9 was not designed to yield specific data on martian wind velocities, but from its overall findings some scientists have estimated that winds on the planet at times may reach speeds of several hundred miles per hour.[29]

The martian atmosphere was extremely thin, Lowell was sure, and his attempts to determine the pressure it exerted on the planet's surface from light-scattering data were quite original for the time.[30] From observations of limblight, the twilight arc, and particularly the planet's albedo or reflectivity, he estimated the surface pressure initially at less than one-seventh that of the earth's atmosphere, adding that because of the smaller gravitational factor on Mars, the density of its atmosphere "must diminish less rapidly than our own" with increasing altitude.[31] A few years later, Lowell refined this, setting the surface pressure at 87 millibars, or only about 1/12 of earth's. More recent determinations of this important datum, from occultation experiments made by Mariner 9, indicate a surface pressure range from 2.9 to 8.3 millibars with a mean of 5.5, or about 1/180 that of the earth.[32]

The extreme thinness of the martian atmosphere was not a barrier to the existence of life, to Lowell's mind at least:

One deduction from this thin air we must be careful not to make—that because it is thin it is incapable of supporting intelligent life. ... That beings constituted physically as we are would find it a most uncomfortable habitat is pretty certain. But lungs are not wedded to logic, as public speeches show, and there is nothing in the world or beyond it to prevent, so far as we know, a being with gills, for example, from being a most superior person. ... To argue that life of an order as high as our own, or higher, is impossible because of less air to breathe than that to which we are locally accustomed, is, as Flammarion happily expresses it, to argue not as a philosopher, but as a fish. [33]

Nor was the problem of the composition of the martian "air" particularly difficult for Lowell. He argued the presence of "aqueous vapor" from the haze occasionally observed over the edges of the waning polar cap, and from the existence of the caps themselves. Observations of what were taken to be "clouds" were also consistent with the idea of water vapor in the atmosphere.

But further, in confutation of arguments by the English scientist G. Johnstone Stoney and others, he contended that the kinetic theory of gases did not bar the presence in the martian atmosphere of such life-essential molecules as oxygen, carbon dioxide, or water vapor. For the critical velocity of 3.1 miles per second which molecules of gas must reach to escape the gravitational pull of the planet, he declared, was above that of any gas except hydrogen. And, as water vapor could be shown to be present in the atmosphere from other phenomena, these other gases must also be present, for their molecular weights were greater than that of water. [34]

Thus, he said in summary, "we have proof positive that Mars has an atmosphere; we have reason to believe this atmosphere to be very thin . . . and in constitution not to differ greatly from our own." [35]

The composition of Mars' atmosphere, of course, does "differ greatly from our own" as astronomers of later generations began to discover and as the Mariner missions still later have demonstrated. Mariner 9 results indicate a martian atmosphere composed predominantly of carbon dioxide, with small amounts of carbon monoxide and water vapor, along with traces of oxygen and hydrogen in the upper levels as the photo-dissociated products of water vapor. Nitrogen, the predominant ingredient of earth's "air," is very rare if present at all.[36] But in Lowell's time, the assumption that the atmospheres of earth and Mars were of similar composition was not uncommon in astronomy in lieu of any direct observational data on the subject and in view of the then prevailing technological incapability of obtaining such data. Lowell, as will be seen, soon attacked the problem in an innovative but unsuccessful attempt to put what was at best a plausible analogy on a sounder scientific basis.

A second major physical requirement for the existence of life on a planet, Lowell not unreasonably assumed, was the presence of at least some water on its surface. The polar caps here again held the key, for "just as the fact of change in the polar caps proves the existence of air, so it implies the presence of water also."[37] The dark blue band encircling the waning south polar cap like a collar, observed from Flagstaff in 1894, was a major basis of his argument. This feature, he noted, was the darkest marking on the planet, and it varied in width, becoming widest "at the martian season when melting should have been at a maximum." It also followed the retreating perimeter of the cap poleward and thus, as "an associated detail" of the cap, "instantly suggested its character, namely, that it was water at the edge of the cap due to the melting of the polar snow. ... That the blue at the edge of the melting snow was water seems unquestionable."[38]

Lowell often cited Pickering's polarization work in 1894 as confirmation of this in his early expositions of his Mars theory.[39] Later, however, he expressed some reservations. In 1906 for instance, he advised science writer Waldemar Kaempffert, who was preparing an article on Mars for *McClure's* national magazine, that "Pickering's polariscope test, although possibly correct, is not of such assurance as you write."[40] Mariner 9's photographs, parenthetically, do indeed show these dark collars around the polar caps, but they unquestionably have nothing to do with water. Rather, astronomers now suggest that the bands are the result of the scouring action of violent winds generated by the sharp temperature gradients between the frozen caps and the warmer surrounding surface areas.[41]

Lowell's belief that the polar caps were frozen water was not shared by all astronomers of his day. Many favored the then newly advanced idea that the white stuff of the caps was more probably frozen carbon dioxide—dry ice. The assumption of polar snow and ice on Mars, in fact, was questioned in print in December 1894 even as Lowell was making it one of the bases of his martian theory.[42] W. W. Campbell of the Lick Observatory, one of Lowell's earliest and most persistent critics, was one of those "who talks seriously of the carbonic acid theory," Douglass early advised Lowell.[43]

The idea that the caps were frozen carbon dioxide was one that continued to plague Lowell's theory, and he strove mightily to combat it. In *Mars*, arguing from the blue band, the polariscope and even early experiments with carbon dioxide under varying temperatures and pressures by nineteenth-century physicist Michael Faraday, he concluded "water to be the most probable solution to the problem."[44] Assuming that the blue band was at least liquid, he made his point quite effectively in the *Annals:*

There is but one substance known to us which under the conditions could be white as a solid, blue as a liquid, and *coexistent in the two states.* That substance is water (The italics negative the whimsical suggestion of carbonic acid.) Unless then, we invoke some unknown substance for which we have absolutely no warrant in the physical premise, the band girdling the polar cap was a polar sea. [45]

A decade later, when English physicist John H. Poynting deduced a mean temperature for the planet of −27 degrees Fahrenheit,[46] Lowell privately chortled that "in this he falls between hay and grass as it is too cold for snow and not cold enough for carbonic acid—which rather hangs him up."[47]

Lowell's phrase "between hay and grass" ironically may turn out to be the answer, figuratively speaking, to this long-debated issue of carbon dioxide versus water as the stuff of the martian polar caps. The 1971 Mariner 9 results seem to indicate that the caps are mainly composed of frozen carbon dioxide. But because summer temperatures of the caps are above the freezing point of carbon dioxide and below that of water, the possibility that frozen water forms a part or all of the lingering residual cap cannot yet be excluded. [48] In addition atmospheric water vapor, clearly confirmed by Mariner 9's infrared spectroscope, appeared to be stronger over the waning south polar cap than over other areas of the planet, suggesting that the cap might be a major source of atmospheric H_2O. [49]

Lowell not unreasonably assumed that life, even extraterrestrial life, could exist only within a narrow temperature range, and frozen carbon dioxide implied temperatures well below the limit. If, however, the caps were water alternately precipitated as snow and melted to liquid as the caps seasonally turned toward the sun, then temperatures on the planet were not any serious obstacle to martian life.

The presence of water vapor, so important to his theory in so many ways, was important here too. For "aqueous vapor is quite specific as a planetary comforter" and probably accounted for "the pleasing amelioration of climate" on Mars. Martian weather seemed to him to be "astonishingly mild."[50] From the seemingly obvious analogy to the melting of terrestrial polar caps, and initially from little else, he concluded that Mars' "temperature is not incomparable with that of the Earth."[51] Later, applying data on solar insolation and reflectivity to methods developed by Samuel P. Langley and Frank W. Very at Allegheny Observatory, he arrived at a mean temperature for Mars of 48 degrees Fahrenheit, somewhat below the earth's, but still amenable to any life there.[52] In this work, the *Scientific American* noted, Lowell introduced "several novel considerations which have not received attention by previous workers on the same problem."[53] Here again, of course, the Mariner series of Mars missions have shown the planet to be far colder than Lowell believed. Although the Mariner 6 and 7 missions in 1969 indicated maximum temperatures of up to 75 degrees F at local noon, the 1971 Mariner 9 maximum was below zero on the Fahrenheit scale. Both the 1969 and 1971–72 results agree in showing great variation in temperatures over the planet and minimums at or near the freezing point of carbon dioxide.[54]

The dark, bluish-green markings, long discerned on the planet and just as long considered "seas," constituted another important physical condition that pointed to its ability to support life. That these were not seas, Lowell declared, was evident first from the "quite unmistakable" changes in color they displayed over the long martian year, and secondly from the discrete features—Pickering's 1892 "streaks" or "river systems" and Douglass' 1894 "canals"—that had been observed within them. Furthermore, Pickering's polariscope work in 1894 gave negative results for these areas. Nor did they return the specular reflection to be expected from a large body of water on a spherical surface.[55]

For Lowell, the most significant characteristic of the dark markings was their reaction to the turning of the martian seasons, a set of phenomena that had been observed and commented on by many earlier observers and which Lowell considered thoroughly confirmed by the extended observations at Flagstaff in 1894. The fact that the markings seemed to darken and deepen in color during the martian spring and fade to shades of gray and brown in autumn was not disputed seriously then or later.[56]

But Lowell discovered a peculiar circumstance about these seasonal changes that had escaped earlier, less systematic observers. There was indeed a "wave of darkening," but curiously this wave swept down over the planet from pole to equator and beyond rather than from the equatorial regions poleward, as spring returns to a terrestrial hemisphere. Thus, he said, all the seasonal phenomena on Mars depend primarily on the alternate melting and redisposition of the polar caps rather than on the relative warmth of the sun as on earth.[57]

Why such general changes? "Water," he declared, "suggests itself; for a vast transference of water from the pole to the equator might account for it." But, he quickly added, "there are facts which seem irreconcilable with the idea of water."[58]

Vegetation, he concluded, was the better answer. The growth of some form of plant life alone seemed to account for the freshening and darkening of the blue-green areas in the martian spring, and for him their gradual fading through late summer and fall reflected "the quiet turning of the leaf under autumnal frosts."[59]

The suggestion that vegetation might cause the long-observed changes was not new, and Lowell repeatedly credited it to a suggestion advanced by Pickering in 1892,[60] although it had been made much earlier—at least by E. L. Trouvelot about 1884 and by A. Liais as far back as 1860.[61] Nevertheless, the belief in martian "seas" was still widespread both in and out of astronomy. Even Schiaparelli, who had recognized the seasonal nature of the changes, apparently felt they were best explained by the annual drying up of shallow seas, a suggestion which Lowell, in rare disagreement with his *cher maître Martien*,[62] rejected because he could not find enough water on Mars to fill such seas even temporarily.[63]

While insisting on the aqueous nature of the polar caps and thus the presence of some water, Lowell stressed the extreme aridity of the planet, and far from considering this a negative factor for its habitability, he turned it to positive account, making it a premise for postulating its actual habitation. There was a scarcity of water, certainly, but:

If, therefore, the planet possesses inhabitants, there is but one course open to them in order to support life. Irrigation, and upon as vast a scale as possible, must be the all-engrossing Martian pursuit. . . . What the physical phenomena assert is this: if there be inhabitants, then irrigation must be the chief material concern of their lives.[64]

Thus he set the stage for his discussion of what he considered "the most startling discovery of modern times—the so-called canals of Mars."[65]

In introducing the canals, Lowell drew an analogy to the checkerboard patterns of terrestrial grainfields. Were we intelligent beings on Mars seeking telescopic evidence of sapient man on earth, he suggested, then "by his crops we should know him."[66] Thus, he said, the very regularity of the planet-wide network of lines—Schiaparelli's *canali*—was alone enough to testify to their irregularity as a work of nature:

The whole system is trigonometric to a degree. If Dame Nature be at the bottom of it, she shows on Mars a genius for civil engineering quite foreign to the disregard for prosaic economy with which she is content to work on our own work-a-day world. Her love for elementary mathematics is evidently greater than is commonly supposed, a private passion which on tenantless Mars she is able to indulge unhampered by fear of unseemly ridicule. [67]

The most conspicuous characteristic of the network, he reported, "is this hopeless lack of irregularity." The lines were "as fine as they are straight," and averaged "about 30 miles wide ... less than a Martian degree," and "generally speaking ... the lines are all of comparable width." Furthermore, the lines almost invariably described arcs of great circles on the planet's spherical surface, extending for hundreds and even thousands of miles in all directions across the martian terrain. [68]

In 1894 a total of one hundred eighty-three of the canals were discerned, described, and drawn. But, Lowell added:

What their number may be is quite beyond the possibility of count at present; for the better our own air, the more of them are visible. About four times as many as are down on Schiaparelli's chart of the same regions have been seen at Flagstaff. [69]

The reality of the canals of course had been strongly questioned, but now Lowell proclaimed them fully confirmed and, commenting on the prevailing skepticism in the *Annals,* declared:

The peculiarities reported of the canals sounded so amazing that the account was, not unnaturally, assumed to exceed the facts. ... But Schiaparelli was, as usual, right. ... The truth is that the facts exceed the account. The better the means of observation, the stranger the whole system appears. [70]

And again, in *Mars:*

Disbelief still makes a desperate stand against their peculiar appearance, dubbing accounts of their straightness and duplication as sensational, whatever they mean in such connection; for that they are both straight and double, as described, is certain,— a statement I make after having seen them, instead of before doing so, as is the case with the gifted objectors. ... Doubt, however, will not wholly cease until more people have seen them, which will not happen until the importance of atmosphere in the study of planetary detail is more generally appreciated than it is today. [71]

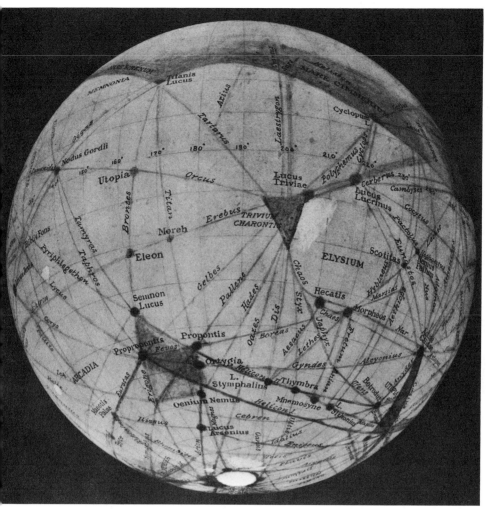

A globe of Mars, prepared by Lowell

But while Schiaparelli had thought the lines, despite their geometrical complexity, to be of geological origin, and others had also advanced natural explanations, Lowell flatly pronounced them to be artificial. There were three major reasons to "negate the idea of natural causation of the lines," he declared: (1) their straightness, (2) their uniform width, and (3) the discovery of their systematic radiation from special points—the small, round, dark spots that Pickering had considered "lakes" in 1892 but which Lowell preferred to call "oases" for the purposes of this theory. For these reasons, he contended, the lines could not be rivers which vary in width and deviate in course, nor could they be geological features such as cracks or canyons in the martian crust, because of the "great improbability" of their observed convergence at the oases. For this same reason, he added, the lines could not represent meteoritic scars on the planet. [72]

Lowell also answered charges by skeptics that the lines were optical illusions, by pointing out that they remained fixed in position regardless of observer and from observation to observation, and that the phenomenon of "gemination" was individual to the particular canal, the distance between the lines and the time that duplication occurred differing in each case. [73] To those who denied the canals on psychological grounds, he argued that they were "not due to a priori prejudice, as they are not what one would expect to see." [74] Schiaparelli, he was to point out a few years later, "started with no preconceived ideas on the subject."

On the contrary it is clear that he shared to begin with the prevailing hesitancy to accept anything out of the ordinary. Nor did he over-come his reluctance except as by degrees as he was compelled. ... In other words the geometrical character of the "canals" was forced upon him by the things themselves instead of being as his critics took for granted, feisted on them by him. [75]

"Not only do the lines look unnatural," Lowell concluded, "but there is no natural explanation, which has so far been advanced, which their very appearance does not negative." Thus, he found himself left with "a supposition which, however startling it may seem at first sight, is not more so than the appearance of the lines themselves—to wit, that they are not natural features at all." [76]

But while the canals appeared to be of artificial origin, they nevertheless participated in the general seasonal changes on Mars, the "wave of darkening" that alternately swept down the planet from its poles. The canals became visible progressively, darkening through the martian spring and summer, not strictly in relation to Mars' position in its orbit but in relation to their proximity to the pole and the bluish-green areas in the spring-summer hemi-sphere.[77] Here again the seemingly obvious explanation of a "direct transference of water" from the melting polar cap to the equatorial regions was not adequate, he declared. First, there was "an insuffi-ciency of water;" secondly there was "a superabundance of time," for months elapsed before a canal reached the point of maximum development, in terms of its visibility vis-à-vis the bright "deserts" and dull "seas" of Mars. The most probable answer, he concluded, was that a canal was not only a canal but "a line of artificially irrigated country."

The explanations advanced ... for the blue-green areas explain also the canals, namely, that what we see in both is, not water, but vegetation. ... If therefore, we suppose what we call a canal to be, not the canal proper, but the vegetation along its banks, the observed phenomena stand accounted for. [78]

Again he credited the suggestion to Pickering.

The double canals he explained similarly but with somewhat greater qualification. The phenomenon of "gemination," he noted, was far more difficult to observe and required "emphatically steady air to see it unmistakably." But when glimpsed under such conditions, the doubles were "more markedly unnatural" and conveyed the appearance of artificiality by their "mathematical precision." [79]

Exactly what takes place ... in this curious process of doubling, I cannot pretend to say. ... From these observations, and those of Schiaparelli, I feel, however, tolerably sure that the phenomenon is not only seasonal but vegetal. [80]

Having thus attempted to explain the puzzling network of *canali,* Lowell moved on to the small dark spots which he called "oases," and to the triangular, checkmark-like markings which he had reported in 1894 at the edges and in the interiors of the blue-green regions.

The "oases" formed a "singularly correlated system of spots" that were intimately associated with the canal network. He had found no instance where an oasis was not connected to another oasis or to a larger dark marking by at least two canals, nor of a junction of two or more canals without its oasis. The spots were generally circular, from 120 to 150 miles in diameter, and appeared to be "temporary phenomena, like the canals." [81] Moreover, they displayed seasonal changes even in the martian tropics and thus "clearly some more definite factor than the seasons enters into the matter. ... That this factor is water seems ... to be pretty certain." [82] The spots were not "lakes" but "areas of verdure."

The spots are oases in the midst of the desert, and oases not innocent of design. ... For here in the oases we have an end and object for the existence of the canals, and the most natural one in the world, namely, that the canals are constructed for the express purpose of fertilizing the oases. ... The canals rendezvous so entirely in defiance of the doctrine of chance because they were constructed to that end. They are not purely natural developments, but cases of assisted nature. [83]

The oases he considered to be as objectively real as the canals, and on at least one occasion he invoked the authority of Schiaparelli on their behalf, even though the Italian astronomer had never reported them. In the first of several letters in which he argued his theory in detail to the U.S. Naval Observatory's Simon Newcomb, at the time the dean of American astronomers, he wrote:

As to the oases or small dark spots connected to the "canals," it may interest you to know that when I spoke of them to Schiaparelli in 1895, he replied, his eyes lighting up: "I suspected them myself but could never see them well enough." [84]

The triangular spots, or "carets," also fell neatly into the overall pattern of the planet-wide irrigation system. These he found both "at the mouths of canals at the edge of the so-called seas," where the lines in the bright and dark regions met and at the intersections of canals in the dark areas. The exact nature and purpose of these strange triangular spots was not clear, "but it would seem to be that of relay stations for the water before it enters the canals." It was significant, he thought, that during the process of change in the martian spring, the carets "develop *before* the canals; the oases *afterwards.*" [85]

His extensive inventory of what was known or believed and of what he had discovered in 1894–95 of both the natural and what he considered the non-natural features of Mars was now complete. "Upon the above results of the observations," he wrote in the conclusion of the first volume of the *Annals*, "is based the deduction I have here put forward—(1) of the general habitability of the planet; (2) of its actual habitation at the present moment by some form of local intelligence." And he added:

Certain as it is to be looked at askance by astronomers trained in the impossibility, from the positions of their observatories, of seeing the phenomena on which it is based, a time is no less sure to come when other observatories shall be put up in favorable places and the phenomena generally recognized. [86]

For readers of *Mars,* however, he ventured somewhat further into the realm of speculation and expressed his conclusions more colorfully. He summed them up in four points: (1) the physical conditions on Mars were "not antagonistic to some form of life;" (2) there is an apparent dearth of water on the planet and therefore "if beings of sufficient intelligence inhabited it, they would have to resort to irrigation to support life;" (3) there "turns out to be a network of markings ... precisely counterparting what a system of irrigation would look like;" and (4) there are "spots placed where we should expect to find the lands thus artificially fertilized, and behaving as such constructed oases should." And he added: "All this, of course, may be a set of coincidences, signifying nothing, but the probability points the other way."[87]

That martians would be physically capable of constructing a planet-wide irrigation system Lowell had little doubt, for "any Martian life must differ markedly from our own." The effect of the size of habitat on the size of its inhabitants was notable. On Mars, a smaller planet where the surface gravity was only 0.38 that of earth, a similar expenditure of energy could accomplish almost three times as much work. Martians also could be nearly three times larger than men, although, he quickly added, he was speaking "to the possibility, not the probability, of such giants."[88]

Mentally, too, martians should be superior to man, for "Mars shows unmistakable signs of being old ... and evolution on its surface must be similarly advanced."

Of its actual state our data are not definite enough to furnish much deduction. But from the fact that our own development has been a comparatively recent thing, and that a long time would be needed to bring even Mars to his present geological condition, we may judge any life he may support to be not only relatively, but really older than our own.[89]

But for Lowell, it was the canal network itself that provided the best testimony that the inhabitants of Mars possessed intelligence:

The evidence of handicraft, if such it be, points to a highly intelligent mind behind it. Irrigation, unscientifically conducted, would not give us such truly wonderful mathematical fitness in the several parts of the whole as we there behold. ... A mind of no mean order would seem to have presided over the system we see—a mind certainly of considerably more comprehensiveness than that which presides over the various departments of our own public works. Party politics, at all events, have had no part in them; for the system is planet-wide. Quite possibly such Martian folk are possessed of inventions of which we have not dreamed, and with them electrophones and kinetoscopes are things of a bygone past, preserved with veneration in museums as relics of the clumsy contrivances of the simple childhood of the race. Certainly we see hints at the existence of beings who are in advance of, not behind us, in the journey of life.[90]

Lowell freely conceded that these conclusions might seem incredible and even repugnant to many people, but he argued that this "strangeness is a purely subjective phenomenon, arising from the instinctive reluctance of man to admit the possibility of peers."[91] And he ended with a plea:

One thing we can do and that speedily: look at things from a standpoint above our local point of view. . . . That we are the sum and substance of the capabilities of the cosmos is something so preposterous as to be exquisitely comic. . . . That man gauges the possibilities of the universe is humorous. He does not, as we can easily foresee, even gauge those of this planet. . . . He merely typifies in an imperfect way what is going on elsewhere, and what, to a mathematical certainty, is in some corners of the cosmos indefinitely excelled.[92]

His appeal, however, was largely lost in the swirl of controversy that his Mars theory stirred anew.

First Reactions

PERCIVAL LOWELL'S PRONOUNCEMENTS of the probable existence of intelligent life on Mars stirred considerable interest in the final years of the nineteenth century, even in an America troubled by political, economic, and social unrest and in an era notable for its excesses of sensation. The fact is one measure of the impact of his thought in intellectual history.

Thousands of ordinary people learned directly of Lowell's plausible postulation of this specific example of extraterrestrial intelligence from his magazine articles, his lectures, or his book *Mars*. Thousands more absorbed the major points of his theory at second hand, often in oversimplified and not always accurate form, from reports, reviews, and commentaries in a press that was already familiar with the lurid techniques of what came to be known as "yellow journalism."

His theme was hardly new. But his clear and logical exposition, bolstered by the unprecedented mass of observational data he presented, gave it a new look and invested it with an aura of scientific authority it had not hitherto possessed. Furthermore, his status as scion of an old and affluent New England family and as a member of Boston's brilliant intelligentsia, as well as his reputation as an erudite writer, scholar, and world traveler, lent added credence to his martian ideas. Certainly this circumstance gained them a more respectful hearing than they might have otherwise received.

The initial public reaction to Lowell's Mars theory ranged from simple uncritical wonder through more or less credulous curiosity

Lowell's favorite portrait.

to skeptical but tolerant amusement. People were clearly interested and intrigued by lengthy and detailed articles that appeared in the newspapers and periodicals of the day, but they also chuckled at humorous or whimsical references to his vigorous astronomical activities in the press. Such items, in the ubiquitous form of the newspaper "filler," showed up with increasing frequency as the Mars furor intensified in subsequent years, and testify to the growing general awareness to his ideas. Typical, perhaps, of the hundreds of examples in the Lowell Observatory archives, is this "filler" appearing in October 1896:

The title of the article, "Mr. Percival Lowell on Mars," in a recent weekly, is misleading. Mr. Lowell hasn't quite got there yet, though he is getting nearer and nearer every night.[1]

Or again:

We read in a contemporary that a certain famous professor "has spent several years in Arizona studying Mars." This will be news to those people who have had doubts in their minds as to where Mars really is located.[2]

Such blurbs were almost always affably irreverent in tone. Only occasionally were they in what Lowell, at least, would have considered bad taste:

Prof. Lowell says the sea-green body of water on Mars changed to a chocolate brown over night. A yen to a kopeck the doctor had a dark brown taste the morning after this transpired.[3]

And as humor sometimes will, occasionally one of these bits of popular whimsy candidly capsuled a precise truth. One anonymous quipster, for instance, penetrated to the essence of the martian furor in two short sentences:

Prof. Percival Lowell is certain that the canals on Mars are artificial. And nobody can contradict him.[4]

Within astronomy itself, reactions were more extreme, running the gamut from polite professional interest to incredulous indignation. These, too, reached the public in lengthy and sometimes opinionated articles which served to further focus popular attention on Lowell and his provocative ideas.

The most vehement attacks on his martian theory were

launched by astronomers at the Lick Observatory on California's Mt. Hamilton where the great 36-inch Clark refractor, the largest telescope of its kind at the time, had failed to reveal anything resembling the elusive Schiaparellian network of *canali,* the existence of which Lowell now confidently proclaimed as confirmed. Less than a year earlier, of course, Lick director Edward S. Holden had denounced Lowell for conjecturing about life on Mars before he had even begun to observe the planet at Flagstaff. Now, as a formal theory bolstered by the extensive data that Lowell and his assistants had gathered during the favorable 1894 opposition, these same ideas evoked an even sharper response. Holden, indeed, resorted to lines by Rudyard Kipling to express his professional consternation over Lowell's unprofessional assumption of astronomical authority. Lowell was, he declared:

> Hanging like a reckless seraphim,
> On the reins of red-maned Mars. [5]

But sarcasm was a far more frequent weapon of Lick astronomers in their sustained attacks on Lowell and his theory. In an editor's note to the Reverend Edward Everett Hale's widely-published account of Lowell's February 1895 lectures on Mars, Holden suggested that Lowell had been prejudiced by his preconceptions. "It is a point to be noted," he wrote, "that the conclusions reached by Mr. Lowell at the end of his work agree remarkably with the facts he set out to prove before his observatory was established at all."[6]

Astronomer W. W. Campbell, Holden's colleague and eventual successor as Lick director, hammered away at this same point subsequently in his formal review of Lowell's *Mars,* noting that "Mr. Lowell went direct from the lecture hall to his observatory, and how well his observations established his pre-observational views is told in his book."

In justice to him it must be said he has written vigorously and at length of the dangers of bias on the part of those having preconceived notions, and in numerous paragraphs throughout the book severely criticises those who write on the subject without having made the observations. So I suppose we shall have to forget his remarkable preliminary lecture. [7]

Campbell also questioned the originality of Lowell's 1894 findings. Lowell, he declared, had merely confirmed in a general way

Schiaparelli's earlier description of the planet and had done so moreover without specific reference to the Italian's work. To show the similarity of Lowell's results, he quoted at length from a translation of Schiaparelli's original martian memoir made by Lowell's 1894 assistant W. H. Pickering.[8]

The Lick astronomer also inquired pointedly how the martian equatorial regions could be irrigated alternately from both poles, "the 'canals' in the two cases of opposite flow being identical." He noted that the canals "apparently do not turn aside from anything" on the planet. "The path of least resistence," he observed dryly, "seems to be unknown."[9]

Campbell's critique hardly touched on Pickering's Flagstaff observations. But he did deal briefly with some of Douglass' work, conceding the possible value of his observations of canals in the dark areas which he considered an extension of work done at the Lick Observatory in 1892, while challenging his measurements and interpretations of irregularities on the martian terminator.[10]

Campbell's principle target, however, was Lowell, whom he bluntly labeled an opportunist. "In my opinion, he has taken the popular side of the most popular scientific question about," he opined. "The world at large is anxious for the discovery of intelligent life on *Mars,* and every advocate gets an instant and large audience."[11]

Campbell was willing to concede that "Mr. Lowell's book is written in a lively entertaining style," and was "printed and illustrated faultlessly." He also agreed that Lowell "is entitled to great credit for devoting his private means so generously to establishing and conducting an observatory, and for his efforts in search of the best, but imperfect atmospheric conditions. He is likewise fully aware of the necessity for making observations continuously and systematically."[12]

But after these brief concessions, he returned to the attack:

The theories advanced are mostly old ones ... but Mr. Lowell has presented them very fully and suggestively. Scientifically, the leading faults of the book are: first that so elaborate an argument for intelligent life on the planet, embracing a complex system of seasonal changes, should be based on observations covering only onefourth of only one Martian year; and second, that there should be so many evidences of apparent lack of familiarity with the literature of the subject.[13]

Lowell's immediate reaction to Campbell's review was to pen a stinging rebuttal which, with additional comments by Douglass, subsequently appeared both in *Science* and *Popular Astronomy* under Douglass' name:

Having sought to throw discredit on Mr. Lowell's work almost before it was begun some two years ago, The Lick Observatory now renews the attack in Prof. Campbell's review of Mr. Lowell's book. Formerly, it decried the work because the theories upon which it was started were too original; now it attempts to seize the credit of the results and calls the theories "mostly old." Such a remarkable act of appropriation cannot be allowed to pass unnoticed.[14]

To Campbell's charge that he lacked familiarity with the literature, Lowell retorted:

To support this statement irrelevant quotations at great length are made from a translation by Prof. W. H. Pickering of Schiaparelli's work to which translation he professed his obligation. Of this it is only necessary to say that the translation in question *was made at the Lowell Observatory*, a fact which Prof. Campbell neglected to mention. ... We are very willing to have the Lick indebted to us for its knowledge of Schiaparelli's work but it must not suppose us ignorant of our own translation to which its knowledge is due.[15]

To Campbell's criticism that the 1894 Flagstaff observations had not been prolonged enough to support a conclusion of seasonal changes, Lowell cited the work of Schiaparelli, adding that "what our observations disclosed was not only the fact of change, which they corroborated, but the character of the changes and the process of their development, thus furnishing an important link in the chain of evidence for Mr. Lowell's theory."[16]

That Lowell wrote this much of the Douglass article, referring to himself in the third person, is revealed by the handwritten draft in the Lowell archives. Douglass, however, added several paragraphs in his own handwriting in which he commented somewhat heavy-handedly on the Lick Observatory and some of its work:

The Lick article asserts that the first irregularity on the terminator was seen at the Lick Observatory in 1890 but it omits to mention that it was a casual visitor who detected it, so that to this visitor and not the Lick staff belongs the discovery. What such an outsider's discovery betokens about the efficiency of the staff is not our purpose to remark.[17]

And his closing shot was aimed at Holden:

As to any knowledge at the Lick Observatory of a Martian atmosphere, it has been purely negative, Professor Holden going so far in an article in the *North American Review* for 1895 entitled Mistakes about Mars as to declare that the opposition of 1894 would be memorable for having proved an absence of atmosphere. We may let Holden's Mistakes about Mars speak for themselves.[18]

Not all the reviews of *Mars,* however, were unfavorable. W. W. Payne of Minnesota's Goodsell Observatory and an editor of *Popular Astronomy,* for instance, wrote that the book was "crowded with an array of facts that are made to signify much that is new by the gifted and ready reasoning power which Mr. Lowell possesses, in a remarkable degree."[19]

Nor were all Lick astronomers as vigorous in their objections to Lowell's martian work as Holden and Campbell. The respected Edward Emerson Barnard, who had seen broad vague streaks on the planet from Mt. Hamilton in 1894 while Lowell was seeing fine sharp lines, noted mildly that his own observations "do not agree with many of those obtained by Mr. Lowell," and suggested that further observations of Mars might "materially change" Lowell's opinion about the planet. "He is a remarkably fine writer," he added, " and with his skill in this line—which so few observers possess— he will be able to present his results in a very acceptable manner to the public."[20] Barnard, parenthetically, resigned from the Lick staff late in 1895, reportedly because of "trouble" with Holden, and moved to Yerkes Observatory then under construction at Williams Bay, Wisconsin.[21] His own astronomical work, however, did not prevent him from remaining on friendly terms with Lowell. He visited the Flagstaff observatory in 1898, and in subsequent years Lowell not infrequently stopped over in Chicago on his trips to and from Arizona to entertain him at dinner and to discuss their respective observations.

Other astronomers too, whose smaller telescopes were located at observatories less favorably situated than the Lick, generally took a more moderate position on Lowell's theory while acknowledging the importance and impact of his work. Prof. Charles A. Young of Princeton University, author of a widely-used textbook on astronomy, was one of these. "The observations of 1894," he wrote in October 1895, for the national readership of *Cosmopolitan* magazine "have made it practically certain that the so-called 'canals' of Mars are real, whatever may be their explanation."

Although the planet has practically withdrawn from observation for a time, the popular interest in it has by no means disappeared, but has been maintained, and perhaps increased, by the bold speculations of Mr. Lowell, presented last season in his captivating lectures, and since then in his charming papers published in the *Atlantic Monthly*.[22]

That the public excitement over matters martian was continuing to grow, Young also noted a year later in an article written for the press:

There seems to be an extreme popular interest in the question of the habitability of "other worlds," and of late it has been greatly intensified by the rather sensational speculations and deliverances of Flammarion, Lowell and others. ...[23]

Flammarion's contribution to the flaring controversy, he noted incidentally, was to suggest "in a caprice of speculation it would seem, that the Martians are winged creatures, but whether bats, birds or butterflies he does not attempt to decide."[24]

Despite such skeptical barbs, however, Young remained circumspect on the issue of martians:

Still, it is always wise to be reticent in denying the possibilities of the future, and no less so to be cautious in accepting as ascertained truth the startling conclusions and unverified discoveries of imaginative observers. It is so easy to see what one expects and wishes to find, especially on a disc so small and delicately marked as that of Mars.[25]

In May 1897, and while *Harper's* magazine was serializing a novel entitled *The Martian* by George du Maurier,[26] the U.S. Naval Observatory's Simon Newcomb had his say about the growing Mars furor that had been rekindled by Lowell's theory:

While every astronomer has entertained the highest admiration for the energy and enthusiasm shown by Mr. Percival Lowell in founding an observatory in regions where planets can be studied under the most favorable conditions, they cannot lose sight of the fact that the ablest and most experienced observers are liable to error when they attempt to delineate the features of a body 50 to 100 million miles away through such a disturbing medium as our atmosphere.
Even on such a subject as the canals of Mars doubts may still be felt. That certain markings to which Schiaparelli gave the name of canals exist, few will question. But it may be questioned whether these markings are the fine sharp uniform lines found on Schiaparelli's map and delineated in Mr. Lowell's beautiful book. It is certainly curious that Barnard at Mount Hamilton, with the most powerful instrument and under the most favorable circumstances, does not see these markings as canals.[27]

By 1901, although the nervous exhaustion which had stricken him in 1897 had virtually suspended his astronomical activities for nearly four years, Lowell and his theory dominated the continuing controversy over Mars. In March of that year, astronomer Henry Norris Russell, soon to win renown for his work in the field of stellar evolution, took public note of the fact. Summarizing the current state of the debate, he concluded:

Perhaps the best of the existing theories, and certainly the most stimulating to the imagination, is that proposed by Mr. Lowell and his fellow workers at his observatory in Arizona.[28]

Some astronomers seemed to be as upset over Lowell's energetic promotion of his theory in the popular media as they were over the theory itself. His *Atlantic Monthly* series in mid-1895, for instance, led an anonymous member of the New York Academy of Science, where Columbia University's Rutherfurd professor of astronomy Harold Jacoby was the leading Lowell critic, to declare publicly that such articles not only "presented fanciful conclusions" but tended to "cheapen the science of astronomy." The anonymous newspaper reporter who recorded this opinion, however, did not entirely agree:

These doubts and difficulties should only spur the astronomers to further and more united efforts. There is no danger that astronomy will be "cheapened" by the freest discussion of its problems. Mistakes and errors recoil only upon their authors. No great advance will ever be made unless the imagination is enlisted in the work. What is purely fanciful will finally vanish of its own accord; but if it leaves a residuum of fact, science will have gained so much.[29]

Lowell certainly must have enjoyed reading this thoroughly Lowellian statement.

The extensive and persistent popularization of his martian theory, which some professional astronomers sternly disapproved, was deliberate on Lowell's part. For he was firmly convinced that the rapid advances of science and technology could be brought within the intellectual range of the general public and discoveries explained simply and clearly so that the nonexpert might not only understand them but gain from them. He made the point repeatedly in his writings:

To set forth science in a popular, that is in a generally understandable, form is as obligatory as to present it in a more technical manner. If men are to benefit by it, it must be expressed to their comprehension. To do this should be feasible for him who is master of his subject. . . . Especially vital is it that the exposition should be done at first hand; for to describe what a man himself has discovered comes as near as possible to making a reader the co-discoverer of it. Not only are thus escaped the mistaken glosses of second-hand knowledge, but an aroma of actuality, which cannot be filtered through another mind without sensible evaporation, clings to the account of the pioneer. Nor is it so hard to make any well-grasped matter comprehensible to a man of good general intelligence as is commonly supposed. The whole object of science is to synthesize, and to simplify; and did we but know the uttermost of a subject, we could make it singularly clear.[30]

Technical phraseology, Lowell insisted, hampered the broad understanding of scientific knowledge and, while "useful to the cult, becomes meaningless jargon to the uninitiate and is paraded by the least profound." And:

Even to the technical student, a popular book, if well done, may yield most valuable results. For nothing in any branch of science is so little known as its articulation.[31]

Here again the consistency, and perhaps a certain naivete, in Lowell's thinking is clearly apparent. More than ten years later he still urged the popularization of science in similar terms:

Discoveries in science have a fatal facility for lying lost in the technical publications which record them. Few persons attend to what is not alluring and columns of figures form but an uninviting portico to the learning within. Yet these very people would take the keenest interest in scientific progress could its beauty and real simplicity be adequately set out for them. For the whole object of science is to explain and make comprehensible the universe around us. Science consists in solving mysteries not, as a layman might imagine, in making them.[32]

Nor should scientists restrict their efforts along these lines to technical books and articles in the more sober journals. "The daily press, indeed with more zeal than discernment, prints everything it can get hold of, but it does so with an impartiality between knowledge and ignorance that leaves nothing to be desired but the truth," he declared. But nonetheless, he added, "it behooves the scientist, in the very interest of science itself, to turn no cloistered ear to lay appeal."[33]

To maintain at least some control over such undiscerning journalistic zeal in the reporting of his own astronomical activities,

Lowell adopted the policy used by many scientific institutions then and since of providing information to the press only on request. Douglass, for example, applied it in 1896 to complain to the *Associated Press* about the handling of news from Lowell Observatory:

Our contributions to the public press are always sent by request. We seriously object to having them appear as if sent on our own responsibility, as was done by you. Nor is it our intention to allow our articles to be copyrighted, thus preventing the free distribution of information which we have taken great trouble to collect and some pains to put in a suitable form for the popular mind. We do not wish our discoveries to be considered a marketable product.[34]

Lowell could quietly bend this policy on occasion to his own purposes, privately volunteering information and suggesting articles on his work to particular writers or editors with whom he had a close rapport. But publicly, he tried to keep up a properly professional appearance of not seeking publicity for himself and his observatory. In 1905, for instance, he berated Carr V. Van Anda, the knowledgeable and influential managing editor of *The New York Times*, for publishing an item from the observatory without labeling it "telegraphed by request," and drew a respectful apology.[35] Years later, incidentally, he found it expedient to consider his "discoveries" as "marketable," arguing to the U.S. Treasury Department and Secretary William Gibbs McAdoo that his observatory was entitled to a business tax exemption because "income has been received from it by me—by the sale of its product."[36]

Lowell was well qualified for the task of popularizing science both by reason of his broad education and his experience in writing about little-known subjects for a more or less general readership. By 1894 he had already acquired a considerable literary reputation from his books and other writings about his travels and investigations in the Orient. His clear and colorful style, his broad humor and pungent wit, in these works had drawn praise from readers and reviewers alike. He was no less lucid and entertaining when his subject was science, a fact also widely noted and appreciated, even by his severest critics in astronomy. Indeed his earliest critic, Holden, had complained that not only were Lowell's pre-Flagstaff conclusions about life on Mars "misleading" but that "they are all the more so because they are very well written."[37] As is evident above, Campbell and Barnard also conceded his literary if not always his astronomical qualifications.

Lowell knew good writing, and he worked hard on his own. The texts and manuscripts of his books, articles, and lectures in the Lowell Observatory archives more often than not are scored with deletions, insertions, and corrections in his bold, slanting handwriting designed to improve the turn of a phrase or clarify a point. It was not at all unusual for reviewers to refer to the flowing if somewhat florid prose of the finished product as poetic.

Lowell had a philosophy about writing, as he had about everything he took seriously, and about some things he didn't. This he expressed long before he took up astronomy in a letter to a Boston friend as he prepared to begin writing about his Far Eastern experiences:

I believe that all writing should be a collection of precious stones of truth which is beauty. Only the arrangement differs with the character of the book. You string them into a necklace for the world at large, you pigeon hole them in drawers for the scientist. In the necklace you have the cutting of your thought, i.e., the expressing of it and the arrangement of the thoughts among themselves.[38]

In his early expositions of his martian theory, the literary quality of his writing was somewhat overshadowed by the sensational nature of his subject matter and of his conclusions thereon. In his later books and articles on Mars, which were essentially elaborations on a still controversial but now familiar theme, the literary aspect of his work increasingly was recognized. Two widely separated examples from reviews of Lowell's third and final book about the planet, *Mars as the Abode of Life* (1908), will serve here to illustrate not only this point, but the wide geographical interest in his work. The St. Louis *Globe-Democrat,* for example noted:

It is not only as an astronomer, but as a writer that Prof. Lowell charms the reader ... the beguilement of the theme is well matched by the grace and literary finish of the style in which it is presented. ... The warmth and earnestness of the true lover of his theme—aye, and the believer in it—shine through the entire work so that in its whole style and illustrations it is a charming production.[39]

And from Lowell's earlier area of activity, the English-language *Japan Daily Mail* commented on both his literary style and his erudition:

The mingling of its advanced thoughts of geologists, paleontologists, biologists, chemists and astronomers, shows very clearly that the author had read most carefully and that he has browsed in many of the greenest fields. .... It would be impossible for Lowell to write anything without imparting to it a touch of poetry, and to make his presentations, argument, and deductions attractive by the manner of his treatment. [40]

Lowell's friends and scientific correspondents on occasion also praised his writing as well as what he was writing about. Lester Frank Ward, the pioneer American sociologist whose own works were quite provocative in their field, was one of these. An avid supporter of Lowell's martian theory, Ward wrote of his *Mars and Its Canals:*

I know it cannot flatter *you* if I say that next to the edifying effect of your epoch-making results I have most enjoyed your brilliant and sparkling style, those sly but pointed plays of humor and dashes of wit, which enliven your pages without the least distracting from the serious and courageous treatment of a profound theme. [41]

Most reviewers of Lowell's martian books commented along the same lines. Most, too, praised them as well-organized, comprehensive summaries of what was known and of what he and his assistants had discovered about the planet from Flagstaff. The review of *Mars and Its Canals* in *Science,* for instance, declared that "while this book is published as a popular exposition of the most recent investigations, it presents practically all that is known, or thus far suspected, presumably, concerning the planet and its inhabitants." [42]

But most reviewers also were cautious about accepting Lowell's more sensational conclusions. Typical were reviews of *Mars and Its Canals* in *The Independent* and the *Catholic World.* The former felt that Lowell's observations "have every claim to acceptance. The theories propounded are by no means so clear." [43] And the *Catholic World's* Father George M. Searles, a frequent Lowell critic, declared that the observed martian phenomena were "explicable enough without any idea of Mars being inhabited. It seems clear that he has let his imagination run away with him." [44] It may be added that most reviews of Lowell's books on Mars, favorable or not, were extensive, some running to thousands of words. Thus they had the incidental effect of promulgating his theory to a far larger audience than the books themselves ever reached. [45]

Lowell was also a highly effective speaker, and his personal dynamism, tempered with an affable good humor and spiced with a sharp penetrating wit, held a strong appeal for those who flocked to hear his lectures. The clarity and confidence with which he explained his ideas on the podium as well as on paper along with his air of earnest sincerity had an almost evangelical effect upon many of his listeners. No less a word-master than the Reverend Edward Everett Hale provides eloquent written and spoken testimony to the fact. In his account of Lowell's February 1895 lectures on Mars at Boston's Huntington Hall, Hale wrote:

It is impossible in print to describe the charm of Mr. Lowell's lectures. His humor, his ready wit, his complete knowledge of the subject with which he deals are such as one has no right to expect in the same public speaker. The most serious considerations are made interesting by analogies with affairs with which we are familiar and in which we are at ease. Everybody knows how light his pen is when he writes of his travels; and his ease as a public speaker and the readiness with which he takes his audience into his confidence give an additional charm to the lectures as he reads, or rather delivers, them.[46]

Hale, author of the patriotic and inspirational *The Man Without a Country* and pastor of Boston's South Congregational Church, was so impressed by the lectures that he made them the subject for his Easter sermon that year:

I was much touched this winter . . . as Mr. Lowell spoke to us simply of the canals on the surface of Mars. He said squarely, "there are not 20 people in the world who have seen them; but the 15 who have seen them have seen them, and we know they are there. We do not ask you to accept this," he said, "on our authority, but we tell you that as soon as you have seen them you will believe in them." When a man speaks in that way, you do not accept what he tells you on his authority; but you find, all the same, that you believe. You cannot resist the conviction which shows itself, not in the language, but in its intensity; not in his words but in the man.[47]

This charismatic quality that Hale described was a hallmark of Lowell's many lectures and one which was to make him almost as famous as a speaker as an astronomer. It was noted not only from the pulpit, but by the more worldly commentators of the daily press. "He has a graphic and vivid style, and his deep ernestness [sic] is now and then relieved by a touch of quiet and delicate humor," one Boston newspaper reporter wrote of his highly successful lecture series for the Lowell Institute* in October 1906.[48] "A literary pro-

---

*The Lowell Institute was founded with a $250,000 bequest by John Lowell Jr., who died in 1836, to provide the public with intellectual and scientific lectures by the leading thinkers of the day. Lowell's brother, A. Lawrence Lowell, was trustee of the Institute at the time of the Mars lectures.

duction of the highest order, abounding in those epigrammatic phrasings which mark Mr. Lowell's productions," another proclaimed.[49]

Not only professional journalists, but many other persons were impressed by these lectures which filled Boston's 1,000-seat Huntington Hall to overflowing at the twice-a-day sessions necessitated by the public demand for tickets, and which proved popular enough to be serialized in *The Century* magazine before being published as the book, *Mars as an Abode of Life*.[50]

"You will perhaps remember a blind lady who spoke to you one afternoon at the close of your lecture," a Mrs. Maria L. Crowell of Cambridge, Massachusetts, wrote Lowell after the series had ended. "I have been interested in Mars for years and have believed it was inhabited so that I was fully prepared to enjoy your lectures thoroughly."[51] And another appreciative, but anonymous listener wrote:

I listen to you with the greatest interest. Ever since I studied astronomy at school, I have longed to know more of the solar system, and of all the heavenly bodies. ... And I look forward with hope to another existence when I can know more with certainty. ... I am a woman of no importance, but I am a grateful and deeply interested listener.[52]

The verdict was not entirely unanimous, however, even on Lowell's platform style:

Is Mr. Lowell aware that he looks at the clock during his lectures on the average of once in three minutes, and wipes his brow with his handkerchief at least once in four minutes to the great distraction of his deeply interested audience? It is a pity that his delivery should be marred by such mannerisms![53]

This complaint was signed only: "One who suffers from them."

Nor was the verdict unanimous about his theory. There were some in his audiences who, despite his forensic fervor, clung to their skepticism about canals and canal-building martians. A Boston *Herald* reporter noted:

Prof. Lowell's Monday lecture did not convince one of his auditors of Mars' habitability, for as she came out with the great crowd she was overheard to say: "Well, I call him a visionary!" All the mathematical calculations of which scientists are capable would be unable to shed light on the probable life on the planet for this incredulous person. Even if a Martian should suddenly appear on the platform of Huntington Hall, some Boston mind would decline to accept him. Therefore, enlightenment can only come through time for such incredulity.[54]

The enlightenment thus so confidently predicted by the *Herald* reporter, no less than by Lowell, never did come, of course, but this is getting ahead of the story. For it was not only through his sensational conclusions about Mars that Lowell inspired incredulity.

CHAPTER

8

New Directions

WITH THE COMPLETION of his 1894–95 observations of Mars and the wide promulgation of his theory of the probable existence of intelligent martians, Percival Lowell bought himself a new and larger telescope and went in search of new worlds to conquer. Along with Mars, he would now consider Venus and Mercury, the two terrestrial planets between the earth and the sun.

He spent the year 1895 in Boston writing about his martian observations and conclusions for popular consumption, publicizing his ideas in lectures, and preparing his book, *Mars,* for December publication. With this done, he sought relaxation in Europe, a voyage that turned out to be "one of the most unpleasant I have ever made."[1] His ship, the *Spree,* ran aground as it entered the English Channel. The passengers were all taken off safely by tugboat, but the incident marked the closest Lowell ever came to shipwreck in a lifetime of traveling over the world.[2]

His companion on the trip was Alvan G. Clark, the surviving member of the family telescope-making firm of Alvan Clark and Sons, with whom Lowell in April 1895 had contracted for a 24-inch refractor. The instrument, its objective lens figured from a blank of fine Mantois optical glass, was to be ready June 1, 1896, for shipment to Arizona where it would be tested before being moved to Mexico for the 1896–97 martian opposition. Lowell agreed to pay $20,000 for the telescope, although press reports at the time put the cost at "nearly $100,000."[3]

The 13-ton refractor was duly completed, with time to spare,[4] and shipped west to be erected in the enlarged wooden dome that the

The Lowell 24-inch refractor.

observatory's man-of-all-work Godfrey Sykes had prepared for it. On July 22, 1896, Lowell, with Clark himself, Clark's daughter, and Lowell's secretary, Miss Wrexie Louise Leonard, arrived by train in Flagstaff, bringing the objective lens with them. The glass was soon installed, and observations began the following night. [5]

For his Mexican expedition, Lowell had assembled a somewhat larger and more diversified group than he had had for his initial martian campaign in 1894. Douglass was on hand, to be sure, but Pickering had departed and subsequently will appear here mainly as a proponent of his own rather imaginative ideas about Mars and as an opponent of Lowell's theory. Newcomers included two young college graduates, Wilbur A. Cogshall and Daniel A. Drew, and more notably the much-publicized Dr. Thomas Jefferson Jackson See, lately of the University of Chicago and George Ellery Hale's staff and a binary star expert of then considerable repute. See, about whom more must be said later, planned not only to pursue his double star studies but to make the first systematic survey of the southern heavens since Sir John Herschel's pioneering work in South Africa in the mid-1830s. Clark and Lowell's new secretary, Miss Wrexie Louise Leonard, went along and made occasional observations. [6]

Douglass handled the logistics of the move and served as the observatory's advance guard, arriving in Mexico City early in November. There, he prospected some sites southwest of the city, finally selecting one at Tacubaya, not far from the National Observatory. [7] See and Cogshall followed later in the month, and work got underway on setting up the 24-inch and its dome.

For Douglass, incidentally, this Mexican sojourn was quietly productive. In addition to observing Mars and continuing his studies of the atmosphere, he turned the telescope on the Galilean satellites of Jupiter, finding linear markings on Ganymede and concluding therefrom that this third jovian satellite rotated in about seven days, a period closely synchronous with its revolution around the planet, and thus kept the same face turned toward Jupiter as the earth's moon does to the earth. [8] Typically, before leaving Mexico the following spring, the adventurous Douglass climbed to the 17,887-foot summit of Popocatapetl, checking on observing conditions at various levels on the famed volcanic peak and later reporting on his ascent in *Popular Astronomy*. [9]

Lowell himself arrived at Tacubaya late in December, well after the martian opposition had passed, and formal observations did not begin until December 30, extending through March 26, 1897. Lowell's work there on Mars was quite routine, involving largely the areographic location of various markings, new measurements of martian diameters, and the addition of nineteen new double canals to his maps.[10]

Mars produced no new sensations. As observations resumed that summer at Flagstaff, its telescopic disk was only eight seconds of arc in diameter—it would grow to only seventeen arc seconds at opposition*—and so for a time the astronomers turned their attention to other celestial objects. Lowell, with Cogshall, Drew, and Miss Leonard, observed Mercury and Venus. See concentrated on double stars, particularly the "dog star" Sirius, managing to stir the first of several non-martian controversies that resulted from the Lowell Observatory's varied activities in 1896–97.

On September 6, 1896, the New York *Herald,* in a widely-copied dispatch, reported See's August 30-31 "rediscovery" of the faint companion star of Sirius that Clark, incidentally, had first discovered in 1862 while testing an 18½-inch refractor he had produced for the Dearborn Observatory in Chicago.[11] The companion had been "lost," from view at least, since 1890 when it had disappeared behind the larger, far more brilliant component that makes Sirius the brightest star in the night sky. Its reemergence in 1896 was not observed by other astronomers until November when it was seen at the Lick Observatory in California, although in a different position from that reported by See.

This circumstance provided Lick director Edward S. Holden with still another opportunity to exercise his talent for sarcasm at the expense of Lowell and his new observatory. Observations by Lick double star specialist Robert G. Aitken, Holden informed the press in October, "show conclusively that there is no object in the place reported by the astronomers of the Lowell Observatory."

It is remarked on several sides to be a strange thing that this discovery was made at the Lowell Observatory and not at Lick. It was, however, not possible for the observers at Mount Hamilton to report any object which they had not seen.[12]

See continued to insist on his rediscovery, however, declaring somewhat later to Douglass that the Lick positions for Sirius' companion "are so much out that I doubt the validity of their work."[13] The validity of See's own double star work, as will be seen, would

---

*At "most favorable" oppositions, the martian telescopic disk reaches 24 or 25 arc seconds in diameter.

Miss W. Louise Leonard at the 24-inch telescope

soon come in question and he would leave the Lowell Observatory staff under a quiet cloud.

See both opened and closed the 1896–97 observations on a note of controversy. With the return of the Lowell expedition to the United States, the Chicago *Tribune,* on April 17, 1897, reported that See had measured "300,000 double and triple stars in Mexico . . . in the first three months of the year." The figure may have been a typographical error, yet it is also possible that it was a result of See's own extravagant claims. The actual number was closer to three hundred, but the *Tribune* report was picked up uncritically by the *Associated Press* and distributed throughout the country. One mathematically inclined editor made the calculated comment that See must have been measuring double and triple stars continuously at the rate of two and one-third stars per minute to have achieved such an astronomical total. [14]

But Lowell himself still contrived to provide the major news that emerged from his 1896–97 campaign through his observations of Mercury, and particularly of Venus. Coming on the background of the renewed furor over Mars that he had stirred in 1894–95, these revelations caused a flurry of excitement in the press and considerable serious interest among astronomers. The general public, however, accepted them quite casually, for they were somewhat technical in nature and, in any case, lacked the compelling fascination of the life-on-Mars theme.

In October 1896 while observations were still underway at Flagstaff, the Boston *Transcript* published an "official dispatch from Mr. Lowell for distribution to astronomers" announcing that both Mercury and Venus rotated on their axes in the same time in which they revolved around the sun and thus, again like the earth's moon, kept the same face turned toward the larger body. [15] Mercurian and Cytherean "days," in short, were just as long as their "years"—88 and 225 earth days respectively. The finding, Lowell pointed out, was not a discovery but merely confirmed the determinations of isochronous motion for the two planets made in 1889 and 1890 by Schiaparelli.

In the case of Mercury, the idea was not even particularly controversial. Before and long after Lowell's time, some astronomers discerned vague streaks on the planet which seemed to remain fixed in position from one observation to another rather than to move across its visible disk as, for example, in the case of a relatively rapidly rotating Mars. Such observations, which are quite difficult, were usually made when Mercury was at "greatest elongation"—that is, when it was most favorably situated at its greatest angular distance from the sun—and were considered presumptive evidence of the planet's synchronous rotation and revolution.

It was not until 1965, indeed, and the application of radar technology that this interpretation was shown to be wrong. Radar measurements then revealed a rotation of about 59 days, or two-thirds of Mercury's 88-day orbital period.[16] It thus also appears that the sun has captured the planet in a stable 3:2 "resonance lock."[17] Astronomers, viewing the planet at greatest elongations and seeing the same markings in the same positions over successive observations, assumed that with four revolutions Mercury had rotated four times, when in fact it had rotated six. Telescopic observers had erred, ironically, largely because they had been careful to select only the most favorable occasions for viewing the planet.

But the problem of the rotation of Venus was something else again. In Lowell's day there were two schools of thought on the matter. One contended for a "short" Cytherean "day" on the order of about twenty-three hours that Domenico Cassini had first suggested as long ago as 1666–67 and which, in the absence of any better estimate, had stood alone until Schiaparelli's challenge in 1890. The other, following the Italian's lead, argued for a synchronous, or at least a very slow rotation. As in the case of Mercury, the "synchronous school" rested its tenuous case on observations of apparently fixed markings. But in contrast to the difficult but fairly well-defined features of Mercury, the only markings even suspected on Venus were nothing more than faint, evanescent shadings generally believed to represent movements of the cloud veil shrouding the planet. Schiaparelli had relied on observations made over a period of four months of diffuse, faintly lustrous spots near the cusps of the planet's sunlighted crescent that appeared to remain stationary in relation to the planet's terminator.[18]

Lowell, however, based his conclusion of isochronous motion on some truly startling observations. In his initial dispatch for astronomers, he flatly declared that "Venus is not cloud-covered, as has been supposed, but is veiled in an atmosphere," and then claimed that he had observed a hitherto unseen system of dark surface markings on the planet. A few days later, in a report of his observations to the Boston Scientific Society, he described these markings in considerable detail. The Boston *Transcript* reported:

"The markings," Mr. Lowell writes, "proved to be surprisingly distinct; in the matter of contrast as accentuated, in good seeing, as the markings on the moon and owing to their character much easier to draw. In the matter of contour, perfectly defined throughout, their edge being well marked, their surface well-differentiated in tone from one another, some being darker than others. A large number of them, but by no means all, radiate like spokes from a certain center. In spite of all this curious system, there is nothing in them of the artificiality observable in the lines of Mars. They have the look of being purely natural."[19]

The markings, he added, "have proved to be not only permanent, but permanently visible unless seeing is very poor, and they are to be seen in the same places. . . . The steadfastness of position is evidence of the slowness of rotation."[20]

This "curious system" in fact was clear enough to Lowell's eye to enable him to produce a chart of the planet and numerous drawings of its purported surface features. These he published along with articles about his Venus work in *Popular Astronomy* in December 1896 and in the *Monthly Notices of the Royal Astronomical Society* in March 1897.[21] A non-astronomical public read of his surprising findings concerning Venus in the newspapers and in his own words in the March 1897 *Atlantic Monthly*.

Initially and briefly, these reports apparently won some favor. Certainly England's knowledgeable historian of astronomy Agnes M. Clerke was inclined to take Lowell's Venus observations seriously even though they defied the widely-held hypothesis of a cloud-shrouded planet that at least explained its exceptional brightness as a celestial object:

This extraordinary brilliance would be intelligible were it permissible to suppose that we see nothing of the planet but a dense canopy of clouds. But the hypothesis is discountenanced by the Flagstaff observations, and is irreconcilable with the visibility of mountainous elevations and permanent surface markings. To Mr. Lowell these were so distinct and unchanging as to furnish data for a chart of the Cytherean globe, and the peculiar arrangement of the divergent shading exhibited in it cannot offhand be set down as unreal.[22]

But while Lowell could claim that other observers had seen the canals of Mars, no other astronomer had, or indeed has, ever seen anything like the "surprisingly distinct" features he described on Venus. Not even his assistants saw them as he did, although Miss Leonard's drawings of the venusian markings in Lowell's 1896–97

Venus logbook approach his own in boldness and detail. Yet to Douglass they were somewhat more vague and diffuse, while Cogshall, Drew, and See apparently saw only a few faint shadings. [23]

It was Lowell's actual observations of Venus that now disturbed astronomers, more than his conclusions as in the case of Mars; and "disturbed" here is the proper word. Even his critics at the Lick Observatory seem to have wondered privately if their eyes might be deceiving them, although they publicly challenged his Venus work. Not having seen what Lowell declared with such great assurance was there, their confidence in their own observations appears to have been briefly shaken. Compare, for example, the tone of Holden's public attacks on Lowell's martian ideas with the plaintive, private argument regarding Venus he made in 1897 in a letter to Douglass:

I have been rather skeptical, as you know, about the markings on Venus, and I still am (you must not, should not, take criticisms as personal which are not meant so. We are all after the truth and nothing else).

How do you personally explain [E. E.] Barnard's and [J. M.] Schaeberle's and my inability to see any of your *class* of markings on Venus? Barnard and Schaeberle are skilled observers. I began to observe Venus in 1873 and missed no chance for several years.

Not once, as far as I know, have any of us even suspected your *class* of markings. I took every precaution in observing. I don't see how your *class* of markings could have escaped me in all those years. [24]

Somewhat later, the respected Barnard, accepting an invitation from Douglass to visit the Lowell Observatory, expressed his own self-doubts:

I am intensely interested in this matter. I do not have any ill feelings—why should I have? It is simply to satisfy my own eyesight that I want to come. Dr. See and Mr. Lowell have always been friendly towards me and if for nothing else, I should want to come so that I shall cease to do them an injustice, for I doubt not that my failure at L. O. [Lick Observatory] to see these things so easily seen from Flagstaff may have had some sort of influence against the observations. [25]

Publicly, Lick astronomers, unable to see what Lowell described or to believe that a smaller telescope could resolve what the Mt. Hamilton 36-inch refractor had not, claimed the disputed markings were merely the result of observational or instrumental flaws—the "spoke-like" markings, for example, being caused by strains on the Lowell 24-inch objective. One newspaper article, headlined "The Strife of the Telescopes," commented irreverently on this charge:

Astronomer Holden, being jealous of the implied reproach upon his great telescope and the "glorious climate of California," denies the conclusions of the Harvard astronomer [Lowell]. Astronomer Holden intimates that the significant markings were "all in the eye" of the observer, or were due to defects in the glass itself.[26]

Clark, who had built both telescopes and figured their lenses, quickly defended the quality of the Lowell refractor. In a letter to *Science,* he declared:

Having worked both of these objectives myself, and expended as much artistic ability on the one as on the other, there can be no impropriety in my saying that the performance of the Lowell glass is equal to that of the Lick or any of our large telescopes.[27]

The "all in the eye" charge, however, was not so easily answered; indeed, five years later, Lowell quietly conceded at least a part of the argument. In a brief notice in *Astronomische Nachrichten* in 1902, he reported that the strange "spoke-like" markings on Venus apparently were "optical effects of a curious and—astronomically speaking—of a hitherto unobserved character." He did not elaborate, but he stressed that this conclusion did not apply to the other kinds of markings he had seen on Venus nor did it negate his confirmation of its synchronous rotation and revolution.[28]

Lowell's 1896–97 observations of Mercury were not particularly contested—he saw by and large what others had seen—and indeed they became the subject of a handsome memoir, with his drawings of Mercurian markings, published in 1898 by the American Academy of Arts and Sciences, of which he was a Fellow. In this, his findings included a determination that Mercury was slightly larger than previously had been supposed, a circumstance he claimed now explained a discrepancy in the planet's calculated motion that had been detected some years earlier by the U.S. Naval Observatory's Simon Newcomb.[29]

His Venus observations, on the other hand, quickly stirred controversy. But almost as soon as opposition began to gather

VENUS, 1896.

Oct. 13, 2ʰ 55ᵐ to 3ʰ 4ᵐ
λ = 37°

Oct. 14, 1ʰ 59ᵐ to 2ʰ 12ᵐ
λ = 37°

Oct. 14, 3ʰ 13ᵐ to 18ᵐ
λ = 37°

Oct. 14, 3ʰ 50ᵐ
λ = 37°

Oct. 15, 0ʰ 5ᵐ to 14ᵐ
λ = 37°

Oct. 15, 1ʰ
λ = 37°

Oct. 15, 2ʰ 10ᵐ to 21ᵐ
λ = 37°

Oct. 15, 3ʰ 48ᵐ to 54ᵐ
λ = 37°

Oct. 15, 4ʰ 37ᵐ
λ = 37°

Percival Lowell, del.

How Venus appeared to Lowell in 1896 (from *Monthly Notices of Royal Astronomical Society,* **March 1897).**

within the astronomical community, Lowell's ability to defend his
Venus work was suddenly suspended for nearly four years by the
onset of nervous exhaustion in April 1897, shortly after his return
from Mexico.[30] His condition was not severe at first and, although
shocked by the death of his friend Clark early in June, he felt he was
on the way to recovery by mid-summer.[31] "I get better myself slowly," he wrote Douglass in July. "My quantity increases and they tell
me that my quality will come after it."[32]

There was relapse, however, a "case of evil-ution," Lowell
quipped.[33] By the end of September, his brother-in-law William L.
Putnam took over the full supervision of his affairs, including his
observatory, advising See and Cogshall in Flagstaff that Lowell "has
been feeling poorly for some days and we are not going to bother him
with any more business."[34] In October, See wrote Douglass, who
was then in California, that Lowell "seems to be worse. I feel awfully
sorry for Lowell—the poor man wanted so badly to continue his
work, and must give it up for a considerable time."[35]

By December, Putnam was only mildly optimistic, noting that
Lowell was "steadily but slowly gaining strength, but there is no
likelihood of our being able to discuss any business with him for six
weeks at the earliest."[36] The enervating illness was to linger far
longer. In June 1899 Miss Leonard warned Douglass: "Do not
allow yourself to break down. Mr. Lowell's condition is a bad example. I can see that he is better but not well yet."[37]

Lowell's long convalescence was carried out variously at his
father's home in Boston, Bermuda, Virginia, Maine, New York, and
by early 1900 on the French Riviera where he held lengthy and
presumably philosophical talks with fellow Bostonian and neuras-
thenic William James. During this sojourn in Europe, he also visited with Sir William Huggins, the pioneer spectroscopist, and other
astronomers; he attended astronomical meetings and prepared to
observe a solar eclipse on May 28, 1900, from Tripoli in North
Africa with Professor David Todd of Amherst College. He had, in
fact, made quite elaborate plans to coordinate his observations with
those Douglass was to make from Washington, Georgia, in the
southeastern United States.[38]

When early in 1901 he had finally regained his former vigor, the
resumption of his astronomical work was welcomed by at least
some astronomers. H. H. Turner, director of the Oxford University
observatory and editor of its journal, *The Observatory*, wrote that "we
are all glad to learn of your returning health. Astronomy cannot

CHART OF VENUS.

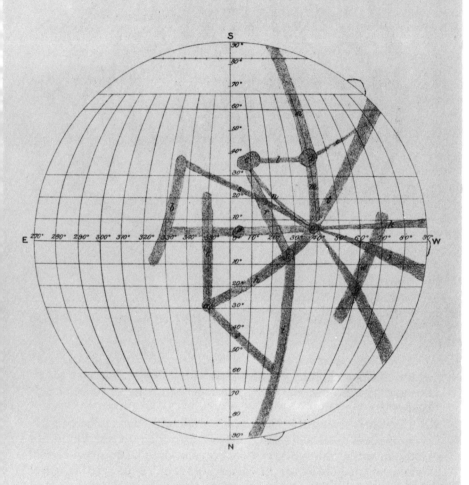

Chart of Venus, drawn by Lowell on October 19, 1896 (from *Monthly Notices of Royal Astronomical Society,* **March 1897).**

MAP OF MERCURY

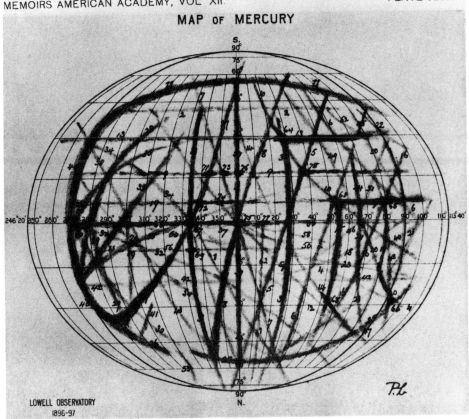

Map of Mercury, drawn by Lowell in 1897 (from *Memoirs, American Academy of Arts and Sciences,* **Vol 12).**

spare its most vigorous workers—they are too few. Don't overdo it, though," he cautioned, "until you feel quite sure of full strength again."[39]

Toward the end of 1899 when he gradually began to take up his astronomical interests again, Lowell's first concern was Venus. In October, while visiting in Boston, Douglass wrote Drew that Lowell "wants to work away on the observations of Venus. ... He is not working very hard but is nervous enough to have it worry him if all material is not at hand."[40] The need to bolster his 1896–97 conclusions about Venus with better evidence, indeed, became urgent a few months later when astronomer A. A. Belopolsky of the Russian Imperial Observatory at Pulkova, in April 1900, reported spectrographic observations supporting a "short" Cytherean day.[41]

"The sentence of Belopolsky," Lowell confided somewhat later, "is indeed a hard nut to crack."[42]

But Lowell intended to crack it, and to crack it moreover with spectrographic evidence of his own. To this end, late in 1900 he ordered what would prove to be an exceptionally fine spectrograph for his Flagstaff observatory from the renowned instrument-maker John A. Brashear of Pittsburgh's Allegheny Observatory.[43] And in the late summer of 1901, all his early assistants having departed for various reasons, he brought in a new assistant—Indiana University graduate Vesto Melvin Slipher—to operate it. The combination turned out to be highly successful, not only for Lowell's work on the rotation of Venus and other planetary problems, but for Slipher's subsequent career in which he distinguished himself by his pioneering work on the radial velocities of spiral nebulae.[44] How quickly and well young Slipher learned to use the new spectrograph with the Lowell 24-inch refractor is indicated in September 1902 in a letter from Lowell to President Henry S. Pritchett of the Massachusetts Institute of Technology where Lowell had just received an appointment as nonresident professor of astronomy:

Enclosed are some prints of Jupiter's spectrum showing the planet's rotation. They are the first fruits of the new spectroscope, made by Brashear, in the hands of V. M. Slipher.

The spectroscope Mr. Brashear considers "the finest and best instrument we have ever made" under the date of Sept. 21, 1901. It was made so that I might be sure of the best possible instrument in the matter of Venus' rotation period.[45]

Some six months later, Lowell advised Pritchett:

You will be interested to know that Mr. Slipher has just finished a spectroscopic research on the rotation period of Venus and that the result, which was got without any bias from me and with every precaution of his own to prevent unconscious bias on his part and to eliminate systematic errors, is completely confirmatory of Schiaparelli's period and of my visual work here in 1896–97—the planet undoubtedly rotates in the time it revolves.[46]

Slipher, however, in formally reporting his findings some months later in a *Lowell Observatory Bulletin,* jumped to no such conclusion, declaring only that his spectrograms revealed "no evidence that Venus has a short period of rotation," and adding that "so fast a spin" as 24 hours could not have escaped detection.[47]

Some six years later, interestingly enough, Lowell flatly claimed that Slipher's Venus spectrograms "yielded, indeed, testimony to a negative rotation of three months which, interpreted, means so slow a spin as this was beyond their power to precise."[48] The statement is as curious as it is gratuitous, for ignoring the matter of how something beyond "power to precise" can yield testimony to anything, Lowell was certainly aware that a rotation of three months, whether positive or negative, was well beyond the practical resolution of the spectrograph. He did apparently persuade Slipher to accept his interpretation, however, although the only documentary hint of such persuasion in the Lowell archives is this passage in a letter from Slipher later in 1903:

Was very glad to get your letter concerning the Venus measurements. You were right (and I wrong as I usually am). I am given to making mountains out of molehills, whenever the latter are unfavorable.[49]

In 1903 while Slipher was examining Venus with the spectrograph, Lowell made another extensive series of visual observations of the planet and again found the "curious system" of dark markings he had described in 1896–97, reporting his findings in another observatory *Bulletin*.[50]

For the remainder of his life, apparently, Lowell continued to believe that he had seen actual features on the surface of Venus and that the markings he described were not spurious as almost everyone else in astronomy then and since has insisted. But in light of the total absence of confirmatory observational evidence, he no longer made much of the point in public. Privately, however, he stood his ground. Late in 1903, for example, he defended the reality of the Venus markings to English astronomer E. Walter Maunder who, intentionally or otherwise, had confused them with the martian "canals" which he considered to be equally illusory.[51] And in 1907, when Lowell's attention was called to a reference to veils of cloud on both Mercury and Venus in a critical article by the Reverend George M. Searles in *Catholic World,* he replied confidently:

The statement you were struck by about Mercury and Venus being covered with uniform cloud is merely a survival of opinion of past eras of thought in the case of Venus, while as to Mercury I do not know where it originated. Observations at Flagstaff are perfectly conclusive however as to the universal character of the surface of Mercury. ... As to Venus they are also practically conclusive. There are absolutely no grounds for Fr. Searles' statement, but you know how difficult it is to change a preconceived opinion.[52]

It is indeed. In later years Lowell continued to respond quickly to published implications that he had changed his mind about the markings on Venus. In 1908, he sought to delete a reference to his 1902 *Astronomische Nachrichten* note from Charles A. Young's *Manual of Astronomy*,[53] and in 1913 he tersely advised the editor of the *Observatory* that "I see in No. 461 . . . that I am said to have rejected some of my observations of Venus. This is in error. I was at one time tempted to doubt them as a matter of opinion, an apprehension which has since been removed by my further observations."[54]

Nor did Lowell ever abandon his conclusion on the rotation of Venus, and here he was on sounder ground, as it turns out. For after the announcement of Slipher's careful spectrographic work, astronomers increasingly began to favor a slow if not entirely synchronous spin for the planet, although for many years textbooks on astronomy would continue to list its rotation period with a question mark.

Again the answer to the problem of the rotation of Venus did not come until the advent of radar techniques in the 1960s, and again the answer was a surprising one. Venus, it is now known rotates very slowly indeed, in about 243 earth days, and in a retrograde, that is to say a "negative," direction. Moreover, this rotation is such that each time the planet arrives at inferior conjunction, that is, when it is closest to the earth, it presents the same hemisphere to the observer.[55]

Even though Lowell was largely out of astronomical action, as it were, from late summer 1897 to early 1901, there was nonetheless some activity at his Flagstaff observatory. In September 1897 a fire, which See claimed to be of "incendiary origin," razed the Grand Canyon Hotel where See and Cogshall were living and destroyed their personal possessions. Lowell, See noted, "with characteristic grace offered to replace our losses and our library."[56]

During this period too, there were extensive comings and goings. Drew left in June 1897.[57] Douglass, becoming ill himself, spent several months that summer and fall recuperating in California, leaving See in charge at Flagstaff. Cogshall departed in October 1897 and See appointed Samuel Boothroyd, a former science student at the University of Chicago, as his replacement.[58] In November, with the return of Douglass, See left on an extended trip

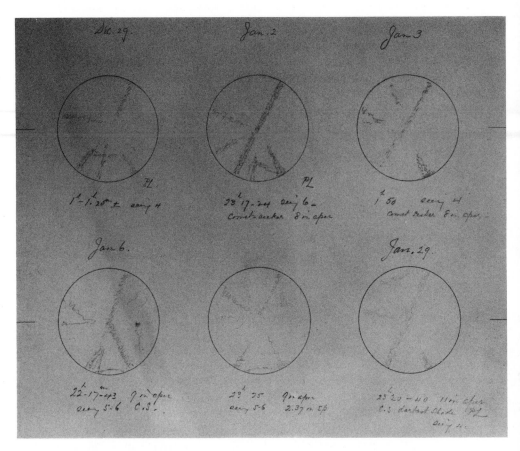

How Venus appeared to Lowell in 1903 (from observatory log, 1903).

through the east to prepare his double star catalog for publication and otherwise to promote his own interests and his grandiose astronomical ideas. Clearly no one on Mars Hill was unhappy over his prolonged absence, and indeed the prospect of his return caused considerable consternation. See, to say the least, was not at all popular with his colleagues on the Lowell staff. Boothroyd, in fact, cited his deep dislike of See as the reason for his reluctant resignation in June 1898 in the midst of Barnard's visit to Flagstaff.[59]

The case of Dr. Thomas Jefferson Jackson See is a strange and fascinating one in the annals of astronomy, but it can be only

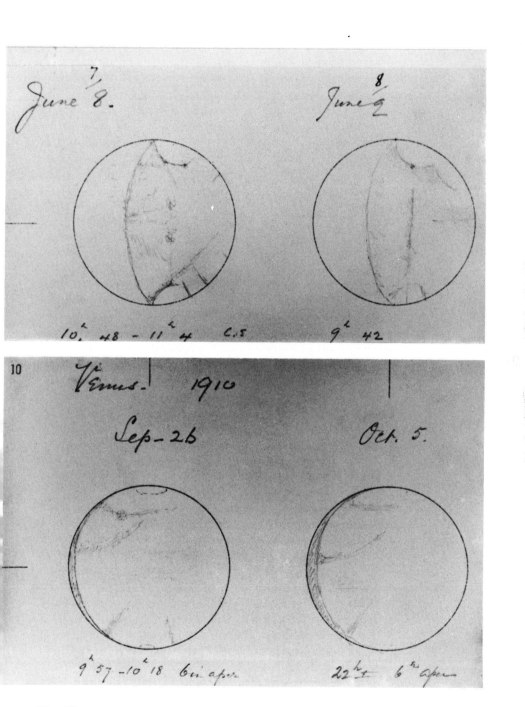

How Venus appeared to Lowell in 1910 (from observatory log, 1910).

touched on here as it bears on his short, troubled tenure at the Lowell Observatory. It is obvious that See's repellent, egotistical personality caused the general exodus of Lowell's assistants during this period. The complaints against him, documented in lengthy letters to Putnam in June 1898 from Cogshall, Boothroyd, and Douglass, comprise devastating indictments of his arrogance, deviousness, incompetence, instability, lack of principle, and propensity for appropriating the work of others.

Boothroyd's charges are typical. See, he wrote, talked to him in a "disgraceful manner," cursed at him, and blamed others for his own mistakes. He was not only "jealous" of the discoveries of others but he "has not scrupled to print such discoveries as his own." Citing See's claims to have measured a hundred thousand or more double stars, Boothroyd declared that these were "not the only instances where I can vouch for his departure from the truth," and added that "having been closely associated with him in his work, I have had occasion to doubt the accuracy of some of his observations. ..."[60]

Cogshall detailed a similar bill of particulars, although with a greater emphasis on the defects in See's character, calling him "the most cowardly man I have ever known. ... He says the meanest things he can think of behind a persons [sic] back and acts as smooth and suave as possible to his face. ..."[61] In a letter written to Douglass the same day, Cogshall also reported that he had discussed See with astronomer Forest Ray Moulton,* who had known See at the University of Chicago, and suggested to Douglass that See's erratic and unprofessional behavior might be an effect of his use of "morphine or other strong drug."[62]

In his own deposition, Douglass concluded from a long list of personal and professional grievances that See displayed "evidences of mental degeneration"; he ended his letter with a peculiarly bitter denunciation. "Personally," he wrote Putnam, "I have never had such aversion to a man or beast or reptile or anything disgusting as I have had to him. The moment he leaves town will be one of vast and intense relief and I never want to see him again. If he comes back, I will have him kicked out of town."[63]

---

*Moulton, soon to coauthor the Chamberlin-Moulton "planetesimal" hypothesis of solar system origins, became a frequent future source of See's "original" theories. See Chapter 13, *infra*, for example, or Moulton, "Capture Theory and Capture Practice," *Popular Astronomy*, 20:67 (1912).

These letters, submitted by Douglass as evidence for the prosecution as it were, as well as the questions some others were raising about See's observations, moved Putnam to sever See's connection with the observatory in early July. He cautioned Douglass, however, to keep the reasons for See's discharge as quiet as possible so that there would be no reflection on the observatory and its work.[64]

The case of Dr. See was thus ended as far as the Lowell Observatory was concerned. Early in 1899 Douglass, apparently on his own initiative, mounted an unsuccessful letter-writing campaign to block See's contemplated appointment to an U.S. Naval Observatory post, advising several prominent scientists with influence in government circles that See was "untrustworthy."[65] Presumably Putnam subsequently brought Lowell up to date on the See situation, for in later years Lowell remained cordial but coldly unencouraging to See's occasional efforts to strike up a correspondence.

The trauma that See inspired within the Lowell staff had by mid-1898 reduced the number of astronomers on Mars Hill to one. By July 1901 there were no assistants at all left at the observatory, Douglass himself having been abruptly "discharged" for "untrustworthiness," Lowell later informed the editor of the short-lived journal the *American Astronomer* who had published a critical article by Douglass entitled "Illusions of Vision."[66] The judgment, appearing suddenly and belatedly in the record, seems harsh and even unfair in view of the seven years Douglass had devoted to Lowell and his observatory. Furthermore, it cannot be otherwise documented in the Lowell archives where, indeed, references to Douglass' departure are noticeably absent. Yet, because Douglass went on to a long and distinguished career in science, some attempt must be made here to explain Lowell's statement.

It would appear that Lowell acted primarily on the undeniable fact that by 1901 Douglass had developed strong doubts about the validity of many of Lowell's planetary observations. Certainly, as noted earlier, he had questioned the objective reality of the so-called "gemination" of the martian canals. In raising the question in the second volume of the observatory's *Annals,* he had even implied that Lowell had doubts of his own about the phenomenon. Also he had expressed reservations about the relative excellence of the location of the Flagstaff observatory for astronomical observations. Moreover, late in 1900, and apparently against Lowell's expressed wish, he had undertaken an investigation to determine whether the disputed planetary details reported by Schiaparelli, Lowell, and others might be psychological in origin—might, in short, be all in the eye or mind of the observer, as Holden and other critics had charged. In this connection, early in 1901 he submitted a long list of pointed questions to Joseph Jastrow, an eminent psychologist at the University of Wisconsin; his accompanying letter reveals the extent to which his astronomical apostasy, at least from Lowell's standpoint, had progressed:

I would have written you long before but for Mr. Lowell's indifference to taking up the psychological question involved in astronomical work. . . . I have made some experiments myself bearing on these questions by means of artificial planets which I have placed at a distance of nearly a mile from the telescope and observed as if they were really planets. I found at once that some well known planetary appearances could, in part at least, be regarded as very doubtful. . . .[67]

Thus there seems to be no doubt that by 1901 Douglass had come to the point of fundamental disagreement with what Lowell had been seeing and saying, and had communicated this disagreement outside the observatory's circle, an action which Lowell most certainly would not have approved. Beyond this, however, Douglass on March 12, 1901, composed a lengthy, confidential letter to Putnam complaining about Lowell's "unscientific methods" and his over-hasty publication of sweeping conclusions on the basis of only a few, selected observations. "His method is not the scientific method and much of what he has written has done him harm rather than good," he advised Lowell's brother-in-law. "I fear it will not be possible to turn him into a scientific man."[68]

Douglass' letter seems to have been well-intentioned, for he also wrote of his "intense loyalty" to his employer. But it certainly was ill-considered in light of Putnam's close relationship with Lowell. For despite his plea to Putnam to keep the letter "strictly between ourselves," Putnam apparently showed it to Lowell, and Douglass promptly found himself out of a job, a circumstance he not only later regretted but considered "a gross injustice."[69]

At least two other factors may have contributed to the dismissal of Douglass. First Lowell may have disapproved of Douglass' vigorous role in See's ouster and his vain efforts to block See's appointment to the U.S. Naval Observatory. For Lowell never seems to have developed the strong antipathy to See displayed by his early assistants and later by astronomers in general. Secondly, Cogshall may have claimed at this time that Douglass had earlier appropriated some of his work on the photography of the zodiacal light. At least in 1903 Cogshall made such a claim to Lowell with some documentation.[70]

Three months after his departure, Douglass appealed for reinstatement, a plea which Lowell tersely rejected by declaring "I have at present all the assistants I want at Flagstaff."[71] Still, in his future writings and lectures, Lowell never denied Douglass full credit for the work he had done from 1894 to 1901.

Douglass disappears from the record in the archives after 1901. He did, however, serve a stint as a probate judge in Prescott (Arizona Territory), returning to Flagstaff in 1905 to teach science and Spanish at the Northern Arizona Normal School (later to become Northern Arizona University). In 1906 he joined the faculty of the University of Arizona where in 1916, just weeks before Lowell's death, he was named the first director of the university's new Steward Observatory, Arizona's second astronomical facility. He noted at the time, incidentally, that observing conditions at Tucson "approach the ideal."[72] In later years, Douglass won renown by developing the science of dendrochronology—the absolute dating of prehistoric and historic materials from the annual growth of tree rings—which has proven of great value in the fields of archaeology and ethnohistory.

While the years of Lowell's illness were not particularly productive astronomically in Flagstaff, they did not come to an end without one final sensation which, appropriately enough, related to Mars. This was the sighting, on December 7-8, 1900, by Douglass, of a prominent "projection" on the edge of the Scarium Mare, which set off another flurry of reports that martians were attempting to signal the earth. The reaction was brief but intense. A press syndicate in Philadelphia, Pennsylvania, wrote Douglass that it was "very much interested in recent announcements concerning the message which it is reported you received from the planet Mars sometime since."[73] And an attorney in Denver, Colorado, inquired:

I observe from the papers that . . . you observed a projection of some kind on the planet Mars. Will you kindly advise me what the projection appeared to you to be, whether a dark material object, or a shaft of light? Suppose the people of Mars have built a monument 10 miles square at its base and a hundred miles high, covered exteriorly with polished marble (which is the fact). Would not the monument reflect a shaft of light? If you saw a shaft of light what was its color, and did the light scintillate?[74]

Mars-and martians still had a strong fascination for the lay mind, as Garrett P. Serviss, science writer and columnist for the Hearst newspapers, pointed out in commenting on Douglass' "projection."

This will probably lead to a renewal of the suggestions made during the last preceding opposition of Mars, that the inhabitants of that planet are signalling the earth. . . . How deeply this idea has sunk in the popular imagination is indicated by a proposal . . . that special researches be undertaken for the purpose of solving the problem of interplanetary communication.[75]

And in a somewhat more general vein, he added:

It is useless to try to disregard the problems of Mars. Every time the planet arrives at opposition something is seen that piques the curiosity and obscurely hints at more interesting revelations behind. The time is past when these things could be dismissed as a mere illusion and speculation. The accumulation of observations is too great to admit of indifference on the part of those who have not actually seen the phenomena described. The most skeptical must now believe that there are on Mars the things called "canals" and the other singular appearances from time to time noted by observers. The only question now is, what do they mean?[76]

Lowell and his theory were gaining ground.

CHAPTER
9

New Techniques

THE YEAR 1901 marked the beginning of a new period in the history of the Lowell Observatory and in the turbulent astronomical career of its founder and director, Percival Lowell.

Restored to full health and his former vigor after nearly four years of enervating illness, Lowell began to assert a far more rigorous direction over the work of his observatory at Flagstaff and over the new staff of young assistants he would assemble on Mars Hill over the next few years.

Now he would attack the problems of Mars, the solar system, and planetary evolution with a new intensity and with the new tools and techniques of spectrography and photography in a sustained and determined attempt to provide incontrovertible proofs for his controversial theories. This unprecedented effort, ironically, fell short of its goal, although it left the science of astronomy in his debt. It also brought the curious intellectual phenomenon of the Mars furor to its frenzied peak and assured Lowell, as its principal *agent provocateur,* of worldwide fame.

Astronomically, the year began auspiciously enough with the discovery on February 22 of a brilliant nova near the familiar eclipsing binary star system of Algol in the constellation of Perseus. The exploding star flared through twelve magnitudes to temporarily surpass Capella in brightness.[1] Lowell, in England at the time, observed the nova a few nights later "through a hole in the clouds" with his friend, the eminent British astronomer Sir Robert Ball, and reacted with typical enthusiasm. "Isn't this new star a marvel for magnitude," he exclaimed in a letter to Flagstaff. "There has been nothing like it since Tycho Brahe!"[2]

[127]

A few years hence, Lowell would cite this "new star" as evidence for his concept of planetary evolution.[3] But in 1901, Nova Persei held only passing interest, and within a month Lowell was on his way back to Arizona to turn his thoughts and his telescope once again toward Mars.[4]

The martian opposition of 1901 was not at all favorable from an astronomical point of view—the planet's disk was just over 13 seconds of arc in diameter at its closest approach to earth and only about half its apparent size at most favorable apparitions. Lowell did not begin his observations until March 31, more than a month after Mars had come to opposition, but nonetheless he managed to find something to report:

> Bright spots made their appearance . . . in parts of the planet and at times which render explanation difficult. . . . The peculiarity of these spots consisted in their long continuance . . . for months. In splendor, they nearly rivalled the sheen of the polar caps. . . . They are clearly local phenomena dependent upon local conditions, for just these parts of the planet have a way of appearing bright.[5]

By early July, with the departure of Douglass, the last of his early assistants, Lowell was left briefly alone at his observatory—a circumstance which does not appear to have concerned him. Indeed, he seems to have soured on the idea of having assistants at Flagstaff in the future, perhaps because those who had been there during the period of his illness-enforced absence had been notably unproductive of significant astronomical observations.

Earlier that spring, however, Lowell had already agreed to employ on a temporary basis a young graduate of Indiana University named Vesto Melvin Slipher who came with high recommendations from astronomer W. A. Cogshall of Indiana's Kirkwood Observatory, himself a former Lowell assistant at Flagstaff and in Mexico.[6] Lowell fulfilled his commitment with considerable reluctance:

> As regards Mr. Slipher: I shall be happy to have him come when he is ready. I have decided, however, that I shall not want another permanent assistant and take him only because I promised to do so; and for the term suggested. What it was escapes my memory. If, owing to this decision, he prefers not to come, let him please himself.[7]

Slipher, of course, did come, arriving in Flagstaff late that August, and the temporary nature of his employment most certainly did escape Lowell's memory. Slipher stayed fifty-three years, serv-

Vesto Melvin Slipher

ing as chief assistant until Lowell's death in November 1916, and then in accordance with Lowell's wishes as the observatory's second director until his retirement in 1954 after a distinguished career. Slipher maintained his interest in the observatory's work and in astronomy until his death on November 8, 1969—three days before his ninety-fourth birthday. [8]

Cogshall was also responsible for the subsequent addition of two other Hoosier astronomers who became longtime members of the observatory's staff.

The first of these was Carl Otto Lampland who took up residence on Mars Hill late in October 1902, remaining as an assistant there until his death in 1951. [9] Under Lowell he pioneered techniques of planetary photography; later, with W. W. Coblenz, he made the first attempts to determine planetary temperatures with the aid of the device known as a thermocouple. [10]

The second addition was Slipher's younger brother, Earl C.

Slipher, who came to Flagstaff in July 1906 under the short-lived Lawrence Fellowship program which had been established the previous year to provide selected Indiana graduates with a year's practical experience in astronomy at the Lowell Observatory. The younger Slipher, second of only three Lawrence Fellows during Lowell's lifetime, became a permanent member of the staff in 1907, and remained associated with the observatory until his death in 1964. For more than half a century he distinguished himself through systematic photography of the brighter planets, most notably by his recognition of the value of multiple-image printing to enhance the quality of planetary photographs and as a leading observer of Mars.[11]

While Lowell was alive, these three young men worked almost entirely under his direct supervision and largely on those problems of planetary astronomy in which Lowell himself was primarily interested. Mars held the top priority, for the planet's 1903, 1905, and 1907 oppositions would be progressively more favorable, providing a series of observational opportunities for bolstering his embattled martian theory that would not occur again until the decade of the 1920s. The other members of the solar system, excepting only the sun itself and earth's moon, also came in for methodical study, for Lowell now began to formulate the ideas on planetary evolution that he would claim as the basis of a new and special branch of astronomy he called "planetology." Sporadically from 1905, and intensively from 1910, he kept the staff and principally Lampland busy on a photographic search for a long-suspected ninth planet beyond Neptune, whose existence he confidently and publicly assumed as early as 1902 and whose discovery in 1930 at the Lowell Observatory posthumously added to his earlier fame.[12]

In most of their work at Flagstaff, the two Sliphers and Lampland functioned essentially as skilled astronomical technicians carrying out Lowell's instructions and devising practical ways of testing experimentally and observationally his theoretical ideas. Lowell did not, however, discourage them from pursuing research of their own, provided it did not conflict with tasks he had assigned them. But only V. M. Slipher in Lowell's lifetime looked effectively beyond Mars and the solar system toward the stars, the nebulae, and the other wonders of deep space. In noting Slipher's early interest in the discovery of new double star systems by determining spectroscopically the varying velocities of their components in the line of sight, Lowell put his position politely but firmly:

With regard to yourself, by all means make your star measures for velocity—whenever there is no pressing planetary work—and good luck to you in the result. [13]

Luck certainly played a part in Slipher's subsequent career, but luck alone cannot account entirely for his many significant achievements in astronomy. Slipher's arrival in Flagstaff coincided with the completion of the spectrograph that Lowell had ordered the previous year from instrument-maker John A. Brashear and which Brashear had designed as an improved copy of the famous Mills spectrograph that had been in use at the Lick Observatory since the mid-1890s. It also coincided with the emergence of spectrography as an important new tool for astronomical research and, on a broader level, a general technological advance in astronomy and its related fields, including photography. Slipher also was fortunate to be associated with Lowell and his Flagstaff observatory, for Lowell had both the sense and the means to provide the finest available instruments and equipment which could be used, moreover, at a location that was especially favorable for astronomical observations.

With the Brashear instrument Slipher pioneered the spectrographic study of planetary rotations and atmospheres, discovered the existence of interstellar gas and dust, found that nebulae rotate, and, through his painstaking determinations of the enormous radial velocities of spiral nebulae, provided the first evidence for the modern theory of an expanding universe.

"I know of no observer in the 20th century who has made a comparable number of fundamental discoveries," John S. Hall, who became director of the Lowell Observatory in 1958, wrote in his obituary review of Slipher's work. "Many of these were basic to present-day studies of the planets, the interstellar medium, diffuse nebulae, galaxies, globular clusters, the light of the night sky, auroras and comets."[14]

Slipher's observational abilities, however, were not especially in evidence when in the fall of 1901 he began to put the Brashear instrument, attached to the 24-inch refractor, into "perfect running order" as Lowell had directed. [15] His early attempts at spectrography drew a series of patient criticisms from Lowell in Boston:

Your prints of spectra I have examined and notice that not only Jupiter but Capella shows a slight shift, not as great as Jupiter, but comparable with Venus, suggesting some other [agent] at work.

Will you look into this. They are all shifted as lines should be were mere rotation at work.[16]

The problem, Lowell advised his youthful assistant a week later after consulting with James McDowell of the Brashear firm, "is due to lack of adjustment."

Either the star is not on the center of the slit or the refracting edge of the prism was not parallel with the slit opening. See that these errors are corrected and then compare Capella's spectrum with Venus and send me prints. . . .[17]

Slipher's problems persisted, and months later, in March 1902 Lowell was still sending him elementary advice:

Incidentally, let me call your attention to the ease with which the two ends of the spectrum can get misnamed. I do not know which print is rightly oriented, red or violet, but the other is evidently wrong as comparison at once shows. . . . Of course you know the two side spectra should be absolutely symmetrical with regard to the central space; the shift or tilt of the lines can then be much more nearly measured.[18]

A few weeks later, he wrote:

When you get some new plates untilted send them on. . . . Stars of course they will be to begin with; then *per astra ad aspera* to the planets.[19]

For a time, Slipher seems to have despaired of solving his difficulties with the spectrograph without some direct instruction. Early in December 1901 he suggested to Lowell that he visit the Lick Observatory to learn the techniques used there with the Mills instrument. Lowell quickly vetoed the idea:

I think it would be inadvisable for you to go to the Lick at present. When you shall have learnt [sic] all about the spectroscope and can give as much as you take it will be another matter and at that time I shall be very glad to have you go.[20]

Slipher's acquiescence apparently was not all that was desired, for Lowell acknowledged it with some acerbity:

I am glad the postponing of your desired trip to the Lick commends itself to you. You are quite right in supposing that everybody encounters the same snags; the only difference being the clever ones

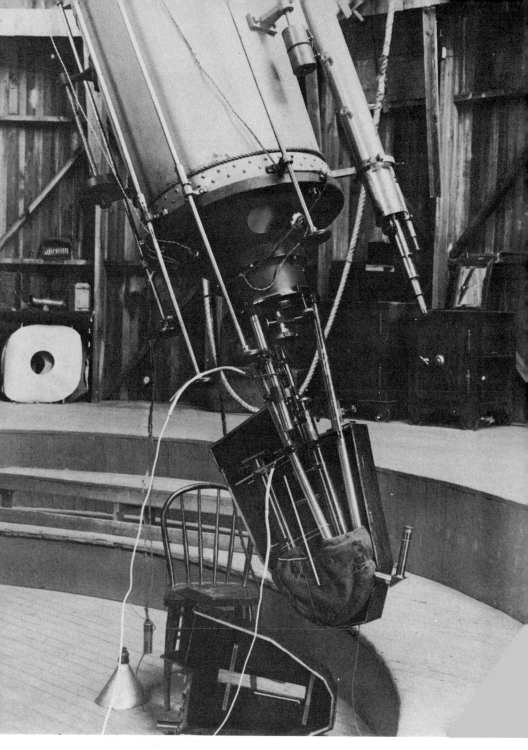

Brashear spectrograph attached to 24-inch telescope.

contrive to get over them themselves whereas the stupid ones have to have recourse to others. [21]

Thereafter Slipher did contrive to resolve most of his problems with the spectrograph and by mid-summer of 1902 produced spectrograms of Jupiter and Saturn which Lowell considered to be good enough to send to a number of eminent scientists. [22] To Brashear, he wrote enthusiastically:

You will be glad to know how your spectroscope is doing. The enclosed spectrogram of Saturn—which is ten times the breadth of [J. E.] Keeler's—will show you better than words what we hoped to get. [23]

Young Slipher, however, was not allowed to rest on his laurels. Even as letters of congratulations for the Jupiter and Saturn spectra came in, Lowell kept a critical eye on the work of his assistant in Flagstaff:

I have been looking over your prints and cannot tell whether these are the "better spectrograms of Saturn" you refer to or not. I feel these spectrograms are not as good as you can get so please take some more at once. [24]

As Slipher tried a hesitant hand at writing up these first results, Lowell offered some practical advice concerning the most effective approach to the publication of scientific data:

Apropos your article about Jupiter; the only criticism I have to make is: that apology for a performance not being better is a mistake. Equally is a suggestion that it may be better in the future. To state the facts themselves is the best way. [25]

Along with demonstrating his new-found competence, and the excellence of the Brashear instrument, Slipher had provided new spectrographic confirmation of the rotation periods of Jupiter, Saturn and Mars, a circumstance which Lowell duly reported to the December meeting of the American Association for the Advancement of Science in Washington, D. C. [26]

The autumn of 1902 was a particularly important time in the history of the Lowell Observatory. That fall, Lowell set Slipher to work on two intensive spectrographic studies that would occupy much of his attention for the next few years and lead to several significant discoveries.

The first of these broad investigations involved the difficult problems of determining the rotation periods of Venus and

Uranus.[27] In the case of rapidly revolving planets with clearly defined equatorial planes, such as Jupiter, Saturn, and Mars, it is a relatively simple matter to detect rotation spectrographically from the so-called Doppler effect in which the varying radial velocities of the opposite limbs of the rotating planet cause a measurable tilt of the absorption lines in its light spectrum—wavelengths from the planet's approaching limb being shortened and their spectral lines shifted toward the violet, and those from the receding limb being lengthened and shifted toward the red.

On cloud-shrouded Venus, however, astronomers could only guess at the position of its equator where, of course, rotational velocity is greatest and best measured, while the equatorial plane of Uranus is inclined some 82 degrees to the plane of the ecliptic and for extended periods during its 84-year orbit the planet's axis is pointed almost directly toward the earth.

As previously noted, Slipher presented Lowell with spectrographic evidence for a "slow" rotation of Venus in March of 1903, only some five months after he took up the problem. These observations seemed so conclusive to Lowell that he felt it "unnecessary" to ever have them repeated.[28]

The problem of the length of the Uranian "day," however, required considerably more time, for during the early years of the twentieth century the planet's aspect was such that its south pole was turned almost directly toward the earth-based observer. Yet Slipher, in a remarkably accurate observation, finally resolved it spectrographically in the summer of 1911, determining a retrograde rotation for the planet of 10¾ hours. Subsequent observations over the years have not appreciably changed this result; indeed, the modern refinements on the datum vary by only a few minutes at the most from Slipher's original value.[29]

The second major investigation Slipher undertook at Lowell's direction in the fall of 1902 involved the determination of the compositions of planetary atmospheres. This also proved to be highly productive, although some of the results did not come in until many years later.

In this study, Slipher reported in August 1903 the presence of free hydrogen in the atmosphere of Uranus as a parenthetical note in a letter to Lowell devoted primarily to a discussion of spectroscopic binary stars.[30] The following April, on the basis of observations late in 1903 and early in 1904, he declared that he had no doubt of the existence of hydrogen in the atmospheres of both Uranus and Neptune, adding his belief that both planets also "must be at a rather high intrinsic temperature."[31] The results of this investigation were published in June in a Lowell *Bulletin*.[32]

By 1905 Slipher had accumulated a series of exceptional spectrograms of the major planets which showed many conspicuous absorption bands that neither he, nor anyone else for that matter, could immediately identify. The source of these lines, faintest for Jupiter and increasing in intensity for the outer planets with their distance from the sun, remained a major astronomical mystery for more than a quarter of a century, until in 1931 Rupert Wildt in Göttingen, Germany, suggested that some of them might be caused by methane and ammonia gas in concentrations well above those normally used by astronomers to produce comparison spectra. Wildt's work was confirmed and extended by Theodore Dunham Jr. at the Mt. Wilson Observatory in 1933 and later by others, including Slipher and Arthur Adel at the Lowell Observatory in 1934.[33]

But the most urgent thrust of this investigation of planetary atmospheres was on Mars. For Lowell realized that if the spectrograph could reveal oxygen and water vapor there, a major point in his Mars theory would be confirmed. The problem here, however, is that the light received from a planet is simply reflected sunlight which must penetrate earth's atmosphere to reach terrestrial telescopes. Oxygen, water vapor, and other gases in the earth's atmosphere produce strong absorption lines in planetary spectra which mask any corresponding lines that might reveal the existence of these gases in the atmosphere of another planet.

In 1902 the only known method for investigating the composition of planetary atmospheres spectroscopically was to compare the spectrum of a planet with that of the moon, observed at the same altitude in the night sky so that the effects of the earth's atmospheric absorption would be the same in both spectra. If the telluric lines appeared to be enhanced or intensified in the planetary spectrum, then it was assumed that this was evidence of their reinforcement by constituents in the planet's own atmosphere.

This method posed serious problems. For one thing, the degree of intensification of a given line was a matter of subjective judgment on the part of the individual observer. For another, the significant lines in planetary work, such as the so-called A and B lines of oxygen and the "little a" band of water vapor are in the faint, almost invisible, red end of the spectrum and were extremely difficult even to see, much less to measure accurately with the rudimentary means available at the time.

The approach, moreover, had produced inconclusive or sharply conflicting results with pioneer spectroscopists such as England's

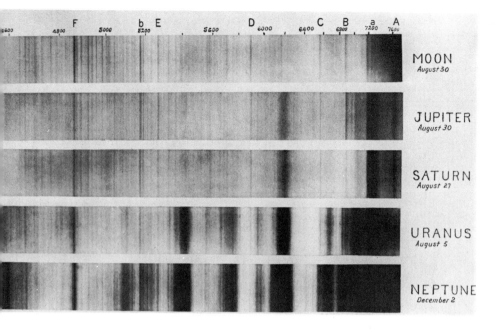

F   b E   D   C B a A
4800   5000 5200   5600   6000   6400 6800 7200 7600

MOON
August 30

JUPITER
August 30

SATURN
August 27

URANUS
August 5

NEPTUNE
December 2

V. M. Slipher's spectra of the major planets in 1907.

Sir William Huggins in 1867 and Germany's H. C. Vogel in 1873 opting for traces of oxygen and water vapor in the martian atmosphere, and Lowell's persistent critic W. W. Campbell as well as J. E. Keeler and others disputing these claims with visual spectroscopy in 1894 and spectrographically in 1896.[34]

In the autumn of 1902 Lowell came up with a better idea. Late in September, while scanning prints of Slipher's Jupiter and Saturn spectra, he found one in which the absorption lines not only were tilted by Jupiter's rapid rotation but also were shifted slightly toward the red, apparently the result of the planet's overall velocity of recession in the line of sight relative to the earth.[35]

A few days later, he wrote Slipher:

There has come into my head a new way of detecting the spectral lines due to a planet's own atmospheric absorption, and I beg you to apply it to Mars so soon as the moon shall be in position to make a comparison spectrum.

It is this. At quadrature* of an exterior planet we are traveling toward that planet at the rate of 18.5 miles a second and we are

---

*"Quadrature" refers to the two points in an orbit midway between conjunction and opposition. A half moon, for example, is at quadrature in its orbit around the earth.

carrying of course our own atmosphere with us. Our motion short-
ens all the wavelengths sent us from the planet, including those
which have suffered absorption in *its* atmosphere. When the waves
reach *our* atmosphere, those with a suitable wavelength are absorbed by it and these wavelengths are unaffected by our motion
since it is at rest as regards us. Even were the two atmospheres alike
the absorbed wavelengths reaching us would thus be different since
the one set, the planet's, have been shifted by our motion toward it
while the other set, our own, are such as they would be at rest. We
thus have a criterion for differentiating the two. And the difference
should be perceptible in your photographs. ... [36]

In the next few weeks, Lowell vigorously pressed this new line
of investigation, firing off letters to Slipher, enclosing charts and
drawings of Mars, advice on observing the planet, and calculations
of its line-of-sight velocity relative to the earth on specific future
dates.[37] Implementation of his new "velocity-shift" method had
clear priority:

It would be well to observe the planet's spectra visually and examine
the lines about which I wrote you ... for increased breadth or
duplicity.
     I should put all spectrographic observations now before the
visual ones as the drawing of Mars from time to time will be often
enough.[38]

Almost immediately Lowell sought to obtain photographic
plates that would be particularly sensitive in the far red of the
spectrum where the elusive oxygen and water vapor lines occur.
This search, which extended to England, was not to delay the new
project, for he was impatient for Slipher to put his idea to a test.
"Your red-sensitive plates are started from London," he wrote a
week after outlining the new method, "but of course you won't wait
for them to begin."[39]
     Lowell's sense of urgency is understandable enough from the fact
that Mars was already approaching quadrature preceding its 1903
opposition, an optimum time for such an experiment. But he had
another reason as well, as Slipher soon discovered, and one quite
typical of the competitive zeal with which he pursued his astronomical interests:

Of course, you have not mentioned to anyone what you want your
red plates for; it would be unpleasant to have someone else put the
idea in operation before you got round to it yourself.[40]

The problem of obtaining plates sufficiently sensitive in the far red to record the sought-after lines, that is, above the 6900-Ångstrom line of the spectrum, proved to be insoluble and was, indeed, quite beyond the technological capabilities of photography at the time. Neither the first batch of "red-sensitive" plates from London nor subsequent ones obtained over the next two years from plate-makers both in Europe and the United States, including some made up to Lowell's special order, were satisfactory for his purpose.[41]

Because of the lack of suitable plates, Slipher made no attempt to test Lowell's new method at the 1902–03 quadratures of Mars. And in 1904–05, with what he considered to be "improved" plates,[42] his spectrographic observations not only of Mars but of Venus turned out to be at best inconclusive. Pessimistic over the possibility of getting results, Lowell nevertheless decided to publish his new method in an observatory *Bulletin*. In this, after outlining his "new way of attacking the problem" and explaining that "the quantity at issue is . . . so small as to offer scant promise of revelation by spectroscopic methods," he reported that Slipher's plates had shown no evidence of any shift in the spectral lines, and that there had been "no perceptible difference in their density either," adding his conclusion that "water vapor is probably non-existent on the illuminated side of *Venus* and extremely scarce on *Mars*. . . ."[43]

Though the result was negative it seems proper to publish it, both because of the novelty of the method and because a negative result serves to fix a value of precision in such spectrographic work."[44]

Slipher, however, writing in the same *Bulletin,* was willing to concede only technological defeat, and this only temporarily:

Although this attempt has failed to detect aqueous vapor in *Mars* and *Venus*, the conclusion should not be drawn that it does not exist in their atmospheres, nor will it always remain impossible to discover it spectroscopically; for it may be present in both, and in the future, the red-sensitive plate may be greatly improved in speed and fineness of grain, enabling the velocity-shift method to be applied with more promise of success.[45]

Slipher was right, although it would be thirty years before Walter S. Adams and Dunham, working with the 100-inch Hooker reflector on Mt. Wilson in 1933, could establish an upper limit of

0.0015 for the ratio of oxygen in the martian atmosphere to that of the earth's with the "velocity-shift" method.[46]

And it would be another thirty years before Hyron Spinrad, Guido Munch, and Lewis Kaplan, again working with the Hooker telescope, in 1963 confirmed spectrographically the existence of minute amounts of water vapor in the martian atmosphere, finding absorption lines in the infrared region of the spectrum shifted 142 Ångstroms due to the relative velocities of the earth and Mars. On the basis of these observations, which incidentally showed the pertinent spectral lines to be strongest over the poles, the water vapor in the atmosphere of Mars was estimated to be between 1,000 and 2,000 times less than the amount in the terrestrial atmosphere—"extremely scarce" indeed.[47]

Slipher's initial failure did not discourage him from pursuing the problem, and three years later his persistence plunged him briefly into the swirling controversy of the Mars furor. From 1905 on he continued to experiment with the spectrograph, trying various new prisms and at one point even urging Lowell to obtain a new spectrograph.[48] He also sought to enhance the red-sensitivity of his plates by bathing them in various chemical solutions, working closely on this with R. James Wallace of Yerkes Observatory, a Lowellian as far as Mars was concerned who had supplied the "improved" plates in 1904 and who on occasion furnished special color filters for the observatory.[49]

For a time these efforts proved fruitless. But in January 1908, during a period when the air over Mars Hill was apparently exceptionally dry, Slipher obtained some spectrograms that he considered to be of more than usual interest. He immediately forwarded them to Lowell in Boston:

I am sending... the negatives of the spectrum of Mars. ... There is a question as to the "a" band being stronger in Mars than in the Lunar spectrum at equal altitude. ... I do not know what conclusion your examination of these plates will lead to but I shall of course be interested in hearing.[50]

After receiving Lowell's telegraphic answer, he reported at greater length:

Glad to have your telegram giving the result of your examination of the Mars spectra. I was not, of course, surprised that you found "a" stronger in Mars than in the moon. I said nothing in my letter as to my opinion or rather the result of my examination of them, for I knew you would prefer using your own eyes on the plates before reading of what I saw. Of course, I was pretty thoroughly convinced there was very good evidence pointing to additional absorption in the spectrum of Mars before sending you the plates. Since the moon has got round again I have photographed the two spectra on 3 nights—two of the plates are good. They agree with those sent you in showing small "a" changes in Mars. I showed these plates to Mr. Lampland and my brother, individually, and had arranged the spectra so that it was impossible to infer which was Mars and which moon. They both said "a" stronger in the spectra which were those of Mars. . . .[51]

With the receipt of this letter, Lowell released to "the world" the news that Slipher had succeeded in obtaining spectrographic evidence of water vapor on Mars, claiming the confirmation of an important point in his martian theory.[52] Newspapers and periodicals throughout the United States and Europe carried the announcement.[53] Almost immediately, the respected British scientific journal *Nature* conceded that the presence of water vapor on Mars "now seems certain," and added a few weeks later, on the basis of prints of Slipher's spectra that Lowell had sent to English astronomer Sir Norman Lockyer, that Slipher's new result "leaves little doubt" on the point.[54]

As fast as prints could be made from Slipher's Mars plates, Lowell dispatched copies to eminent scientists and friends on both sides of the Atlantic.[55] One set went to the *Encyclopaedia Britannica* which was then preparing a new edition and had already sought prints of Slipher's earlier planetary spectra through astronomer Simon Newcomb.[56] In the *Britannica,* Lowell exuberantly informed Slipher, "they are sure to go 'thundering down the ages'—as Prof. [E. S.] Morse would say—bumping at every obstruction."[57]

Slipher's spectrograms and Lowell's claims for them, after the initial flurry of attention in the press, seem to have stirred little public interest. Indeed, some editors, perhaps wearied by the sustained intensity of the Mars controversy over the preceding years, greeted this latest revelation about the red planet somewhat irreverently, not to say frivolously. *The New York Times* for example, which usually received Lowell's astronomical pronouncements respectfully, delivered itself of a lightly satirical essay on the subject of martian humidity which it headlined, "A Steamheated Planet."[58]

Astromers for the most part, while acknowledging the technical quality of Slipher's spectra, remained polite but noncommittal, an attitude that can be explained at least in part by the fact that as Slipher's work involved a then virtually unknown region of the spectrum they were hardly in a position to comment on it critically.

But there was inevitably some active opposition, led primarily by Campbell, whose negative spectroscopic findings for martian water vapor at the Lick Observatory in 1894 and 1896 were now, of course, brought into question. Campbell, after requesting and finally obtaining prints of Slipher's comparison spectra, advised him that he could "see no evidence that any lines or bands are stronger in Mars," and that indeed, "any excess of strength appears to be in the moon's spectrum."[59] This conclusion Slipher confided to John A. Miller, his former astronomy professor at Indiana, who brushed it aside by noting that "of course Campbell will not readily admit what he has contended against for a long time."[60]

Curiously enough, Slipher himself seems to have had some reservations of his own about the implications of his martian water vapor observations. In describing his conclusions to Lowell in February, he had closed his letter by noting that "this same band [a] or at least the band in this position [7200 Ångstroms] in the spectra of Jupiter and Saturn is very strong."[61] And in submitting a carefully-worded draft of his conclusions to Lowell in April he made this circumstance a parenthetical condition. His Mars spectra, he wrote:

... establish the conclusion that the "a" band is reinforced by absorption in the atmosphere of Mars. ... Then as this band in the terrestrial spectrum has been found to be due to water vapor in the earth's atmosphere, we are justified in concluding that the spectrograph has demonstrated the presence of an atmosphere about Mars, and that this atmosphere almost certainly contains water vapor ("almost" because of the strong bands in Saturn and Jupiter spectra coinciding with the "a" band).[62]

The qualification hardly satisfied Lowell,[63] who considered the spectra "the first certain or definite spectrographic proof of the presence of water vapor on Mars,"[64] and who already had planned to insert Slipher's results, along with a plate of his martian spectrograms, into the book version of his popular 1906 lectures *Mars as the Abode of Life,* then being readied for publication.[65]

Not satisfied himself, Slipher that summer packed spectrographic equipment to the top of the 12,600-foot San Francisco Peaks just north of Flagstaff, hoping to get above as much of the obscuring veil of water vapor in the earth's atmosphere as possible. These new spectrograms, however, were disappointing:

Although the elevation of the station on the Peaks is almost 13,000 feet the strength of the water vapor absorption was greater than that on some of the days last winter when I was photographing the spectrum of Mars. I am more firmly convinced than ever of the extremely favorable conditions under which some of those Mars tests are made.[66]

After Campbell's devastating evaluation of the earlier findings, Slipher also solicited opinions from several other astronomers, sending prints of his spectra to Forest Ray Moulton at the University of Chicago, co-author in 1905 of the so-called Chamberlin-Moulton "planetesimal" theory of solar system origins, and to the venerable Newcomb at the Naval Observatory in Washington, D.C. Both men declined to draw any conclusions from the prints. Moulton explained apologetically that he was "sorry that I do not know more about such things," adding that "my information is barely enough to make me thoroughly appreciative of good work along such lines."[67] Newcomb, conceding that martian water vapor was "intrinsically very probable," replied that Slipher's spectra showed "varieties of shade which render it impossible to speak decisively from their examination." Then he proceeded to read Slipher a solemn lecture on the desirability of obtaining impersonal, quantitative measurements of his plates.[68]

Lowell too, aware that the conclusions drawn from Slipher's Mars spectra were qualitative, was anxious to quantify them; even while Newcomb pontificated, Lowell was claiming that such quantification had been achieved. Within days of his public announcement of Slipher's work, he had accepted an offer by Frank W. Very, a former associate of Samuel P. Langley, to measure the intensity of the pertinent absorption bands in the Mars plates with a new instrument Very had devised which he called a "spectral band comparator."[69] It was some months before Slipher got around to sending some plates to Very at his small Westwood Astrophysical Observatory near Boston,[70] and some months before Very came up with a report. When he did, however, his findings startled even Lowell:

As I telegraphed you today, Professor Very has brought in the quantitative results of his new spectral band comparator and they are really surprising. ... The results are so good that I think we must have all plates so measured in the future for intensity.[71]

Very's results, which Lowell quoted to Slipher, showed the "little a" water vapor band 22% stronger in the spectrum of Mars than in that of the moon.[72] A few months later, Very reported that the B band of oxygen was intensified by fifteen percent.[73]

Despite Very's "quantitative" measurements, however, the

general undercurrent of skepticism became, if anything, stronger. Miller for example, reading a paper by Slipher on martian water vapor to a scientific meeting soon after Very's findings were announced, encountered "objections" which he attributed to "the critical temper of men towards Lowell's work."[74]

Campbell's response came late in the summer of 1909 when, in the company of Smithsonian Institution secretary Charles Greeley Abbott, he made new spectrographic observations of Mars from the summit of California's 14,495-foot Mt. Whitney, once again claiming negative results.[75] A garbled telegraphic report of his findings served to briefly confuse the issue further, but Campbell personally advised Slipher that even in the best of his spectrograms, his "colleagues and others ... all have difficulty in seeing 'a' in the Martian spectrum until I tell them exactly how to find it."[76]

Lowell, in turn, pointed out tersely that Campbell's observations were not only made during the summer—a "most unfavorable time ... because the Earth's moisture is then at a maximum"—but, from Abbott's own account of the Mt. Whitney expedition, at a time when the air above and around the peak was saturated with moisture from a regional storm system and thus could not possibly yield any significant results.[77]

For Lowell, this was apparently the end of it. His conclusion that water vapor existed on Mars rested, in any case, upon other criteria as well, and he was confident it would stand.[78] Moreover, he was now busy with a number of other intriguing astronomical matters. Very, certain that his spectral band comparator would eventually produce incontestable proof, continued to measure Mars plates, reporting the intensification of oxygen and water vapor bands in subsequent Lowell *Bulletins* in 1910 and 1914.[79] By this latter date, however, Lowell could list the accomplishments of his observatory in the field of planetary astronomy without any reference to either Slipher's martian water vapor spectra or to Very's repeated attempts to quantify them.[80]

From mid-1905 to 1907, Slipher's work also involved a Lowell-initiated spectrographic study of chlorophyll, the ubiquitous substance in green plants which carries on the life-essential process of photosynthesis, in the hope that it might be detectable in the blue-green areas of Mars, thus confirming Lowell's contention that these were areas of vegetation.[81] This investigation, Slipher confided to Cogshall, proved "quite interesting," although not immediately productive.[82] Lowell later explained that "the light was too weak to permit any answer."[83] In 1924, however, Slipher reported that the reflection spectrum from the dark areas of Mars did not resemble that of chlorophyll, a negative result confirmed in 1952 by astronomer Gerard P. Kuiper who suggested, instead, a similarity to the spectrum of terrestrial lichens. [84]

The martian water vapor issue marks Slipher's only public appearance as a participant in the Mars furor. In only one other instance did he allow himself, reluctantly, to be drawn into Lowell's astronomical imbroglios. This dispute, which paralleled the water vapor debate in time, concerned rival claims by Lowell and Campbell over the relative efficiencies of the Lowell 24-inch and the Lick 36-inch refractors at their respective locations on Mars Hill and Mt. Hamilton as determined by the number of stars visible in a given field. Campbell's comments on his Mars spectra may have goaded Slipher into this uncharacteristic action. Slipher confided both his distaste for controversy, and his reason for involving himself, to Miller:

This observatory has been getting it so generally that I have come to the conclusion that where we can defend ourselves that we shall have to do it or otherwise everything we publish will be discredited. [85]

Miller assured Slipher that he was "more than justified" in answering Campbell:

Good seeing anywhere but at Lick is a sore point for Lick observers, and Lowell's charting of the Lick starfields was a good argument for good seeing at Flagstaff. ... Lowell Observatory has been criticized but no one observatory has been gaining in scientific favor recently more than Lowell Observatory. [86]

Slipher's subsequent criticism of the performance of the Lick 36-inch, in a signed article in the *Outlook,* drew a half-reproachful, half-scornful answer from Campbell which marks the end of Slipher's brief, and presumably unhappy, venture into the field of astronomical forensics. [87] After this time, correspondence between the two astronomers was carried on with formal cordiality. In the future, Slipher was content to leave the polemics to Lowell whose mastery of the art by now was widely acknowledged.

From 1908 on, Slipher concerned himself primarily, although not entirely, with investigating objects far beyond the solar system. A major exception, along with his work in 1910–11 on the rotation of Uranus, involved his spectrographic studies of Halley's Comet, which passed perihelion in May 1910. These observations, Lowell contended, proved that carbon monoxide and cyanogen molecules in the comet's tail were "repelled" from its head by "light pressure" from the sun, a process which the prominent Swedish scientist

Svante Arrhenius had recently [1908] declared to be impossible on theoretical grounds. Wrote Lowell in an observatory *Bulletin:*

As the incompetency of light pressure to repel molecules in a comet's tail has been widely published (Arrhenius: *Worlds in the Making*), this observational proof that molecules in such a tail are repelled—whether they can be or not theoretically—is of considerable interest. [88]

Lowell also forwarded a paper on his Halley's Comet conclusions to Lockyer in England and to Henri Deslandres in France, requesting them to present it respectively to the Royal Society and the French Academy because, as he remarked to Lockyer, "it seems to me to be of cosmic importance."[89]

But it was Slipher's investigations outside the solar system that led to his most significant discoveries and won him the respect and approbation of his fellow astronomers in later years. As these are somewhat beyond the scope of the present study, they can only be briefly summarized here.

From the time of his arrival in Flagstaff in the late summer of 1901, Slipher had been interested in the subject of spectroscopic binaries, that is, star systems whose dual nature is not discernible telescopically but is revealed by the Doppler shift of spectral lines that results from the varying velocities of their components in the line of sight as they revolve around the system's center of gravity. The field was quite new in 1901, having opened up only a dozen years before with the discovery of the first such objects by Harvard College Observatory,[90] and Slipher pursued it sporadically whenever he was not working on Lowell's planetary assignments.[91] In 1904, German astronomer J. F. Hartmann, observing the double star Delta Orionis, had noted that the conspicuous H and K lines of calcium in its spectrum failed to participate in the general shift of spectral lines and concluded that these particular lines were caused by absorption of light by a cloud of calcium vapor at some point in space between the earth and the star.[92]

In the summer of 1908, Slipher discovered a similar situation for the binary Beta Scorpii; during the next year he found that the calcium H and K lines remained sharp and unshifted among the otherwise blurred and broadened spectral bands in the spectra of a number of double stars, as well as some single stars, not only in Scorpio, but in Perseus and Orion as well. Slipher's work not only provided confirmation of Hartmann's single earlier observation but extended it by providing evidence for the existence of interstellar gas in three widely-separated regions of the sky, each, moreover, "in or near branches of the Milky Way."[93] This work brought Slipher congratulatory letters from many eminent astronomers, including Hartmann himself, Holland's J. C. Kapteyn, Germany's Karl Schwartzschild, and the Danish astrophysicist Einar Hertzsprung.[94] But his suggestion of what Lowell called an "interfering veil"[95] in space did not win immediate general acceptance. Some astronomers, among them Campbell, postulated other origins for the stationary calcium lines, and the issue was not settled on the side of the Hartmann-Slipher hypothesis until 1926 with the work of the English theoretical physicist Sir Arthur Stanley Eddington and subsequently by other astronomers.[96]

Slipher was also the first to offer evidence of the existence of dust, or "pulverulent matter" as Lowell termed it,[97] in interstellar space. This involved a spectrographic study of nebulous material in the Pleiades near the star Merope, which was begun in 1910 and culminated in December 1912 with the discovery that the Pleiades nebula shines not because of any intrinsic luminosity but by light reflected from nearby stars.[98] This "great piece of work"[99] again brought praise from many astronomers, and Lowell, ever alert to an opportunity to promote his observatory, urged Slipher to distribute his Pleiades spectra widely, as "we have to conduct a campaign of education."[100]

Slipher's interest in nebulae seems to have been stimulated originally by Lowell. At least early in 1909 Lowell directed him to make spectrographic observations in the far red of what then were known as "green" and "white" nebulae, the latter group including the enigmatic spirals, in the hope of finding out whether these then little-known objects were indeed incipient solar systems, as some had hypothesized.[101] As Slipher recalled shortly after Lowell's death:

The problem was of great interest to Dr. Lowell and he did all he could to advance the work. He thought their study might shed some light upon the evolution of planetary systems. He was most especially interested that it be found how they were rotating with reference to the shape of the arms of the spirals.[102]

Slipher, at first, saw little hope of getting such spectra because of "the high ratio of focal length to aperture of the 24-inch gives a very faint image of a nebula" which would require exposures with the Lowell refractor of some thirty hours or more.[103] By December 1910, however, he had reversed himself and had obtained a plate of the Great Nebula in Andromeda which "seems to me to show faintly peculiarities not commented upon" by the few earlier observers who had attempted the task, notably J. E. Keeler and J. S. Scheiner in the 1890s:

These other observations were made with reflecting telescopes and the idea seems to go undisputed that a long focus telescope, and of course a refractor, is unsuitable for such work. But I convinced myself that I knew of no reason why the focus-to-aperture ratio had the slightest part to play in spectrum work on extended objects and this plate proves the proposition completely to my mind. . . .[104]

In this investigation, Slipher again experimented with lenses and prisms, finally devising a single-prism instrument with a commercial Voightlander f2.5 lens which, although providing low dispersion, gave "something like 200 times the speed of the usual 3-prism spectrograph."[105] The low dispersion of this instrument, as it turned out, made little difference in his work on radial velocities. For, as he later explained to Hertzsprung, "in consequence of the extraordinary velocities the displacements [of the spectral lines] are often quite well measurable in spite of the low dispersion it is necessary to employ."[106]

Over the last four months of 1912, Slipher obtained four spectrograms of Andromeda which, surprisingly, all revealed a considerable displacement of lines toward the violet indicating an unusual velocity of approach. The final plate in this series required exposures over three consecutive nights, the last on New Year's Eve. "I feel it safe to say here," he advised Lowell as the year 1913 dawned, "that the velocity bids fair to come out unusually high."[107] A month later, after making careful measurements of his plates, he reported that the Andromeda nebula was approaching the solar system at the then unheard-of speed of 300 kilometers per second.[108] "It looks as if you had made a great discovery," Lowell replied, adding: "Try some more spiral nebulae for confirmation."[109]

This Slipher did, starting with a "peculiar nebula" in Virgo, the edge-on, spindle-shaped spiral of NGC 4594. Almost immediately he discovered "a great displacement of spectral lines"—this time toward the red—corresponding to a velocity of recession of 1,000 kilometers per second, an astounding figure at the time.[110] Over the next three years, his observations of twenty-two other spirals turned up similar velocities which, he informed Campbell shortly after Lowell's death, "continue preponderantly positive," that is, away from the earth.[111] By 1925, Slipher had provided thirty-nine of the forty-five known radial velocities of spiral nebulae.[112] Since then, of course, velocities representing significant fractions of the speed of light have been recorded. But nonetheless, Slipher's work provided the first step toward the modern theory of an expanding universe.[113]

Slipher's work on Virgo also brought a second major discovery. On a 1913 plate of the nebula he had found that not only were the lines shifted toward the red but also were slightly inclined. It was a full year before he was able to get a second satisfactory plate to confirm this tilt, but on May 24, 1914, he wired Lowell: "Spectrograms show Virgo nebula rotating."[114] A few months later he found similar evidence for the rotation of the Andromeda nebula.[115]

These discoveries of extraordinarily high nebular velocities and of nebular rotation stirred great interest within astronomy and resulted in a considerable correspondence between Slipher and prominent astronomers both in Europe and America. To a few of these Slipher allowed himself to speculate informally on the possible significance of his findings, although he told Miller "that I do not wish to theorize, not until I have more observations at least."[116]

For instance, he did not think at first that the evidence he had begun to accumulate was sufficient to show that the spiral nebulae were other galaxies rather than nebulous stars or perhaps incipient solar systems within the Milky Way. This, as he explained to Hertzsprung among others, was because they did not show the composite spectrum to be expected from such objects and observed in globular clusters of stars. Yet he conceded the difficulty in reconciling the "enormous volume of a nebula like the great one in Andromeda with a mass no greater than that of a massive star."[117] As his observations continued, however, his thinking here changed, and in April 1917 he advised a meeting of the American Philosophical Society that his work seemed to support the long-debated theory that the spiral nebulae were "island universes," vast and distant systems of stars far beyond the limits of the Milky Way Galaxy.[118] Very early he noted "a suggestion" of group motion among the nebulae and globular clusters which his later observations of radial velocity tended to sustain.[119]

But he also wondered whether such high radial velocities might not make it worthwhile to study the proper motion of spirals, that is, their angular movement in the heavens, a suggestion to which Lampland, not then realizing the great distances of these objects, later devoted much time and effort.[120] Slipher even thought that the Virgo nebula might be traveling faster than the Andromeda spiral because of its edge-on aspect to the line of sight and flight, "as any disk meeting resistance."[121] Time and later work—his own and others—modified or mooted such early tentative speculations.

The major obstacle to the spectrographic study of the spiral nebulae was their extreme faintness. Exposures extending over exceptionally long periods of time were required to accumulate enough light to record their spectra on a photographic plate. This circumstance had deterred other astronomers from pursuing any extensive research along this line, and consequently, as Miller noted, Slipher found himself "in a rich virgin field."[122] But plates were still slow, and even with his "fast" single-prism spectrograph, exposures of twenty or more hours proved to be necessary and did not always result in readable spectra.[123] Of all the many problems involved, Slipher noted in 1914, "the slowness of the accumulation of data is worst."[124] A few years later he remarked ruefully that he should have included some of the bright planetary nebulae in his studies as "it would be a real recreation to be able to secure a satisfactory spectrogram of a nebula in one night's exposure."[125]

Slipher's work on nebulae won him recognition and wide respect within astronomy, and within a few years such high honors as the Lalande Prize of the Paris Academy of Science, the Henry Draper Medal of the National Academy of Sciences, the gold medal of the Royal Astronomical Society, and the Bruce Medal of the Astronomical Society of the Pacific.[126]

He wore his laurels modestly. Late in 1914, after his discoveries began to become generally known, he was extended the privilege of sending contributions regarding his work to the prestigious National Academy of Sciences. He graciously acknowledged this honor, adding: "It only remains for me to do something worth sending."[127]

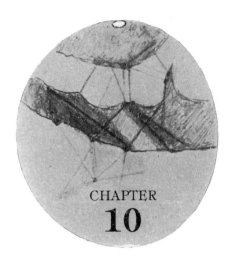

CHAPTER
10

More on Mars

FROM 1901 the Mars furor began to grow gradually in intensity. Over the next eight years, Percival Lowell would expend enormous amounts of energy defending the reality of Schiaparelli's *canali* and promoting his own provocative postulate of their artificial construction by intelligent martians struggling valiantly if vainly to stave off the final desiccation of their dying planet. Because the Newtonian dictum that action always brings reaction applies in principle to men as well as motion, critical attacks on his Mars theory also progressively increased, becoming not only more frequent but more effective.

For the oppositions of 1903, 1905, and 1907, Lowell spent the better part of each year in Flagstaff, not only carrying out his own rigorous observations but directing the work of his assistants and engaging in a variety of other activities ranging from picnicking and politics to paleontology. In Boston or on his trips to Europe in off-opposition years, he lectured and lobbied tirelessly, wrote articles and papers, and kept up an extensive correspondence with astronomers and others on behalf of his observatory's work and his controversial martian theory.

In this period, too, he published four books—*The Solar System* (1903), *Mars and Its Canals* (1906), *Mars as the Abode of Life* (1908) and *The Evolution of Worlds* (1909). In the first and last of these he provided an early brief outline and then the full development of his theory of planetary evolution.[1] Both books were based on lectures he had presented at the Massachusetts Institute of Technology

[151]

Percival Lowell in his study in the Baronial Mansion, 1905.

where, since 1902, he had held an appointment as nonresident professor of astronomy. *Mars as the Abode of Life* also deals extensively with planetary evolution, or what he called "planetology," although his martian theory is expounded therein at length to illustrate his argument.

In *The Solar System*, incidentally, he flatly declared his belief in the presence of a ninth planet beyond Neptune.[2] This small book received only limited distribution and caused little stir, although Lowell had ordered that it be "extensively advertised."[3] His confident prediction that a trans-Neptunian planet eventually would be found, apparently was noted without particular comment, where it was noted at all.[4]

Mars and Its Canals, on the other hand, is Lowell's *magnum opus* on the red planet. Appearing at a time when the public interest in Mars was intense, it reached a far greater audience and drew lengthy, detailed reviews and commentaries in newspapers and periodicals which, although frequently critical, further spread his thought and its rationale in the literate world. Crammed with facts

Lowell Observatory staff in 1905. Left to right are Harry Hussey (porter), W. L. Leonard, V. M. Slipher, P. Lowell, C. O. Lampland, and J. C. Duncan.

as well as what many considered fantasies, it was and so remained for some years the best available summary of martian lore in or out of science. Reprinted a number of times, the book was translated into French, German, and Swedish. The French edition, at least, was a failure from a financial standpoint, a circumstance Lowell considered "a great pity."[5] And one publisher in Leipzig advised against bringing out a German edition "because a great deal of your investigations are amongst some German astronomers considered not to be on a scientific basis." The publisher was willing to go ahead, however, if Lowell would put up 3,000 marks.[6]

Parenthetically, the peaking Mars furor in 1907 also spurred sales of his earlier book, *Mars*.[7] And that year from Peking, Professor Isaac Taylor Headland wrote asking permission to publish a Chinese edition of *Mars*, presumably in a translation by Sir Robert Hart that Lowell had authorized in 1898,[8] and declaring "I think we could sell 100,000 copies."[9] Lowell was delighted and granted the request, waiving all claims to royalties and telling Headland: "It will be gratifying to have the book widely disseminated in China. ... I only hope it will reach the number of ciphers you predict."[10]

Mars and Its Canals is by far Lowell's most comprehensive exposition of his martian theory. Although he restates the basic points he had made eleven years earlier in *Mars* and expounds them in greater detail and lusher prose, his thesis now is firmly set in the matrix of his evolutionary ideas, planetary and otherwise, and is buttressed not only by the mass of observational data accumulated at Flagstaff after 1901, but by new, logical arguments, often insightful and occasionally far ahead of their time, drawn from such other fields as chemistry, biology, geology and geography. These skillful combinations of essentially nonastronomical fact and theory are further developed in his *Mars as the Abode of Life* where they are applied not alone to Mars but to the fundamental question of extraterrestrial life.

To bolster his hypothesis, for instance, he argues from biological and paleontological evidence the chemical evolution of the organic from the inorganic to show that life is an inevitable result of evolution and thus could arise not only on Mars but on any planet in the universe possessing the requisite physical conditions. This idea, of course, is widely accepted by scientists today as the result of the 1952 work of Nobel laureate Harold C. Urey and his collaborator Stanley L. Miller, who achieved the successful laboratory synthesis of complex amino acids, the building blocks of proteins, from water vapor, methane, ammonia, and hydrogen that Urey had postulated as ingredients of the earth's primordial atmosphere.[11]

"That evolution is nothing else than such a gradually increasing chemical synthesis is forced on one by the study of the facts," Lowell declares in his *Mars and Its Canals*. "That organic springs from inorganic is not only shown by what has taken place on earth, but is the necessary logical deduction from its decay back into the inorganic again."[12] He uses the "cousinly closeness of the lowest organic with the highest inorganic substances," combined with the interdependence of terrestrial plants and animals, to argue the mutual origin of the two life forms. Citing taxonomic studies of the primitive organism, *Chromacea*, by the German biologist Ernst Haeckel,[13] he reasons:

At the time when inorganic chemical compounds first passed by evolution into organic ones, the change was of so general a character that even such tardy representatives of it as survive today tax erudition to tell of which of the two great kingdoms they belong. ... If ancestors of two great kingdoms were thus simultaneously produced here, we are warranted in assuming that they would similarly be produced elsewhere.[14]

The point applies to Mars, he notes:

The assurance that plant life exists on Mars ... introduces us at once to the probability of life there of a higher and more immediately appealing kind. ... The presence of flora is itself ground for suspecting a fauna.[15]

Then he carries his logic a step further:

All that we know of the physical state of the planet points to the possibility of both plant and animal life existing there, and furthermore, that this life should be of a relatively high order is *possible*. Nothing contradicts this.[16]

The evidence for an atmosphere and for former seas on Mars shows that the ingredients for chemical evolution are present and "that life had there as here the wherewithal to begin."[17]

In *Mars as the Abode of Life* Lowell restates his premise of chemical evolution in somewhat more detail and in even more modern-sounding terms, pointing out that "virtually only six elements go to make up the molecule of life" and that it "is the number of its constituent atoms, and the intricacy of their binding together" that make it viable:

By it, starting in a simple, lowly way and growing in complexity with time, all vegetable and animal forms have since been gradually built

up. .... In itself the organic molecule is only a more intricate chemical combination of the same elements of which the inorganic substances which preceded it are composed.[18]

In the two books, too, Lowell develops another quite novel argument for the time that has often been used by subsequent proponents of the existence of extraterrestrial life. This is the idea
that relatively extreme physical conditions do not necessarily pre-
clude the origin and survival of life on a planet. Here, he draws
primarily on the pioneering work of biologist C. Hart Merriam in
1889 on the San Francisco Peaks just north of Flagstaff to show that
life could exist on Mars despite the planet's depressurized oxygen-
thin atmosphere and its greater range of temperatures.

"The paucity of air," he writes in his *Mars and Its Canals,* citing
Merriam's survey of the life forms, "is nothing like the barrier to life
we ordinarily suppose ... and not for a moment to be compared to a
dearth of water." And he adds:

If animal or plant life can in a comparatively short time become
accustomed to 30 inches of barometric pressure, it would be rash to
set limits to what time may not do.[19]

He cites further examples of this "indifference" of animal life to
extremes of pressure in *Mars as the Abode of Life,* pointing to the
ancient thriving city of Quito at an elevation of over 10,000 feet in
the Equadorean Andes and to the presence of marine forms in the
abyssal oceanic deeps where life was thought impossible until the
1870s when the historic voyages of the British research vessel
Challenger proved otherwise.[20]

Merriam's study also provided him with an additional argument
to counter the contention of his critics that Mars was simply too
cold to support life. In *Mars and Its Canals* he concedes that the
average temperature on Mars is below that of the earth, although he
insists it is still "above the freezing point of water," a conclusion he
draws primarily from the behavior of the planet's polar caps.[21] In
Mars as the Abode of Life he sets the mean annual temperature on
Mars at 48 degrees Fahrenheit, applying insolation data for the
planet to earlier bolometric studies of the moon by Langley and
Very, which contrary to prevailing belief "indicate thin air is compatible with great surface heat." He arrives at substantially the same
result by using a formula devised by the Swedish scientist Svante
Arrhenius for determining the effect of atmospheric carbon dioxide
on planetary temperatures.[22]

But also "man does not live by mean annual temperatures

alone." The "instructive fact" here is Merriam's finding that "the existence of a species was not determined by the mean temperature of its habitat but by the maximum temperature during the time of procreation."

A short warm season in summer alone decides whether the species shall survive and flourish; that it has afterwards to hibernate for six months at a time does not in the least negative the result.[23]

The point is developed at greater length in *Mars as the Abode of Life* where he stresses its "immediate bearing upon the habitability of Mars; for the martian summer is twice as long as ours, and . . . the probable acme of warmth attained in it is by no means small."[24] Thus, he argues:

It is by these attributes of its climate, and not by its mean annual temperature, or by the great cold its surface very possibly experiences in winter, that its ability to support life must be judged.[25]

A particularly revealing example of his eclectic use of material from outside astronomy is his reference in *Mars as the Abode of Life* to the work of Alexander von Humboldt, the nineteenth-century scientist and explorer. From studies in the Himalayas and the adjacent Tibetan Plateau, he notes, Humboldt found that temperatures on a table land generally exceed those at the same elevation on surrounding mountains. This conclusion Lowell tested for himself by observing comparative temperatures and the differing plant forms at various altitudes in and around the San Francisco Peaks on several camping trips during the summer of 1907. And because he believed the martian surface to be "more like some vast plateau," Humboldt's point, along with an illustrated description of his own confirmatory experiments, becomes an additional argument for a life-sustaining warmth on Mars.[26]

One further example of this kind of reasoning, drawn now from geology and geography, is of interest here. For while Lowell develops the point in greater detail within the framework of planetary evolution, he introduces it initially to explain why his postulated martians were forced to build their vast and intricate system of irrigation canals in the first place.

This is Lowell's concept of "desertism"—the idea of the gradual encroachment of land on sea and of arid land on fertile land which is "symptomatic" of the inevitable drying up of a planet, of its transition from what he calls the "terracqueous" to the "terrestrial" stages in its evolution. The earth, he says, is in the terracqueous stage because, unlike Mars and the moon, it still retains its oceans. But the expanding deserts that belt the earth athwart the Tropics of Cancer and Capricorn, and its slowly shrinking seas, reveal that the fatal planetary disease has already set in:

Desertism, the state into which every planetary body must eventually come and for which, therefore, it becomes necessary to coin a word, has there made its first appearance upon the earth. Standing as it does for the approach of age in planetary existence, it may be likened to the first gray hairs in man.[27]

As evidence of this, in *Mars and Its Canals,* he cites the historic desiccation of Palestine[28] and presents maps taken from James Dwight Dana's *Manual of Geology* to argue the progressive withdrawal of the seas from the continental area of North America in geologic time.[29] In *Mars as the Abode of Life,* to show the continuing depletion of earth's seas, he seized on a contemporary report by a member of the Sladen scientific expedition to the Maldives who inferred that the emergence of coral reefs from the Indian Ocean was due to a lowering of the sea level.[30] The Caspian Sea, he also declared citing studies by the young geographer Ellsworth Huntington, "is disappearing before our eyes."[31] Closer to home, he pointed to the lost lushness of Arizona's petrified forest, drawing on the investigations made in 1899 by paleontologist-sociologist Lester Frank Ward which led eventually to the designation of the area as a national monument.[32]

But if desertism has only just begun to infest the earth, it has already ravaged Mars where its progress would be accelerated by the relatively rapid evolutionary course run by a smaller, less massive planet. Three-fifths of its oceanless surface is desert, and "the picture the planet offers us is thus arid beyond present analogue on earth."

Pitiless as our deserts are, they are but faint forecasts of the state of things existent on Mars at the present time. Only those who as travelers have had experience of our own Saharas can adequately picture what Mars is like and what so waterless a condition means. Only such can understand what is implied in having the local and avoidable thus extended into the unescapable and world-wide; and what a terrible significance for everything Martian lies in that single word: *desert.*[33]

This then is the condition that the great martian canal network is designed to alleviate, for only by fully exploiting the last, dwindling sources of water, that is, the melting snows of the polar caps, could life there be able to prolong its survival. Given an advanced stage of evolution on Mars, this life would be intelligent enough for the task:

As a planet ages, any organisms upon it would share in its develop-ment. They must evolve with it, indeed, or perish. At first they change only as environment offers opportunity, in a lowly, uncon-scious way. But, as brain develops, they rise superior to such occasioning. Originally, the organism is the creature of its sur-roundings; later it learns to make them subservient to itself. In this way the organism avoids unfavorableness in the environment, or turns unpropitious fortune to good use. [34]

Martians must have long ago foreseen the challenge to survival present in the progressive desiccation of their planet:

Before it was upon the denizens of the globe preparations would have been made to meet it. ... A planet's water-supply does not depart in a moment. Long previous to any wholesale imminence of default, local necessity must have begun reaching out to distant supply. Just as all our large cities today go far to tap a stream or a lake, so it must have been on Mars. [35]

Thus the martian response to the relentless advance of des-ertism was to create an irrigation system which, from presumably localized beginnings, gradually grew in size and complexity to even-tually become planetwide in scope. Lowell minimized the labor and engineering skill required for such a vast project, citing the weaker gravity and relatively level surface of Mars compared to the earth, and emphasized the compelling necessity for such a system of canals: "Terrain offers the least of objections, terror the greatest of spurs, to their construction." [36]

And yet, he concluded, a "sadder interest" attaches to this desperate stand by the martian canal-builders, for, as the disease of desertism is terminal, their struggle for survival cannot succeed. Inevitably, Mars is dying:

The process that brought it to its present pass must go on to the bitter end, until the last spark of Martian life goes out. The drying up of the planet is certain to proceed until its surface can support no life at all. ... When the last ember is thus extinguished, the planet will roll a dead world in space, its evolutionary career forever ended. [37]

Such imaginative and even ingenious arguments were only supplementary, at least as far as they relate to his martian theory. For in *Mars and Its Canals,* as in *Mars,* he relies primarily on massed observational evidence to make his case. Again, he begins by setting forth the physical characteristics of the planet to show its inherent

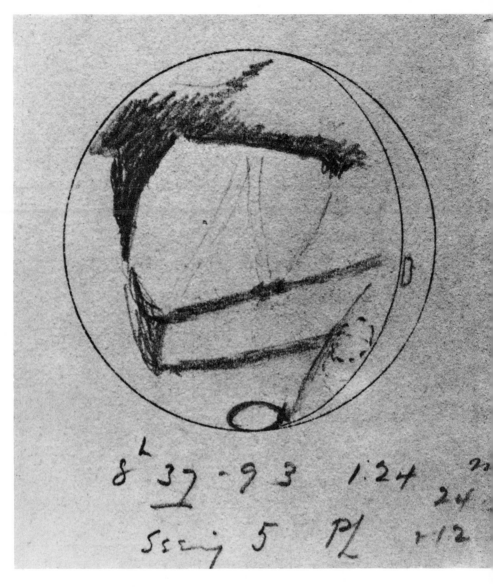

"Projection" on Mars, May 1903

habitability. And again, he follows this with descriptions, often in exhaustive detail, of what he considers to be the "non-natural" features of Mars, that is, the canal network, to show its actual habitation. What is new here as far as the evidence goes, and what has been observed at Flagstaff in recent years, proves to be "entirely corroborative" of his 1894 observations "and of the theory then deduced."[38]

Most of the new material in *Mars and Its Canals* was gathered during the martian opposition of 1903 when Lowell spent a particularly productive six months at his observatory in Arizona. Most of it, too, was his own work, for he continued to make visual observations while his two young assistants V. M. Slipher and Lampland struggled with the knotty problems of spectrography and planetary photography. Lowell's observations ran from January 21 through July 26[39] and included a period during April and May when he was able to observe Mars for forty-six consecutive nights, and to make hundreds of drawings of the planet.[40] He turned up only one mild sensation during this time—the sighting on May 26 of "an enormous cloud" as a "projection" on the planet's terminator. Lowell reported the phenomenon to the press and then discussed it in the first of the observatory's *Bulletins*, the publication of which he initiated in June of that year.[41]

The "projection" set off a new round of speculative comments about signals from Mars, some of which reflect the degree to which scientific fact and fiction had become intermingled in the martian controversy. In noting Lowell's report, the London *Daily Mail*, for example, remarked:

The brilliant imaginations of H. G. Wells have familiarized us all with the possibilities of life on Mars and no one who has read *The War of the Worlds* can help shuddering slightly when he remembers that just such a projection indicated the commencement of that terrifying invasion.[42]

From the standpoint of his martian theory, his 1903 work at Flagstaff provided two major thrusts.

One of these involved his extensive observations of seasonal phenomena on Mars. This voluminous data, recorded systematically in his drawings of the planet and reduced after his return to Boston that summer, yielded what Lowell called the "cartouches," or "sign manuals," of the canals of Mars.[43] In essence, these were graphs on which Lowell plotted the progressive development of an individual canal in terms of its relative visibility on Mars as a combined function of martian dates and latitudes. To obtain the cartouches of 109 canals, he says, he used 10,900 separate determinations of their visibility between the martian dates of June 6 and September 1 and between the martian latitudes of 87 degrees north and 35 degrees south.[44] The resultant curves not only confirmed his earlier descriptions of a vernal "wave of darkening" sweeping alternately from the planet's poles to its equator and beyond but demonstrated quantitatively that the canals participate in this semiannual seasonal change. "The evidence is direct," he declared, "and independent of any theory."[45]

His cartouches also showed that this seasonal wave spread down the planet to its equator, and even beyond, at a uniform rate.[46] This he estimated to be about fifty-one miles per day or 2.1 miles per hour, a rate he considered to be wholly consistent with the idea of a spring quickening of vegetative growth along the banks of a canal.[47] And in first reporting the revelations of his cartouches in an observatory *Bulletin*, he argued that this unterrestrial behavior of the canals, that is, their pole-to-equator development, was also proof of their artificial nature:

The above deduction of the artificiality of the canals springs from their mode of development; it is, therefore, kinematic in character. In the Lowell Observatory *Annals*, Vol. I., a like conclusion of artificiality followed a discussion upon the form and position of these markings; the arguments there being essentially static. Thus the behavior of the canals in action leads to the same view of their nature as does their appearance at rest.[48]

Lowell based his point here on the not unreasonable assumption that water on Mars, if it flowed at all, flowed downhill. The unidirectional flow implied by reversed seasonal development of the canals, then, was hardly credible unless some unnatural agency was called into being to account for it. "A few canals," he pointed out in *Mars and Its Canals*, "might presumably be so situated that their flow could, by inequality of terrane [sic], lie equatorward, but not all."

Now the water which quickens the verdure of the canals moves from the neighborhood of the pole down to the equator as the season advances. This it does, then, irrespective of gravity. No natural force propels it, and the inference is forthright and inevitable that it is artificially helped to its end.[49]

The fact that development of the canals not only proceeded toward the equator but continued across it into the other hemisphere "is more unnatural yet," he added. "For even if we assume, for the sake of argument, that natural forces took the water down to the equator, their action must there be certainly reversed and the equator prove a deadline to pass which were impossible."[50]

Lowell outlined these arguments and their implications for the reality of the martian canals late in 1903 to Slipher and Lampland in lengthy detailed letters. Slipher, at least, was duly impressed. "How can anyone doubt their artificial origin?" he wondered.[51]

Many did, of course. Certainly Lowell's arguments failed to convince such prominent astronomers as W. H. Pickering and the influential Simon Newcomb, who now contended that what Lowell was seeing on Mars was analogous to the craterlet-spattered cracks, rills, and escarpments on the moon, the apparent linearity of such features on Mars being a function of psycho-physiological phenomena connected with the observation of a far more distant object through the earth's shimmering atmosphere.[52]

At the 1903 opposition Lowell developed other observational evidence to bolster his contention that the seasonal changes on Mars were due to the presence there of vegetation. For one thing, as in 1901, he again observed white spots on the planet, finding that at one point its northern, summer hemisphere appeared to be speckled with them, even in the equatorial regions. These spots, he concluded, must be transient patches of frost, and he carried his reasoning on from there:

Such an explanation accords well with the distance of the planet from the sun and the thinness of its atmosphere. At the same time it shows that the mean temperature over the greater part of the planet the greater part of the time is above the freezing-point and that consequently it is no bar to vegetation of a suitable sort.[53]

Even more to this point were his studies of the changes in color of Mare Erythraeum, largest of the dark martian markings, which he pursued systematically over a span of six terrestrial, or three martian, months. This huge southern hemisphere feature he found turning from blue-green to a chocolate brown as the martian autumn gave way to deep winter and gradually greening again with the approach of the martian spring:[54]

As the change in hue at just the season of the planet's year when vegetation, if it exists, would presumably die out and leave the bare ground visible, great countenance is afforded to the vegetation theory by this phenomenon.[55]

Lowell repeated these Mare Erythraeum observations at the opposition of 1905 and, finding similar changes, concluded that "from the recurrence of the phenomenon on two successive [martian] years, it is likely that it annually takes place."

Unlike the ochre of the light regions generally, which suggest desert pure and simple, the chocolate-brown precisely mimicked the complexion of fallow ground. When we consider the vegetal-like blue-green that it replaced, and remember further the time of year at which it occurred on both these Martian years, we can hardly resist

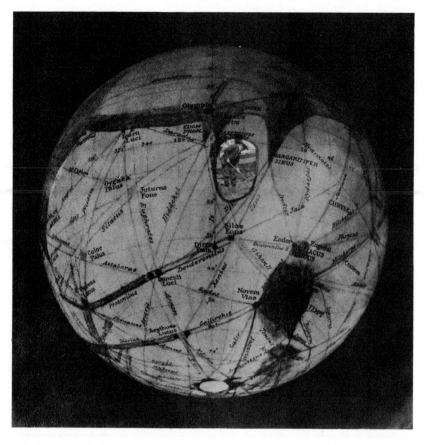

Lowell, "gardening" on Mars. One of his charts, with the small photo pasted on it, was a birthday gift from his staff in 1905.

the conclusion that it was something very like fallow field that was there uncovered to our view.[56]

Lowell considered this work on Mare Erythraeum to be of great significance to his martian theory. He published his 1903 observations first as an observatory *Bulletin* and then, combined with the confirmatory 1905 data, as a full chapter, entitled "Vegetation," with charts and drawings in color in *Mars and Its Canals*.[57]

The second major thrust of the work at Flagstaff in 1903 was the accumulation of both observational and experimental data which Lowell subsequently used to counter persistant criticisms that the canals of Mars were simply optical illusions. This key issue was brought to a head in June of that year by English astronomer E. Walter Maunder who, in a report to the Royal Astronomical Society, revealed that when a group of boys from the Royal Hospital School at Greenwich were asked to copy a canal-expurgated picture of Mars, they had drawn the planet criss-crossed with lines.

Thus, Maunder declared after presenting a lengthy description and photographic evidence of the tests, the apparent lines on Mars "are simply the integration by the eye of minute details too small to be separately and distinctly defined."

It would not therefore be in the least correct to say that the numerous observers who have drawn canals on *Mars* during the last twenty-five years have drawn what they did not see. On the contrary they have drawn, and drawn truthfully, that which they saw; yet for all that, the canals which they have drawn have no more objective existence than those which our Greenwich boys imagined they saw on the drawings submitted to them. ... It seems a thousand pities that all those magnificent theories of human habitation, canal construction, planetary crystallisation and the like are based upon lines which our experiments compel us to declare non-existent.[58]

Maunder's report, which provided proponents of the illusion theory with a strong new argument, had both an immediate and lasting impact on the controversy over Mars. Almost at once, the Mars section of the British Astronomical Association reversed itself on the subject of Mars' canals, an action which Garrett P. Serviss, science columnist for the Hearst newspapers in the United States, condemned as "an attempted revolution in the accepted view ... so to speak a revolution backwards." The association, he wrote, "brushes off all the network of canals from the chart of Mars on the ground that it is based on illusion, leaving out certain features which were known to all astronomers before Schiaparelli."

The members ... have themselves seen more than a hundred of the so-called canals during their observations since 1900, yet E. M. Antoniadi, their director and editor, apparently has their concurrence in holding that their eyes were probably deceived and that they really saw something very different from the straight lines they imagined they were looking at.[59]

Maunder's revelation also prompted Flammarion to run a series of similar visual tests on a group of French school boys to find, as Lowell later remarked, that "not a boy of them drew an illusory line."[60] And to Newcomb, Lowell quoted Sir Norman Lockyer as saying that "it looks as if some leading questions had been put to the English school boys."[61] This same statement, without quotation marks or attribution, also appears in *Mars and Its Canals.*[62]

Lowell himself ordered a series of experiments in 1903 that were designed to combat the optical illusion explanation of the martian canals, or the "Small Boy Theory," as he now began to call it. These, involving systematic observations of telegraph wire of known thickness at varying distances, were carried out on Mars Hill by Lowell, Slipher, and Lampland and reported in two Lowell *Bulletins,* and as a lengthy note in *Mars as the Abode of Life.* [63] In general, they showed that while imaginary wires did indeed appear at great distances and near the limit of visibility, the observer was almost always able to distinguish the real wire from the illusory ones as long as he could see anything at all. From these experiments, Lowell calculated that the eye, under optimum conditions, should be able to discern a line 3/16 of a mile wide on Mars, although because of the loss of light and definition inherent in telescopic observations, one-half mile would probably be a more practical figure as the "limiting perceptible width."[64] Thus, because he considered the martian canals to be strips of vegetation several miles in width, they were not at or near the limit of visibility as the optical illusionists claimed, but well above it. [65] As a matter of fact, he concluded the canals could be seen even below this limit because of their length. For the wire had remained visible when its diameter subtended only 0.69" of arc which, he noted, was only 1/96 of the value of one minute of arc he considered to be the usual figure for the minimum resolution of the normal human eye.

Why a line can be seen when its width is but 1/96 of the *minimum visible* seems to be due to summation of sensations. What would be far too minute an effect upon any one retinal rod to produce an impression becomes quite recognizable in consciousness when many in a row are similarly excited. [66]

Lowell's telescopic observations of Mars during the 1903 opposition also provided him with arguments for the non-illusory nature of the canals, and more specifically against the charges of diplopia, the condition of seeing double, and optical interference that critics directed particularly to the double canals.

While the lines on Mars clearly displayed an unnatural uniformity and interdependence in their structure, their kinematic behavior, their visibility or invisibility in relation to the changing martian seasons, "is an individual characteristic of the particular canal" and thus must be intrinsic to the canal itself rather than an extrinsic effect of observation. [67] The cartouches provided evidence for this for if the canals were only illusory the graphs of their seasonal development would be uniform straight lines whereas in all but three instances they were irregular curves. [68]

Nor, Lowell declared, could the martian canal network simply be an illusion of the mind, a matter of suggestion or preconception, and here he philosophized from the history of science. "The accusation of illusion is a very convenient epithet for jaculatory purposes but as a missile it partakes of the nature of the boomerang. For the moment one evokes psychic phenomena he summons a power that works both ways."

It is just as easy from a preconception not to see what exists as to see what does not; as has been time and time again exemplified in science. . . . Galileo's contemporaries refused to credit the existence of the satellites of Jupiter; Huyghens' explanation of Saturn's ring was derided. Newton's deduction of the law of gravitation was for long refused belief on the continent; Adams could get no hearing in his search for Neptune until Leverrier's parallel studies alarmed Airy to action; lastly Schiaparelli's marvelous discoveries on Mars, Mercury and Venus have been puerilely treated not only with disbelief but scorn. [69]

Thus, he added, "while each generation marvels how its forerunners can have been so blind, it then proceeds to be equally dense in its own day—so prone is poor humanity to buy and not borrow its experience."[70]

The double canals, which stirred the greatest incredulity, gave Lowell his strongest arguments for the reality of the whole canal network. For if their gemination was caused by diplopia, all of the canals, rather than just the same few, should appear double at one time or another. "Diplopia might be a respecter of persons but it certainly could not be one of the canals. . . . For gemination is an attribute of certain canals and never of others."[71] Also the distance between the parallel components of the doubles differed from canal to canal, and this could not be explained by diplopia, interference, or any other kind of illusory phenomena, optical or otherwise. "Each has its individual width . . . and this individual width differs" regardless of the alignment of the doubles on the planet's disk, the telescope or aperture used, or the observer, if the latter could see them at all. Were the doubles caused by optical interference, he also noted, the area between their twin dark lines should appear brighter than other parts of the planet, but this had not been observed. [72]

Finally, he declared, the argument that the doubles result from a failure to focus "is here specifically out of the question." For focusing the eye is a "reflex action," and "so automatic" that "one is commonly conscious when an object is out of focus, always so when the object presents detail."

Consequently, that an experienced observer should not know his business in so primary a matter is preposterous. One may or may not believe that "the undevout astronomer is mad," but that the perpetually unfocused one would be is beyond debate. [73]

It must be noted here that Lowell's arguments for the objective reality of the canals, however persuasive to some, did not prevail even in his own lifetime. Progressively, as the tide of the Mars controversy came to flood and then subsided, the explanation advanced by Maunder, Antoniadi, and their followers gained ascendancy and came to be favored by astronomers who concerned themselves with the observational problems of Mars at all.

Lowell, these critics contended, simply did not see all there was to see on the planet, and what he did see he did not see clearly enough. Ironically, they cited Lowellian scripture to their purpose. They conceded, for example, Lowell's claims that the canals could be seen only with smaller apertures and under certain seeing conditions. But, they argued, if Lowell would use a larger telescope he would find that in the very best seeing, the purported lines were resolved into a mass of diffuse but discrete detail. These dots and blotches when imperfectly seen did indeed produce an impression on the retinal rods of the eye "recognizable in consciousness" as lines. Maunder, Serviss noted, "insists that instead of continuous independent lines they [the canals] are only rows of spots and edges of dark-shaded areas." [74]

Antoniadi in particular pursued the point in later years, notably at the 1909 opposition when he had his first opportunity to observe Mars through a larger telescope—the 33-inch at Meudon Observatory near Paris. [75] "I am glad you are to use the Meudon refractor," Lowell wrote to him, "but remember that you will have to diaphragmed [sic] it down to get the finest possible details. Even here we find 12 or 18 inches the best sizes. ...." [76]

Lowell's advice went for naught, and a few weeks later Antoniadi sent him four of his "best" drawings of Mars, made using the full aperture of the Meudon instrument, which showed many of the planet's features, including some of the canals, as composites of small, dark patches and spots. And he wrote:

The tremendous difficulty was not to see the detail but accurately to represent it. Here, my experience in drawing proved of immense assistance, as after my excitement, at the bewildering amount of detail visible, was over, I sat down and drew correctly ... all the markings visible. ... My drawings ... speak for themselves to an areographer like yourself who knows the planet better than anybody else. [77]

Lowell thanked Antoniadi politely for the drawings, noting "the one you marked tremulous definition strikes me as the best. It is capital."

The others seem not so well defined and this is the great danger with a large aperture—a seeming superbness of image when in fact there is a fine imperceptible blurring which transforms the detail really continuous into apparent patches. On the other hand, a bodily movement often coincides with the revelation of fine detail. This subject we have carefully invesitgated here and all of our observers recognize it. [78]

Lowell enclosed some of his own Mars drawings, showing canals as continuous lines, but Antoniadi nonetheless remained firmly convinced that his own interpretation of the markings as groupings of discrete features was the true one. "There *are* many points of similitude in our drawings, proving that we have seen the same thing," he replied, "but our interpretations of said drawings are different."

You have a great advantage in your *splendid* atmosphere, an advantage which requires no proofs. In 1909 I have the advantage of aperture and *separating* power.
I base all my ideas of Mars on what I saw myself at Meudon; and as I have not seen any geometrical canal network, I am inclined to consider it as an optical symbol of a more complex structure of the Martian deserts, whose appearance is quite irregular to my eye. Such a position differs from yours. . . . A great many canals discovered by Schiaparelli and some discovered by yourself I have held steadily, and they are *quite* real; probably the same holds good of many others; and there are interlacing canals—a sort of vague network—but all irregular and knotted or diffuse. [79]

Despite the respectful tone of his correspondence with Lowell, Antoniadi expressed his conclusions in much stronger terms in reporting his 1909 observations to the British Astronomical Association. Scoffing at drawings which showed the canals as lines, he declared that "it is high time that evidence of this kind should be pronounced worthless; and but for such uncertain data, usually obtained with inferior telescopes, there would never have been a question of the canals on Mars." And while conceding that "the fleeting apparition of hideous straight lines" had been observed from time to time even with the Meudon refractor, he added flatly that this occurred only when seeing conditions were poor:

However welcome such phenomena may be to the believer in canals, they can have really no other meaning than that of show-

ing a very complex and irregular structure under a fallacious geometric form. ... A fact of the highest import is that, under good
seeing, almost all Schiaparellian "canals" visible either broke up
into most complicated and irregular groups of shadings or became
the mere indented edges of these shadings.[80]

As an addendum to this report, Antoniadi presented a letter
from American astronomer George Ellery Hale, who had observed
Mars in 1909 with the Mt. Wilson 60-inch reflector, then the most
powerful telescope in use in the world. Hale wrote that while he had
been "able to see a vast amount of intricate detail," he found "no
trace of narrow straight lines or geometrical structures." He added
that all other observers at Mt. Wilson—including, incidentally,
Lowell's former assistant Andrew E. Douglass—had viewed Mars in
1909 with the 60-inch and were in agreement on the point. "The
frail testimony of small refractors," Hale thus concluded, "has vanished before the decisive evidence of giant instruments."[81]

The question of whether the martian canals were real or illusory, however, was not to be settled so easily or so soon. Over the
next dozen years, as Mars passed through a series of less favorable
oppositions—and the earth passed through its first world war—the
issue, although subdued, was kept alive in the astronomical journals and even surfaced sporadically in the public press.[82] In 1924,
at the most favorable martian opposition of the twentieth century,
not only observers at Flagstaff but astronomer Robert J. Trumpler,
observing with the 36-inch refractor at the Lick Observatory, saw
and sketched networks of lines on his charts of Mars. "The so-
called canals," he wrote, "were not found to be the sharp fine lines
as which they have sometimes been described. Even under the
best atmospheric conditions the canals appeared somewhat diffuse and the narrowest ones not less than 25 miles wide." But,
he added:

It can, however, not be denied that the better defined canals show a
strange directness of line; mostly, but not always straight nor following a great circle, they are at least always smooth continuous
lines without break unless an oasis or intersection with another
canal is met.[83]

V. M. Slipher would subsequently cite Trumpler's work to Sir
James Jeans in calling the English astrophysicist's attention to
"what appears to us to be a blemish" in his popular book, *The
Universe Around Us*. After quoting Jeans' statement that "the supposed canals of Mars disappear when looked at with a really large
telescope and have not survived the test of being photographed,"[84]
Slipher declared:

The observational evidence is strongly contradictory to these statements. But you might think us here an interested witness. ... In that case, you might want to look into the work done at Lick Observatory on Mars, where they used a powerful telescope under good conditions, and have been, if anything, opposed to what may be called the Lowell view of Mars. As long ago as 1894 and as late as 1924, their observations, visual first and lately also photographic, have shown these pencil-like markings. ... The views you have expressed appear not in accord with the prevailing interpretation of Martian observations.[85]

Slipher's loyal defense of the Lowellian network of canals, however, was in a lost cause. In that same year—1930—Antoniadi published a book, *La Planète Mars,* in which he summed up his work from 1909 through 1926, presenting a series of careful, comparative drawings of the planet which showed the canals resolved into vague, variously shaded, but distinctly separate features.[86] This, combined with later confirmatory work with large telescopes at prime locations tended to reduce the canal controversy, if not the question of martian life, to the status of an historical curiosity.

It may be said, in fact, that not even the advent of the modern space age and the success of the Mariner series of Mars probes in photographing the planet from near space has appreciably modified the Maunder-Antoniadi hypothesis of the canals as illusory optical effects resulting from the observations of discrete markings imperfectly seen.

"In place of canals," astronomer Robert B. Leighton reported after studying more than two hundred photographs taken in August 1969 by the Mariner 6 and 7 spacecraft, "the new pictures record only fairly large, irregular, very low-contrast splotches, with no physiographic details."

Far-encounter photographs have been examined for evidence of martian canals, defined here as those dark, diffuse, more or less linear features generally of low contrast, which have been recorded by visual observers and telescopic photographs. Although the Mariner pictures are still in relatively rough form, several previously identified canals appear as well-defined features. ... Other canals appear to be resolved into a sequence of dark patches of varying size and contrast. In some cases the individual dark areas seem unrelated, suggesting that many canals involve a chance alignment of randomly distributed dark patches. Variegated shading has been noted in some of the well-defined canals, but the true physical nature of these features is still unknown.[87]

Their nature is better known now, however, with the successful completion of the 1971 Mariner 9 mission to Mars. Photographs taken by the spacecraft from orbit around the planet reveal an incredibly varied topography in which at least a few of the telescopically observed features are marked by towering mountains, huge chasms, rugged highlands, and vast basins. An enormous canyon near the equator, for example, closely coincides with the classic martian canal known as Coprates, and the large circular feature called Nix Olympica turns out to be a vast volcanic cone more than 335 miles in diameter at its base and perhaps seventeen miles high.[88]

The year 1903 from the standpoint of the amount of observational work accomplished and reported was the most productive one up to that time in the brief history of the Lowell Observatory. In addition to making the observations and experiments described above and working on Venus, Lowell also recorded extensive measurements of the size and position of the north polar cap of Mars, the one turned toward the earth in 1903; made new determinations of martian longitudes and the tilt of the planet's axis; analyzed in detail the development of the canals Thoth and Amenthes; and studied the relationship between the double canals and oases, finding that in every case, "the oasis stood exactly embraced by the two arms of the double."[89]

Yet, productive as the year was in quantity of data, the 1903 Flagstaff observations did not have that dramatic quality that so often, before and after, piqued the public interest in Mars. Graphs and charts, micrometric measurements and repetitious descriptive detail made dull copy for the popular press and the mass circulation magazines, despite Lowell's insistence on their scientific importance and their significance as confirmation of his martian theory. Mars itself failed to cooperate too, for aside from the May "projection," the distant planet revealed nothing of a more sensational nature. The observatory's work did indeed win mention in some journals, particularly in Europe where the activities of Maunder, Antoniadi, Flammarion, and others kept a fire burning under the martian controversy. And it helped bring Lowell the French Astronomical Society's prestigious Janssen Medal in 1904.[90]

Still, in November, while discussing Lampland's photographic efforts, Lowell could remark somewhat wistfully to Slipher that "it would be encouraging to have an undisputed canal show in a plate."[91]

CHAPTER
11

Canals and Cameras

C ARL OTTO LAMPLAND was, if possible, even more of a scientific conservative than his senior Lowell Observatory colleague Vesto Melvin Slipher. Yet, ironically, within a few years of his arrival at Flagstaff in October 1902 he gave the peaking Mars furor its ultimate sensation.

In May 1905 Percival Lowell announced to the world that Lampland had succeeded in photographing canals on Mars, a feat generally considered beyond practical reach, even assuming the reality of the canals, because of the formidable technical difficulties inherent in planetary photography. Schiaparelli, the 1877 discoverer of the canals, reflected the widespread incredulity at the announcement when he confided wonderingly to Lowell: "I should not have believed it possible."[1]

Two years later, during the favorable opposition of 1907, Lowell reported that not only Lampland but he himself as well as the younger Slipher, Earl, had found scores of single and double canals on plates exposed both on Mars Hill and on the arid Andean desert at Alianza, Chile.

These claims, widely publicized in the popular press, brought the long-simmering canal controversy to a full boil. Yet when in the course of time the much-heralded pictures generally failed to appear in print, or failed to reveal indubitable canals when they did appear, the pot boiled over, so to speak, and all but put out the fire.

In the early years of the twentieth century, photography was not considered a viable technique for planetary work although its value for deep space research, which derives from the ability of the photo-

[173]

graphic plate to record light cumulatively, had been clearly recognized and was being exploited by increasing numbers of astronomers. In planetary studies, however, definition of detail is usually more important than illumination, and in this regard the eye was and still is unsurpassed. Indeed up until the arrival of Mariner 9 at Mars in November 1971, astronomers still observed the planet visually for detail that could not be resolved by telescopic cameras, and recorded what they saw in drawings much as Lowell did, and indeed as observers have done since Huygens' first sketch of the Syrtis Major in 1659.[2]

The pre-Lowellian history of martian photography is brief and quite unspectacular. The first photograph of the planet was taken by Benjamin Apthorp Gould of Argentina's Cordoba Observatory in 1879, a few years after the "dry-plate" process became available, but it showed almost nothing of the profuse detail visible to the eye on the martian disk. Subsequent attempts were hardly more successful, and indeed it was hailed as a considerable achievement when Harvard's W. H. Pickering, observing the 1892 opposition at Arequipa, Peru, was able to record photographically a gross increase in the size and brightness of the south polar cap, by far the most prominent feature on the planet at the time.[3]

Pickering did not pursue the photography of Mars at the 1894 opposition when he and his colleague at Arequipa, Andrew E. Douglass, became Lowell's first assistants at his newly-established observatory in Arizona. But Douglass did, and he continued his largely unsuccessful efforts to photograph the planet sporadically until his departure from the observatory's staff in 1901.[4]

When Lampland came to Flagstaff he apparently already was quite adept at photography, for his first plates drew quick appreciation from Lowell, along with the seed of an idea that would eventually bear fruit:

Your photographs of the moon do you credit and they suggest to me that you might try your hand on the planets, especially Mars and Venus. If you could devise the mechanism for making practically instantaneous shots on one plate you may catch some detail.[5]

The problem in photographing planets, even one as close to the earth and as well marked as Mars, is that the relatively small image of the planet at the telescope's focus simply does not stand still long enough to have its picture taken. The turbulence of the earth's atmosphere keeps the image in virtually constant motion. The ready human eye is quick enough and sensitive enough to follow some of this movement, but the photographic plate, which requires at least some time to make its record, can neither follow these shimmerings nor stop the action, as it were.

Initially, at least, Lowell was quite pessimistic about the possibility of photographing anything so telescopically elusive as the martian canals, although he was keenly aware of the potential value of photography for astronomical work in a more general way. His new assistant's progress during 1903, however, stirred his innate optimism. Taking Lowell's almost casual suggestion, Lampland had set to work on the problem and had devised a "mechanism" for use with the 24-inch refractor with which in late 1903 he obtained photographs of Jupiter and Saturn that Lowell pronounced "the best I've ever seen. I agree with you," he added, "that some of the canals of Mars are within hope after these results. . . ."[6]

Essentially, Lampland had produced two devices designed to enhance the definition of planetary photographs. First, by inserting a secondary lens tube within the main tube of the Lowell 24-inch, he extended its normal 38-foot focal length to 143 feet, thus gaining greater resolution. Secondly, he devised a two-way sliding plateholder with which twenty or more planetary images could be recorded on a single plate in rapid succession in a matter of minutes. Each individual exposure was limited to an area of the plate just large enough for the planet's image, and by moving the plate at right angles, up or down or to the left or right, a series of consecutive exposures could be obtained as fast as the observer's hand could turn the ratchet and squeeze the shutter bulb. Thus the observer could take better advantage of the brief moments of good seeing.[7]

Lowell's interest was aroused. "What you say about the possibilities of photographing Mars is extremely interesting, and the way you mention the only way to do it," he wrote Lampland early in 1904. "A great many photographs have got to be made and a few chosen."[8] The principle here, of course, was sound, and indeed is fundamental in planetary photography today, although the automatic motor-driven camera replaces the hand-operated double slide plateholder, and fast fine-grained films take the place of the slow undependable emulsions and fragile glass plates of that earlier day. Even the sophisticated modern systems, while quicker than the hand, are still not quicker than the eye.

To supplement these instrumental devices, Lampland embarked on a series of continuing experiments to find the most suitable plates and determine the most compatible color filters to use with them in photographing the planets. Like the elder Slipher he

C. O. Lampland

conducted experiments with various developing solutions and procedures in an effort to improve the contrast and thus the detail on his plates. Together, Lowell would later claim, Lampland's innovations in planetary photography constituted a "mechanical" advance in astronomy equivalent to George Ellery Hale's invention of the spectroheliograph.[9] While modern historians of astronomy eschew such comparisons, they still credit Lampland with the first systematic application of direct photography to planetary studies.[10]

For a while at least, and even after announcing that the martian canals had been repeatedly photographed, Lowell was reluctant to reveal the full details of the techniques that Lampland had devised. "The photography of the canals of Mars," he coldly informed a newspaper writer in 1907 in demanding a correction on the point, "is not done at other observatories because it cannot be done there, the process consisting of one invented by and practiced solely at this observatory."[11] Later, of course, Lampland's innovations became known and were adopted by some astronomers, although Lowell was not always satisfied with the way they were applied in the case of Mars, as he later wrote Pickering:

It is certainly complimentary to have our method of planetary photography so generally copied, but those now using it make the mistake of developing too hard and enlarging too much, so that they lose the delicate canals.[12]

In contrast to the furious activity at Flagstaff in 1903, and particularly during the six months Lowell was in residence on Mars Hill, the year 1904 was unusually quiet, as Lowell complained to Slipher early in April:

I don't quite understand why more has not been accomplished this winter at the observatory. Nothing has been published or got ready ... since I left except what I have sent myself.[13]

And in mid-May, in a letter to Lampland, he inquired somewhat impatiently:

How gets on the photography ... of Mars for the next opposition? We *must* secure some canals to confound the skeptics. I have written to Mr. Slipher to ask you to see what you can some morning about sunrise on the planet and let me know cartographically.[14]

Late in June Lowell sailed for Europe, and even the volume of correspondence between Boston and the observatory dwindled to a trickle. This certainly reflects the fact that his efficient personal secretary Wrexie Louise Leonard, who usually kept a steady stream of instructions, advice, comment, and chit-chat flowing to Flagstaff whenever Lowell was traveling, had herself taken a ten-week "leave" to go abroad that summer.[15] As always when Lowell was in Europe he mended his astronomical fences, as it were. He visited Milan and presumably Schiaparelli and among other things agreed to have Slipher study the star V Puppis spectrographically for Miss Agnes M. Clerke, the eminent English historian of astronomy.[16]

Lowell returned in October and immediately began to prepare

for the favorable opposition of Mars scheduled for May 8, 1905, when the planet would approach to within some fifty million miles of the earth. Again as in 1903 he would spend more than six months at Flagstaff, and again his sojourn on Mars Hill would be highly productive of new observations. Well in advance of the opposition Slipher launched his unsuccessful attempt to determine the presence of oxygen or water vapor in the martian atmosphere spectrographically with Lowell's "velocity-shift" method. Somewhat later, as Mars neared opposition, Lampland began the systematic photography of the planet using the 24-inch and the new appurtenances he had devised for it.

Lowell took a hand in both Slipher's and Lampland's work, but he also pursued a wide variety of astronomical projects of his own. During this period, for instance, he initiated the brief inconclusive spectrographic investigation of chlorophyll that Slipher carried out over the next two years, as well as the long mathematical search for a new planet beyond Neptune that would not end successfully until fourteen years after his own death.

Lowell also continued his visual observations of Mars, following his 1903 measurements of the polar caps; refining his value for the tilt of the martian axis, which he now set at 23 degrees, 59 minutes; and repeating his 1903 studies of the Mare Erythraeum to confirm the nature of the seasonal changes he had recorded there earlier.[17]

In these observations he was joined briefly by his close friend Professor Edward S. Morse, an eminent zoologist, a former president of the National Academy of Sciences, and then director of Boston's Peabody Academy of Science. Morse in October 1906 would lend his lustrous reputation to Lowell's side of the mounting Mars furor by publishing a popular book, *Mars and Its Mystery,* in defense of Lowell's martian theory, and writing a Sunday supplement article about his Flagstaff experiences, headlined "My 34 Nights on Mars," in which he informed readers of the New York *World* that Lowell was "the most conservative astronomer of them all."[18] Another visitor was the American physicist Albert A. Michelson who with Edward W. Morley performed the famous 1887 experiment which showed that the speed of light is independent of the motion of its source, a finding fundamental to subsequent Einsteinian relativity.[19] Michelson wrote no books or Sunday Supplement articles about his Mars Hill experiences, however. Miss Leonard, incidentally, was also on hand in Flagstaff for the 1905 observations.[20]

But Lampland's photographic observations provided the major sensation of the opposition. His plates recorded other phenomena besides canals that Lowell considered newsworthy, including the first "snowfall" of the season around the martian north pole.[21] When this event was announced in the press, B. J. Jenkins, a Fellow of the Royal Astronomical Society, wrote to Lowell to suggest that rather than snow, the bright material observed and photographed in the polar region might be an "incandescent" substance being extruded from the interior of the cooling planet as it contracted within its crust. To support this, he cited an 1862 observtion by Sir Norman Lockyer of "very brilliant matter" that had appeared to be "welling up somewhat in the manner of a fountain" on Mars.[22] Lowell, it seems, was not the only imaginative astronomer of his day.

The canals first appeared photographically, to Lowell and Lampland at least, on a plate exposed May 11, three days after the planet's opposition.[23] Subsequently additional canals showed up on later plates to a total of thirty-eight in all for the opposition, including the broad double known as the Nilokeras.[24] Lowell described the drama of the discovery colorfully, if briefly, in *Mars and Its Canals:*

Many pictures were taken on each plate one after the other, both to vary the exposure and to catch such good moments of seeing as might chance. Seven hundred images were thus got in all; the days of best definition alone being utilized. The eagerness with which the first plate was scanned as it emerged from its last bath may be imagined, and the joy when on it some of the canals could certainly be seen. There were the old configurations of patches, the light areas and the dark, just as they looked through the telescope ... and there more marvelous yet were the grosser of those lines that had so piqued human curiousity, the canals of Mars. ... By Mr. Lampland's thought, assiduity and skill, the seemingly impossible had been done.[25]

"Thus," he added, "did the canals at last speak for their own reality themselves."

They did not speak thusly to everyone, however, and almost immediately Lowell found that his triumph was to be tempered with frustration.

Lowell's telegraphic announcement of Lampland's feat was published initially in the German journal *Astonomische Nachrichten.*[26] Subsequently, through the ensuing summer and autumn, the photography of the canals was described in scores of often repetitious articles in newspapers, magazines, and journals

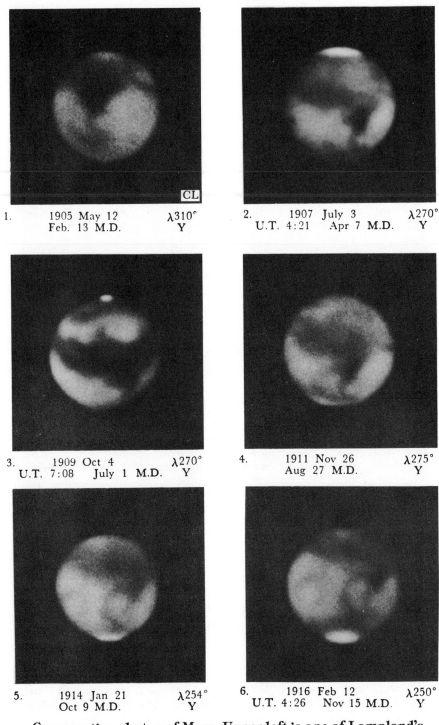

1.   1905 May 12   λ310°
        Feb. 13 M.D.   Y

2.   1907 July 3   λ270°
   U.T. 4:21   Apr 7 M.D.   Y

3.   1909 Oct 4   λ270°
  U.T. 7:08   July 1 M.D.   Y

4.   1911 Nov 26   λ275°
      Aug 27 M.D.   Y

5.   1914 Jan 21   λ254°
      Oct 9 M.D.   Y

6.   1916 Feb 12   λ250°
   U.T. 4:26   Nov 15 M.D.   Y

Comparative photos of Mars. Upper left is one of Lampland's.

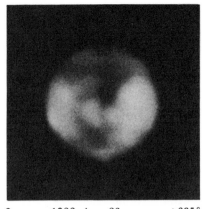

7.   1918 Mar 31   λ284°
   U.T. 3:34   Dec 25 M.D.   Y

8.   1920 Apr 23   λ285°
   U.T. 8:47   Jan 25 M.D.   Y

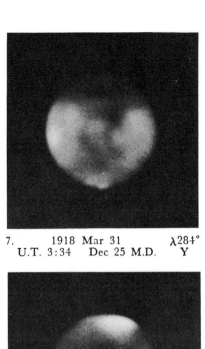

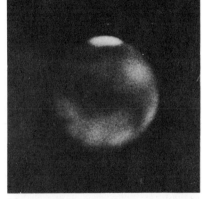

9.   1922 June 18   λ260°
   U.T. 7:25   Mar 16 M.D.   Y

10.   1924 Aug 31   λ250°
   U.T. 9:34   May 17 M.D.   Y

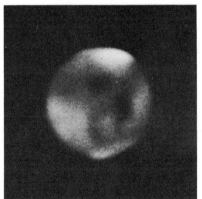

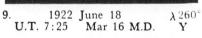

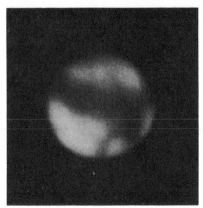

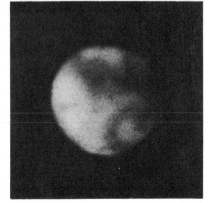

11.   1926 Nov 6   λ255°
   U.T. 5:19   Aug 6 M.D.   R

12.   1928 Dec 29   λ245°
   Sept 28 M. D.   Y

throughout Europe and the United States.[27] Lampland's accomplishment was generally hailed by the press and would win him the 1907 medal of the Royal Photographic Society.[28] Some writers declared, sight unseen, that the photographs had settled once and for all the issue of the reality of the martian canals. Lampland's plates, an "occasional correspondent" for the London *Daily Graphic* opined, "will put an end to the most hotly-debated controversy which in recent years has disturbed the serene lives of celestial observers."[29] In America, Garrett P. Serviss, the knowledgeable science writer for the Hearst newspapers, commented on the "surprising news" from Lowell Observatory with somewhat more restraint:

The photographic demonstration—if it is one—that the canals exist is in itself a great step in advance sufficient to place Mars once more in the forefront of interest. The next step will be to discover what the "canals" are for.[30]

A few days later apparently, and after inspecting direct prints from some of Lampland's plates, Serviss was able to resolve his parenthetical doubt:

On the copies of the photographs which I have received only parts and in some instances mere suggestions of the canals are to be seen, yet the general resemblance to the drawings is most remarkable. . . . We may now look upon the existence of at least the principal Martian "canals" as established.[31]

But Serviss had also discovered a disquieting fact from his examination of the photographs which would shatter Lowell's hopes of having the canals "speak for their own reality themselves" to the world at large:

Unfortunately, the photographs are so small and the shadings on them so delicate, that it would be impossible to reproduce them in a newspaper.[32]

For the canal photographs from the outset proved to be "not generally publishable," as Lowell ruefully conceded somewhat later to Carr V. Van Anda, the astute managing editor of *The New York Times* and others who sought to present them to the public.[33] The image of the planet on the plates was indeed small, less than a quarter of an inch in diameter, and enlargement only increased the already considerable distortion from the coarse grain of the photographic emulsion. While the dark martian "seas" showed up well enough on the tiny images, the lines that Lowell, Lampland, and others identified as the more prominent canals were indeed evanescent. As a practical matter, there was no chance that such faint delicate features could survive the rude process of mass media reproduction.

In the few instances where publication of the photographs was attempted, surely the results better illustrate the problem of reproduction than the canals. In June the *Scientific American* printed five of the pictures, but there are no definitely linear markings in them.[34] In November *Popular Astronomy* published six images with Lowell's paper on the canal photography, and again Mar's face is mottled but unlined.[35] In that same month, however, the *Scottish Review* printed two photographs with an article by British science writer, Hector McPherson, Jr., in which several broad curvilinear markings stand out with startling boldness.[36] Yet when the English journal *Knowledge and Scientific News* published a single image in March 1906, the reproduction was so poor that even the planet's pale disk was barely discernible.[37]

Lowell was particularly concerned over the impact the photographs would have in England where mounting skepticism over the reality of the canals, fostered by the insistence on their illusory nature by Maunder and others had become the most serious challenge to his Mars theory. To this end, on a brief trip abroad late that summer, he took with him a set of direct prints to show personally to a number of influential astronomers. As soon as his *Bulletin* on the photography was off the press, in fact, he had sent a copy, along with some prints from the best plates, to William H. Wesley, assistant secretary of the Royal Astronomical Society and an authority on interpreting astronomical photographs. Wesley, Lowell would later claim to a still-dubious Simon Newcomb, "sees in them as much as Mr. Lampland or I do ... and he judges from the prints alone which of course are far behind visual observations in definition."[38]

Wesley did, in fact, see something in the prints, although certainly not all that Lowell did, and subsequently described what he saw in an article in the British journal the *Observatory*. This, however, was not illustrated with the photographs but only with a composite drawing of Mars. "The small photographs," Wesley noted, "could not be satisfactorily reproduced ... by any photomechanical process and I therefore give a drawing which I have made from them. I have inserted in this *all* the details that I can make out with certainty on *any* of the six pictures."[39]

The Lowell Observatory photographs not only show the seas with much more defined outlines, but also several of the so-called "canals." Of these, Mr. Lowell enumerates eight as being more or less well shown in his photographs, and I certainly think he is justified in his claim. In fact, these photographs go far to remove the skepticism I have always felt in regard to these features. ...[40]

At the same time, Wesley was not willing to go along with all of Lowell's claims for Lampland's pictures:

He says: "The negatives thoroughly confirm the eye in showing not only the existence of the canals but *the fact that they are continuous lines and not a synthesis of other markings."* Does he not go too far in the words I have italicized? Doubtless a photograph has no imagination, but imperfect definition applies equally to photographic and visual observations. That which is seen or photographed imperfectly as a smooth continuous line *may* be full of small irregularities, and may not be strictly continuous. It may, indeed, be a row of dots or discontinuous marks. We can only say that *in the main* these run in straight lines. [41]

The *Observatory's* editor, Oxford Professor H. H. Turner, was even more equivocal about the photographs, as he reported in the same issue:

I do not mean to deny the claim any more than I am prepared to admit it. Personally, I find it extremely difficult to say exactly what is on these prints and friends to whom I have shown them differ a good deal in their interpretations. [42]

But some other British astronomers agreed with Lowell that the new photographs provided proof of the reality of the much-debated linear markings on Mars. A.C.D. Crommelin gave particularly strong support for this view in his presidential address to the British Astronomical Association in October 1905 when he reviewed "some recent advances in astronomy, among which the photographs of the Martian canals secured at the Lowell Observatory have a high place."

The results are a very notable advance in planetary photography; nearly every print shows one or more canals, while the best of them show six or eight. I had the pleasure of meeting Prof. Lowell a few weeks ago ... when he showed me the pick of his prints, on many of which the canals were clear and unmistakable, appearing as continuous narrow, slightly curved lines. These photographs did a great deal to strengthen my faith in the objective reality of the canals which I had previously looked on as probable but not quite certain.[43]

Where there were differences over the interpretation of the photographs within astronomy, the technical points of the debate, such as Wesley's significant reservation, seldom reached the public at large. Inevitably, real canals proved to be more newsworthy than illusory ones, and the popular press in general concluded uncritically that because some astronomers discerned linear markings, however vague, in the pictures, the actual existence of the canals had now been finally established. An easy, but erroneous corollary to this, for some writers at least, was that the optical illusion theories advanced by Maunder, Antoniadi, Newcomb, and others had now become untenable. Shortly after Crommelin's statement, for example, London's *Pall Mall Gazette* declared:

We can no longer accept the explanation of these channels endorsed by Professor Simon Newcomb, when he says that the appearance "grows out of the spontaneous action of the eye in shaping slight and irregular combinations of light and shade too minute to be separately made out, into regular forms."[44]

At least one member of the prestigious Royal Society itself publicly agreed with this assessment. The Honorable R. J. Strutt,* commenting in the London *Tribune* on Maunder's insistence on the optical character of the canals, reported that Lowell "has just read a paper before the Royal Society which effectively refutes this suggestion. For he gives photographs of Mars which exhibit the lines not, indeed, so well as his drawings, but still unmistakably."[45]

The public, not having seen the pictures, was hardly in a position to challenge such statements and tended to accept them at face value, at least for a while. Most mass media articles about the canal photography were either not illustrated at all or were illustrated only by drawings, and their authors usually were careful to explain that

---

*Possibly Lord Rayleigh, the discoverer of argon, whose initials, however, were "J. W." rather than "R. J." Internal evidence in the article would seem to indicate as much, but on the other hand it seems strange that the *Tribune* did not use his title in the byline.

the photographs themselves could not be successfully reproduced in a general publication. In the few instances where some of the photographs were published, they were little more than a "Chinese puzzle" to the layman, as one scientific commentator pointed out in declaring that "the peculiar markings on them are beyond question the so-called waterways of Mars. Astronomers the world over recognize the achievement as one of the most important in recent years."[46]

But at least one small segment of the public did have an opportunity to inspect Lampland's photographs at first hand. Early in 1906 Lowell arranged to place some of the best canal pictures on display at the Massachusetts Institute of Technology. This exhibit, in which slides of Lampland's plates were mounted side by side with drawings of Mars and backlighted by small electric lamps, resulted in a new round of favorable publicity for his canal claims. "Camera Proves Canals on Mars" was the headline announcing the opening of the exhibit not only in the Boston *Herald,* but also in the San Francisco *Call.* [47] The following day the Boston *Post* reported that the photographs were "proving a great drawing card," and the Boston *Transcript* noted that "the amount of detail shown on these photographs will be a revelation to those who are familiar with astronomical photographs of the planets."[48]

In the future, as the problem of photographic reproduction persisted, Lowell would rely more and more on such exhibits to provide public exposure for the photographic and spectrographic work of his observatory, and increasingly the preparation of these displays became a part of the routine responsibilities of his assistants at Flagstaff.[49] "I have concluded," he once explained to the elder Slipher in discussing an exhibit, "that it is well to have the visual representation of our work ... more known to the world. Hale," he added, "totes his around and makes an impression."[50]

If Lowell was keenly disappointed over the fact that the 1905 canal photographs were generally unpublishable, he was still not discouraged. The canals were there, he knew, on Lampland's plates as well as on the planet itself, and given either improved reproduction techniques or better original photographs, he was convinced that they could be made to appear on the printed page. He freely conceded that even the most prominent canals were only faintly discernible on the 1905 plates and that "some acquaintance with photography" was required to see them.[51] But that the plates showed anything at all of the elusive lines was encouraging. He could do little about the reproduction problem, but he was sure that better photographs could be obtained at the highly favorable 1907 martian opposition when the planet would be almost fifteen million miles closer to the earth and its telescopic diameter nearly a fourth again as large as in 1905. To this end he now directed his energies.

In September 1906 Lowell informed the elder Slipher in Flagstaff that he would have to forego observations with the 24-inch refractor for several months because "for the very best photographic work such as Mr. Lampland and I have to do next year on Mars, a refiguring of the glass is advisable."[52] The objective lens was duly removed from the telescope and Lampland brought it to Boston and Alvan Clark and Sons where the refiguring was done late that autumn.[53]

While Lampland was in Boston, he and Lowell attended a reception for Commodore Robert E. Peary, who had just returned from an unsuccessful attempt to reach the North Pole.[54] Lowell had long scoffed at what he considered to be the futility of polar expeditions, contending that they could add little if anything to mankind's knowledge of the earth.[55] Yet after his long conversation with Peary on this occasion, such derogatory references disappeared from his writings and public remarks. Perhaps, in light of his martian theory, he was intrigued to learn from the famous explorer that darkness rather than intense cold was the major problem posed by the extreme conditions of the high Arctic.[56]

Late in 1905 Lowell began to consider the acquisition of a large reflecting telescope for his observatory, primarily for planetary photography. In October he confided to Slipher that he was "reflecting on the reflector,"[57] and in the spring of 1906 he explored the possibility of an 84-inch instrument with astronomer George W. Ritchey, who had completed the 60-inch Mt. Wilson reflector in 1902 and would later build the 100-inch Hooker reflector for that observatory.[58] Lowell informed Ritchey that he would be willing to pay up to $55,000 for such an instrument, pointing out that "with a 7-foot reflector in the air at Flagstaff we ought to get photographic conditions which for many years will be unapproached."[59] And when Ritchey questioned this figure, he answered: "Do not let that worry you. Whatever you think proper . . . I should be only too glad to pay as I consider scientific work very underpaid as it is."[60]

Lowell did eventually contract for a large reflector but not until April 1908 when he ordered a 40-inch instrument from the Clark firm for $10,800, to be completed by June 15, 1909, for use at the martian opposition due the following autumn.[61] Lowell apparently had reflected too long, however, for the telescope was not opera-

Workmen setting up the Lowell 40-inch reflector in 1909.

tional until well after this last really favorable opposition of Lowell's lifetime had come and gone. [62]

The most important, and by far the most publicized of Lowell's efforts to obtain better photographs of martian canals, was the expedition he sent to South America for the 1907 opposition. This, he advised Lampland early in February, was the suggestion of Amherst Professor David P. Todd, his companion for his 1900 eclipse observations in Tripoli, and interested him because it provided an opportunity to test the seeing conditions at high Andean altitudes. "If young Mr. Slipher would like to go," he wrote, "I should propose that you should instruct him as minutely as possible in the photography of Mars."[63]

Later that month Lowell entered into an agreement under which Todd would be director of the expedition and would undertake "every cost . . . for the sum of $3,500."[64] The 18-inch refractor at Amherst would be shipped to South America for the observations, and young Slipher, who had been working with Lampland on photography since taking up his Lawrence Fellowship in July 1906, would be observer-photographer. Todd's wife, Mabel Loomis Todd; A. G. Ilse, engineer for the Clark firm; and R. G. Eaglesfield, one of Todd's students, completed the expedition's roster. [65]

In the context of the Mars furor, this astronomical adventure, this journey to a remote region of one planet to explore the mysteries of another, captured the interest of countless thousands around the world. The imminence of the most favorable 1907 opposition, which so many hoped would reveal so much, had already touched off a flood of articles on Mars and martians in newspapers and other periodicals on both sides of the Atlantic. Some of these were sober summaries of the debate over Mars; others were in a speculative or even fanciful vein; many were argumentative, marshaling the case for or against Lowell's theory of the probable existence of intelligent martian life. The Lowell expedition sailed on this unprecedented high tide of pseudoscientific publicity, set flowing, it should be remembered, by what was essentially a routine astronomical event. [66]

Much of what was written about the expedition in the popular press voiced frank optimism that a final *denouement* in the long, increasingly bitter controversy over Mars was now at hand. This, however, was not the way that Lowell conceived of his expedition. Its purpose, to his thinking, was simply to further confirm and to extend through photography what to him were the already well-established facts of the planet's habitability and present habitation.

"I see your reporters have given queer accounts of the Lowell expedition (of this observatory) to the Andes," he wrote from Flagstaff in May to his friend Edward H. Clement of the Boston *Transcript*. "It is not to *settle* anything about Mars as I need hardly say we can do that better here."[67]

The expedition's publicity preceded it. Arriving at the Panama Canal Zone in mid-May, young Slipher assured his brother in Flagstaff that "Lowell expedition to Andes" was a sufficient address for mail. "We advertise so thoroughly that this ... would find us at any time I think."[68]

By late June, the expedition had chosen a site near Alianza, Chile, some seventy miles inland from the small port of Iquiqui, at an elevation of 4,000 feet. The 18-inch telescope was set up in the open desert with only the sky for a dome, and Slipher began exposing the first of the 13,000 images he would accumulate over the next three months.[69] Almost immediately, canals appeared on his plates, as Todd cabled in code to Lowell on June 28,[70] and within three weeks Todd could report that young Slipher had succeeded in photographing a number of the more prominent doubles.[71] To this, Lowell replied exuberantly:

Bravo! ... The world, to judge from the English and American papers, is on the *qui vive* about the expedition, as well as about Mars. They send me cables at their own extravagent expense and mention vague but huge (or they won't get 'em) sums for exclusive magazine publication of the photographs.[72]

Slipher considered the Alianza site "exceptionally good," although Todd, or perhaps Mrs. Todd, felt a better one might be found and, despite the exigencies of time and the martian opposition, kept urging that the telescope be moved to a new location. "I don't see how I can follow Mr. Lowell's directions and also please Prof. Todd," Slipher complained to his brother late in July. "I guess he must ask Mrs. Todd. She seems to wear the trousers."[73] That Todd did not see any canals on the plates also distressed the younger man. "As Prof. Todd's eyes are not good and he cannot see them," he confided to the elder Slipher, "I do not like to say so and so is true although it shows to my eyes absolutely and distinctly."[74]

Yet if Todd did not see all that Slipher saw in the martian plates, he apparently was willing to take Slipher's word that the canals were there. In forwarding a new batch of prints to Lowell in August, for instance, he had no difficulty in praising the photographs with a certain verbose grandiloquence:

They seem to me fine and speak volumes for themselves. As they looked at these photographs and then flew over the pampas, even the bats of Alianza screamed *oasis, oasis, oasis–canali, canali, canali!*[75]

Actually, Todd contributed little to the expedition as far as the observation of Mars was concerned, although a somewhat different impression emerged from the reports he and his wife subsequently provided for the press.[76] Indeed his major accomplishment, astronomically at least, seems to have been to photograph an eclipse late in August, and this he did at Newcomb's rather than Lowell's behest. The eclipse was routine, but nonetheless Mrs. Todd undertook to write a lengthy newspaper article about the event.[77]

Lowell, of course, was not idle while his expedition was making headlines. The more than eight months he spent at his observatory in 1907 represents not only his longest but his busiest single sojourn in Flagstaff. Arriving in March, and remaining on into November he threw himself into a bewildering variety of activities without, however, neglecting the rigorous observational routine for Mars that he had set for himself at previous oppositions. Photography was now a part of this, and he and Lampland together, with the refigured 24-inch telescope, obtained some 3,000 images of the planet, many of which, Lowell claimed, showed both single and double canals.[78] The best prints from the best of these plates, Lowell immediately dispatched to his favorite martians, the aging Schiaparelli at his Milan observatory, and to Flammarion at his Observatoire de Juvisy near Paris.[79]

Again, Lowell also made extensive visual observations of the planet, reporting that the progressive changes observed in the appearance of the canals not only confirmed his work in 1903 and 1905 but followed the predictions of his Mars theory.[80] Again, he studied the polar caps, both of which were equally visible in 1907, remeasured the position of the martian axis, and even reported the doubling of the well-known surface feature known as the Solis Lacus.[81]

Nor was the work at Flagstaff confined only to Mars. "We are having interesting observations of Jupiter and finding asteroids *en masse*," he wrote enthusiastically to Todd in April.[82] The Jovian

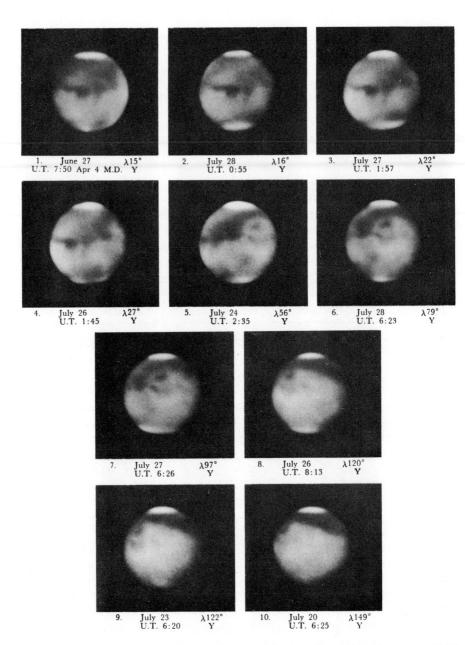

1.   June 27   λ15°
U.T. 7:50 Apr 4 M.D. Y

2.   July 28   λ16°
    U.T. 0:55   Y

3.   July 27   λ22°
    U.T. 1:57   Y

4.   July 26   λ27°
    U.T. 1:45   Y

5.   July 24   λ56°
    U.T. 2:35   Y

6.   July 28   λ79°
    U.T. 6:23   Y

7.   July 27   λ97°
    U.T. 6:26   Y

8.   July 26   λ120°
    U.T. 8:13   Y

9.   July 23   λ122°
    U.T. 6:20   Y

10.   July 20   λ149°
    U.T. 6:25   Y

Photos of Mars made in Chile by E. C. Slipher in 1907

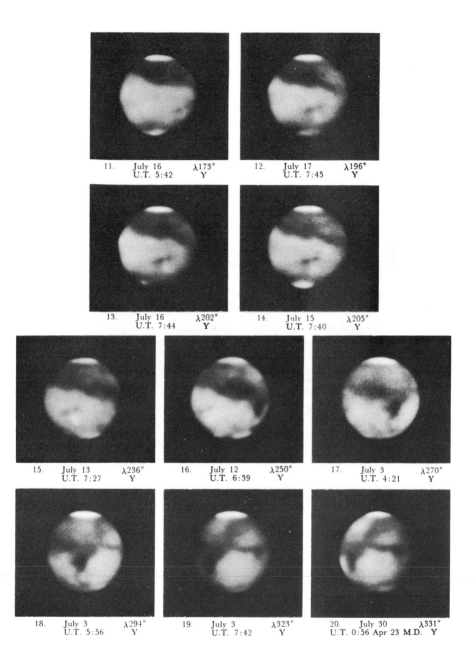

11.   July 16 $\lambda 173°$
        U.T. 5:42   Y

12.   July 17 $\lambda 196°$
        U.T. 7:45   Y

13.   July 16 $\lambda 202°$
        U.T. 7:44   Y

14.   July 15 $\lambda 205°$
        U.T. 7:40   Y

15.   July 13 $\lambda 236°$
        U.T. 7:27   Y

16.   July 12 $\lambda 250°$
        U.T. 6:39   Y

17.   July 3 $\lambda 270°$
        U.T. 4:21   Y

18.   July 3 $\lambda 294°$
        U.T. 5:56   Y

19.   July 3 $\lambda 323°$
        U.T. 7:42   Y

20.   July 30 $\lambda 331°$
        U.T. 0:56 Apr 23 M.D.   Y

observations were indeed of interest to the British astronomer Scriven Bolton who considered them "valuable confirmation" of the faint wispy lines he had seen trailing from some of the giant planet's equatorial belts some eight years before. [83] Lowell usually forwarded any asteroid data gathered at Flagstaff to Max Wolf and A. Berberich who included observations of minor planets among their astronomical interests. [84] That spring, too, the elder Slipher turned the spectrograph on Saturn, finding no evidence for the long-held belief that there was water vapor in its atmosphere, and that the spectrum of its rings differed from that of the planet itself. [85] Lowell also completed his insolation studies of temperatures on Mars, [86] while Slipher pursued the inconclusive spectrographic investigation of chlorophyll and made observations for the line-of-sight velocity of both single and double stars.

Along with this busy astronomical regimen, Lowell played a genial host to a steady stream of guests who visited the observatory and tarried on Mars Hill for varying periods that spring and summer. There were Lowell's fact-finding camping trips as well as less extensive outings of a more social or sightseeing nature, into the scenic country around Flagstaff. In addition, he found times to seek special railroad excursion rates and "stopover" privileges for Flagstaff; to join J. Walter Fewkes as a speaker before the Flagstaff Board of Trade and to entertain the pioneer southwestern archaeologist at dinner; and to lecture about Mars to summer students at Northern Arizona Normal School. [87] Along with all this, late in June he made a quick trip back to Boston where he received his first honorary LL. D. degree from Amherst College. [88]

Despite such varied activity, Lowell also managed to keep up a vigorous promotion and defense of his martian theory in the popular press, in scientific journals, and in his increasingly voluminous correspondence.

But it was the photography of the martian canals by the Lowell expedition to the Andes rather than Lowell's observations at Flagstaff that commanded the attention of the public that summer as Mars passed through opposition. Todd's cabled news of young Slipher's successes did indeed result in a lively if limited bidding for first publication rights to the South American photographs. This began with a request for an illustrated article on the expedition from S. S. McClure, whose national magazine was then at the height of its "muck-raking" popularity. [89] Lowell replied at once, asking for a "definite proposition" in regard to the photographs and pointing out to the renowned publisher "how very valuable they are and what crushing evidence to the critics of the canals."[90] *McClure's* major competitor was the *Century* which had agreed to publish Lowell's October 1906 Lowell Institute lectures as an eight-article series after both *McClure's* and *Harper's* had turned him down.[91] On the same day he answered McClure, he advised the *Century* that as the South American pictures were not part of his lectures, "I should have to have a definite proposition made me for a separate article as other magazines are wiring for them. The photographs are as you can well judge practically priceless."[92]

Within a week, the offers were in and Lowell made his decision. "The magazines have been tumbling over each other to secure publication of the South American photographs," he advised his friend and fellow Boston scientist George R. Agassiz who had been in Flagstaff with him in the spring, "and the *Century* has grasped the prize."[93] To the elder Slipher, then vacationing at his family home in Frankfort, Indiana, he explained that he was giving the photographs to the *Century* because "I was more affiliated with them on account of their publishing my lectures and because they offered the most."[94]

But the problem of reproducing the canal photographs satisfactorily on the printed page could not be solved so easily and loomed as large in 1907 as it had in 1905, as Lowell confided somewhat desperately to Agassiz:

Now to reproduce them *tel quel* would be beyond present processes to accomplish short of great loss of detail. They would have to be retouched under direction, as Wesley was kind enough to do for the Royal Society paper two years ago, and that direction cannot come from me as it would vitiate the very proof they convey. It must be done by some one who is not me and yet knows—You are the best man to do it, if you will. Will you?[95]

Agassiz answered that he would be "delighted to do the retouching," adding, however, that "of course it must be thoroughly understood that the results are not direct reproductions from the negatives."[96] This was evidently satisfactory to Lowell, but at the *Century* the idea was quickly vetoed. Associate editor R. U. Johnson wrote:

There is no time to retouch the photographic plates and we should consider it a calamity to do so, as it would certainly spoil the autographic value of the photographs themselves. There would always be somebody to say that the results were from the brain of the retoucher.[97]

To this, Lowell replied that "if, in your opinion, the canals show well enough without retouching why of course so much the better. If not, Mr. Agassiz is ready to see what he sees comes out."[98]

Lowell's intense desire to have some canals show up in published photographs of Mars also led him to propose to the *Century's* editor, Richard Watson Gilder, that he "make a departure from the usual method [of illustration] for the sake of bringing before the inhabitants of this planet the portraiture of another world." The *Century*, he urged, should paste direct prints from Slipher's plates into each copy of the magazine's December issue.[99]

And further, in his anxiety to insure successful publication of the photographs, he directed young Slipher to deliver his pictures in person to the *Century's* offices in New York City. "My purpose in having you go with the negatives and prints is that you may explain their fine points to them and keep a supervision over them in their workshops," he told his newest assistant.[100]

In the final analysis, however, the reproduction problem proved insoluble, and indeed even today it remains a serious obstacle for astronomers and other scientists who seek to publish photographs in which faint or intricate detail is of overriding importance. It is significant, certainly, that even in Lowell's later books about Mars where presumably particular attention could be given to engraving and printing the illustrations, the canal photographs are conspicuous by their absence, despite the glowing descriptions therein of Lampland's achievements, and despite the fact that they were then of considerable interest simply as unusually good photographs of Mars, if not of martian canals.[101]

Slipher's South American photographs, incidentally, posed one other publication problem for Lowell, but one he could, and did, resolve. "I happen to know," he had warned the *Century* in accepting its offer for the photographs, "that another magazine has already agreed with Prof. Todd to publish an article on the expedition and has asked me to send them prints. These I shall not send but when Todd himself shall ask me I cannot well refuse. You are ahead now and will be provided the article comes out as soon as possible."[102]

The other magazine was *Cosmopolitan*, which claimed a nationwide circulation of over one million and which was then edited by the able and aggressive S. S. Chamberlain. Todd had indeed promised the *Cosmopolitan* an article about the expedition, but he denied to Lowell that he had guaranteed any exclusive publication of the pictures.[103] And Chamberlain had indeed requested some photographs and had been advised by Lowell that he was "welcome to the use of the illustrations in *Mars and Its Canals* for Prof. Todd's article."[104] Whatever Chamberlain's understanding of the situation may have been, and it seems he considered the South American photographs to be under Todd's jurisdiction, Lowell felt obliged subsequently to point out rather bluntly that he was "laboring under a misapprehension."

The South American photographs are my property and therefore not Todd's to dispose of and are not to be published without my express agreement. I shall be very glad later to help Prof. Todd's article if it is not going to appear before January ... as I think it is capital that he is writing a paper for you. I think it is proper to write you this as you might inadvertently get into serious trouble by infringement of property.[105]

Lowell may have had second thoughts about the strong tone of this letter, for within days he wrote soothingly to the *Cosmopolitan's* editor:

With your interest and knowledge of astronomy you might do great work for the proper popularization of the subject by making a special feature of astronomical articles through the columns of your widespread magazine. I think a great many people who read the fairy tales of Earth would read the truer but no less romantic tales of science if they had the chance.[106]

Chamberlain, however, was plainly disappointed and threw Lowell's suggestion back at him in a new appeal. "I had counted upon Prof. Todd's paper as the first of a series on astronomical subjects which would appeal to the million readers of the *Cosmopolitan* and help toward a greater popularizing of the wonders of astronomy," he replied:

The story of the recent expedition, with the first actual photographs of the Martian canals, would constitute so striking a proof in refutation of the skeptics who have persistently derided the canal theory that it seemed to me to furnish the most fitting possible article with which to open a popular series on astronomical subjects. But to be limited to printing a merely general description of the expedition, without a single one of the actual photographs of the canals, would deprive the article of the particular interest upon which I had relied. ... Without it, I shall be compelled to abandon the whole scheme.[107]

When Todd arrived at Panama in October on the expedition's return voyage to the United States, he wrote Lowell to ask if he could promise *Cosmopolitan* a single canal photograph and some drawings of the planet as illustrations of his article.[108] Lowell agreed, but with two conditions. First, as his own article with young Slipher's photographs was scheduled for December publication in the *Century,* the Todd article was not to appear until January. Second there would be no photographs at all for the *Cosmopolitan* unless Todd's article met with Lowell's full and final approval. The first of these conditions provided no problem, but Lowell found that he had to threaten legal action to enforce the second.

Todd's article, as it turned out, was not at all acceptable to Lowell, for the expedition's director had taken unto himself credit that was due solely to the expedition's photographer. Lowell demanded corrections.[109] When the *Cosmopolitan* sent him the press proofs of the article in mid-December, however, the corrections had not been made, and Lowell forthwith wrote an attorney in New York City, A. L. Everett, "in regard to bringing an injunction against the publication of an article by David Todd in the next issue of *Cosmopolitan.*"[110] In a second letter two days later, he told Everett that "I do not wish to make an enemy of the *Cosmopolitan* if I can avoid it and therefore an injunction ought not to be issued if conciliatory measures can effect the same result."[111] Everett's initial efforts along this line were of no avail, and on New Year's Eve, Lowell wrote Todd that "I have decided not to send you the photographs for publication you asked for. I regret very much that the corrections promised in the *Cosmopolitan* article were not made."[112]

Todd, however, had already obtained some prints of the canal photographs from young Slipher and had given them to the *Cosmopolitan* for use with his article.[113] Lowell acted swiftly. Ten days before the magazine's January 20 publication date, he had obtained duplicate prints of the photographs Slipher had given Todd, applied for a copyright on them, and directed Everett to serve an injunction on the *Cosmopolitan* as soon as the copyright papers were in hand. "They will then have to show their hand," he explained to the attorney.[114]

An injunction, as events proved, was not necessary, and early in February Lowell triumphantly advised Everett of the outcome of this legal maneuvering:

I think it will please you to know that that last letter of yours brought down the game! Though they said nothing the *Cosmopolitan* management was so scared that they rewrote the article incorporating the necessary corrections—delaying the issue ten days or so.[115]

And to young Slipher, he sent a copy of the belated January *Cosmopolitan*. "The story read otherwise before the copyright was got," he wrote. "Then the issue was suddenly delayed for ten days at, I fancy, considerable expense to them. I think," he added, "you will like to see the result of holding up a magazine!"[116]

CHAPTER
12

The Mars Furor

HOWEVER SEEN OR CONCEIVED, the elusive lines that Schiaparelli, Lowell, and some others insisted they saw on Mars have been almost universally referred to as "canals," the use of quotation marks usually denoting reservations concerning the scientific propriety of the word.

Schiaparelli, who discovered the lines in 1877, had called them *canali,* or channels, but was reluctant to read any implications into the term, believing them to be of natural origin. He was, however, equally reluctant to deny the implications that others drew from the lines, or from the peculiar polygonal patterns they made on the planet and the phenomenon of their "gemination," or doubling:

Their singular aspect, and their being drawn with absolute geometrical precision, as if they were the work of rule or compass, has led some to see in them the work of intelligent beings, inhabitants of the planet. I am very careful not to combat this supposition, which includes nothing impossible. ... Let us add further that the intervention of intelligent beings might explain the geometrical appearance of gemination, but it is not necessary. ... [1]

Even thirty years after his discovery, the cautious Italian still remained somewhat circumspect, although he could now concede that as the *canali* showed "a marvelous harmony of system," they "cannot be the work of chance." He carefully added, however, that "there is nothing to indicate that there are upon Mars individuals closely resembling human beings."[2]

[200]

Largely as a result of Lowell's well-publicized activities, and perhaps in spite of those of the flamboyant Flammarion in France,[3] the early widespread disbelief in the physical reality of Schiaparelli's *canali* had yielded gradually to what might be called a conditional credulity. As more and more observers reported the lines, more and more people, astronomers included, were willing to admit that at least there were some topographic features on the martian surface that showed up in linear configurations when viewed from the earth. Consequently, the canal controversy turned progressively away from the issue of the objective existence of these markings to the far more intriguing and emotionally charged questions of their nature and possible function. The Lowell Observatory photographs of the "canals," if they accomplished little else, served to assure this subtle shift in emphasis. English science writer W. E. Garrett Fisher summed up the new status of the "canal" debate succinctly as the year 1906 came to an end:

A long disputed matter has finally been determined by the success-ful photography of some of the Martian canals. It is no longer possible to assert that these curious markings are due to some optical defect of the eye or instrument of those who have seen them—though, of course, it does not follow that the anthropomor-phic interpretation put upon them by Mr. Lowell is therefore cor-rect.[4]

In America a few months later, writer Waldemar Kaempffert listed the problems posed by Mars for readers of *McClure's* magazine and noted that the objective reality of the martian "canals" was no longer among them, Lampland's photographs in 1905 "forever dis-posing of their illusory character."[5]

This was not a full triumph, as it still could and would be argued that optical effects produced the unnatural linearity and the geomet-rically precise arrangement of the markings if not the markings themselves. But it was a significant one in terms of the Mars furor. For as long as the "canals" were considered to be merely illusions, there was no real point in inquiring seriously into what they were or how or why they had come into existence on distant Mars, or even in speculating on their possible implications. But once admit that this gossamer network was the manifestation of actual physical features on the planet's surface—that there was indeed *something* there—then such questions became immediately germane, and all possible answers acquired a compelling interest for scientists and laymen alike.

Thus as the martian controversy reached the peak of its intensity, the *canali* or "canals" with or without quotation marks, became
many things to many people.

To Maunder, Antoniadi, and their followers of course, the
"canals" were discrete spots and splotches of shadings probably
roughly aligned and thus appearing as lines when seen with smaller
telescopes and under less than optimum conditions. W. H. Pickering, Newcomb, and others thought the "canals" might be natural
cracks seaming the martian crust, geological faults, or volcanic
fissures imperfectly resolved by the eye and lens and possibly
analogous to some of the linear features that had long been known
on the moon.[6] It may be added here that Pickering also had other
more imaginative explanations, including one that envisioned the
markings as "parts of otherwise destroyed forests left standing in
long lines like those ... on the hills in the Azores."[7]

The "cracks" hypothesis, incidentally, had acquired another
influential advocate as the martian controversy progressed in the
person of the venerable Alfred Russel Wallace, the co-originator
with Darwin of the theory of evolution and one of the giants of
nineteenth-century science. Wallace, aged eighty-four in 1907, had
become the highly articulate spokesman for the sizeable body of
opinion, centered largely in England, which still held to the Whewellian view that life was unique to the earth, and could not exist
anywhere else in the universe. There was simply no provision in this
teleologically anthropocentric cosmos for martians or their "canals."[8]

To Lowell, of course, the markings were irrigation canals pure
and simple—too small to be seen themselves to be sure, but
nonetheless rendered visible at certain times of the martian year by
the progressive darkening of broad bands of vegetation lining their
banks—and as such they formed the foundation of his theory of the
probable existence of intelligent canal-building martians.

These, then, were the positions from which the key antagonists
in the martian controversy were prepared to do battle as the planet
swung toward its highly favorable 1907 opposition. The advantage,
perhaps, was with Lowell, temporarily at least. For one thing, even
astronomers who vehemently disputed his concept of the canals
usually were willing to concede the superior location of his Flagstaff
observatory for astronomical work.[9] For another, the mass of systematic data about Mars he accumulated over the years and since
founding his observatory was in fact unique; Otto Struve would
declare nearly sixty years later that "it is impossible to describe

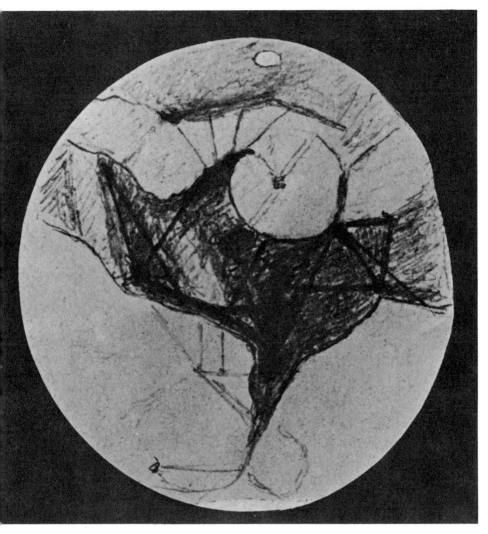

"Canali novae" on Mars, as drawn by Lowell in 1909.

adequately the wealth of observations" included in Lowell's Mars books and in the *Annals* of his observatory.[10]

But two other, broader factors, perhaps, also swung the balance briefly in his favor.

The first of these was the considerable reputation he had now acquired as a planetary observer and as the ranking expert in the world on Mars and matters martian. Although he had been considered something of a *dilettante* when he plunged into astronomy in 1894, his diligent investigations of Mars no less than his vigorous

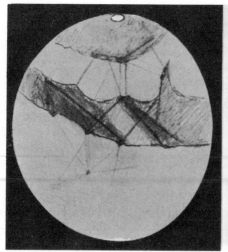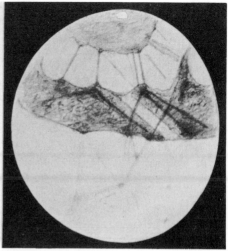

Comparative drawings of Mars, by P. Lowell (left) and E. C. Slipher (right), November 8, 1909.

promulgation of their results had brought him international renown, and in the public mind at least, his name had become virtually synonomous with the red planet.[11] In 1905 British author Hector McPherson Jr. in his book *Astronomers of Today* included Lowell among such eminent astronomers as Lockyer, Huggins, and John Ellard Gore in England; Young, Campbell, Scheiner, and S. W. Burnham in America; Germany's Herman C. Vogel; and Russia's Aristarch Belopolsky.[12]

Those who followed the martian controversy in the popular press knew Lowell as "the foremost living authority on the most interesting of planets,"[13] "the world's chief authority as to the planet Mars,"[14] "our greatest Martian student,"[15] and, in a particularly apt comparison to the ebullient man who was then President of the United States, as "the Roosevelt of astronomy."[16]

Even among his astronomical colleagues, his credentials if not his conclusions were now generally acknowledged; and, as Mars came up to its 1907 opposition, no less a figure than England's Sir Norman Lockyer could declare flatly that "it is to Mr. Percival Lowell of Arizona, to whom we must look for any addition to our present knowledge on this subject. He is the greatest authority on Mars we have. . . . "[17]

Secondly, the Mars furor raged during what was probably the most canal-conscious era in world history.

Canals and canal-building, in fact, were very much in the public eye around the turn of the century, not only as a martian but as a mundane activity. Canals then were still quite commonplace as

regional arteries of transportation and trade and were familiar land-
marks in many parts of the eastern United States and Europe. The
pre-railroad barge canal boom in America in the early 1800s, which
had opened up the vast resources of the continental interior to world
markets, had left its impression not only on the countryside but on
the national consciousness as well in the form of a colorful roman-
ticized body of folklore celebrated in both song and story. By the end
of the nineteenth century, a second great canal boom was
underway—this time to build huge waterways for ocean-going
ships—and this would continue until the advent of World War I.
Indeed, for those of Lowell's generation, the Suez Canal, completed
in 1869, was still a measure of mankind's economic and engineering ingenuity and was hailed as one of the wonders of the modern
world; and the turbulent history of the Panama Canal, which almost
exactly parallels the Mars furor in time, was still being unfolded
event by controversial event in the pages of the same newspapers
that were reporting the running debate over the martian canals.[18]
Even the work-a-day idea of canals for irrigation had acquired fresh
interest for many with the establishment in 1902 of the U.S. Bureau
of Reclamation and the subsequent implementation of a series of
projects to irrigate vast tracts of the arid lands of the American west
and, in the much used phrase, "make the desert bloom."

If any special imagination was required at this time to conjure
up martian no less than terrestrial canals, it was provided by the
press which more often than not explained the one in terms of the
other. The St. Louis *Post-Dispatch*, in a fancifully illustrated Sunday
supplement article published in mid-1907, commented typically:

When Prof. Lowell speaks of "canals," he means huge artificial
waterways built by the Martians and so big as to make the Panama
Canal look "like thirty cents." In this way they have an irrigation
system that would make California green with envy—in fact this
intricate system turns Mars green at certain seasons of the year.[19]

Some newspapers, such as the Providence *Journal*, could relate
martian canals to more localized projects:

Professor Lowell's recent book on "Mars and Its Canals" appears
in a Boston newspaper as "Massachusetts and Its Canal." Coming
at a time when Bostonians are discussing the proposed canal
across Cape Cod, the announcement seems wholly appropriate.[20]

Lowell, of course, never conceived, never claimed that the
canals of Mars were anything like the great ship canals that men had

built or were then building to link the oceans and traverse the peninsulas and isthmuses of the world to expedite commerce. He was well aware that such comparisons were seized upon by skeptics to ridicule his theory and he was careful to deny their validity. "No observer of them has ever considered them canals dug like the Suez Canal or the phoenix-like Panama one," he declared in his *Mars and Its Canals.* "The supposition is exclusively of critic creation."[21]

Lampland's 1905 photography of the canals was, in effect, the opening gun in Lowell's final all-out battle to win acceptance for his Mars theory. This strange engagement, fought on an emotional as well as on an intellectual plane, reached the peak of its intensity in 1907, specifically in the months immediately preceding and following the planet's highly favorable opposition on July 6 of that year. For a while, at least, it appeared almost as if Lowell might carry the day. But as Mars retreated again to the distant reaches of its orbit, it became increasingly clear that the outcome was, at best, inconclusive. The opposing forces had stood their ground well enough, but Mars, whose symbolic image as the god of war literally hovered over this figurative battle, declined to give a sign of his favor to one side or the other, and the issue thus remained in doubt. Generally, astronomers who turned their telescopes on the planet in 1907 found nothing in the way of positive fact that could be used to refute Lowell's provocative ideas, and equally important, nothing that would provide confirmation for any of the various alternative hypotheses that had been advanced against them.[22] But Lowell's own exhaustive observations, as well as those by his expedition to South America, also failed to reveal anything startling or even new about the planet. As a consequence he could only proclaim his now-familiar theory anew and argue that the data on which it was based had been once again confirmed.

This satisfied hardly anyone, for the preliminary barrages laid down by the principal combatants, to say nothing of the spate of speculative articles about Mars that heralded the approach of the 1907 opposition in the popular press, had led many to expect some spectacular revelations. "We look to Prof. Lowell to tell us 'some more' about life [on Mars]," one editor quipped as public excitement mounted, "for having gone so far, it would be cruel to stop at the boiling point of human curiosity."[23] When no revelations were forthcoming, however, a disappointed public soon wearied of the repetitious attacks and counterattacks and progressively lost interest. Lowell fought a few skirmishes in 1909 when the planet again swung close to earth, but these were nothing more than rearguard actions, as it were, and failed to breach the walls of scientific incredulity and public indifference that by then surrounded his martian ideas. Even the press, usually a Lowell ally in the early phases of the Mars furor, began to desert him and by 1910, for example, the editor of *Hampton's Magazine* could hint politely but firmly to the world's acknowledged authority on Mars that he find some other subject to write about. "Personally," this functionary advised Lowell, "I feel so much has been written about Mars within the last few years, that it might be better for you to take a new field."[24]

But this is getting ahead of the story. For in the years just prior to the 1907 opposition, Lowell's cause was far from lost. The 1905 canal photographs, despite the serious reservations that his critics in science held against them, had provided Lowell with a major propaganda victory in the crucial if uncritical public arena where his provocative ideas were now being hotly debated not only by astronomers but by philosophers, moralists, theologians, sociologists, and just about everybody else whose curiosity or imagination had been touched by the mysteries of Mars.[25]

In these years, Lowell launched what amounted to a series of offensives on behalf of his embattled theory, the most effective of which was the publication of his *Mars and Its Canals* in the spring of 1906. Fisher, in his London *Tribune* review, pointed out a major reason for the book's impact:

> The conclusions ... are still, of course, only a working hypothesis; but those who read Mr. Lowell's book without prejudice will admit that up to the present no other explanation has been offered which can for the moment compare with it. ... It is enough to say that no other theory of the seasonal changes in the appearance of Mars has been suggested which has any pretension to explain the observed facts, whereas this theory fits them all and accounts for them in a perfectly satisfactory manner.[26]

Mars and Its Canals, of course, was intended primarily for a popular rather than a strictly scientific readership. But at the time it appeared, Lowell was also busily distributing the third and final volume of his observatory's *Annals* to astronomers, scientific organizations, and journals, for professional perusal. This work too had some impact, for the thick volume was crowded with tables, graphs, maps, and charts detailing the mass of observational data Lowell and his assistants had accumulated over a dozen years and six oppositions of work at Flagstaff. The British journal *Nature* in a lengthy assessment of this exhaustive report, concluded that Lowell's martian observations "are of the first importance."[27]

Lowell's third major thrust in 1906 was his highly successful series of lectures at Boston's Lowell Institute in late October and early November of that year. Although these lectures were heard first hand by only some 2,000 persons, they were reported at length by Boston's newspapers and caused quite a stir, as an anonymous Boston *Herald* writer pointed out somewhat whimsically after the series had ended:

The impression which Prof. Lowell's theories have created are [sic] evidenced by the announcements of ladies lecturing on such subjects as "Is the planet Mars Inhabited?" ... and so on and so on. Once let the woman's clubs get on to Mars, the subject I mean, and the astronomers will find their occupation gone. Prof. Lowell is responsible for considerably more than he imagined.[28]

How widespread was the interest that Lowell's later Mars books created is illustrated by scores of letters preserved in the Lowell Observatory Archives, only a few of which can be quoted here.

A Minneapolis real estate man, for instance, wrote to express "my appreciation of the magnificent work you are doing in your study of Mars and its canals. I have just finished reading your book of that title, and wish to say that for clearness and eloquence of expression and convincingness of argument it is one of the best I have read in recent years."[29] And this letter from a Chicago minister cannot have failed to please Lowell:

It is among the greatest of the books I have read in the past five years. Your style is clear, logical, concise, convincing. You have *seen*—that is certain. ... Pardon me, as a minister, if I add: *And some day theology will be shaken out of its seed-pod present of earth-centrocism and egotism by the study of soul-life on this brother planet.*[30]

The principal of a high school in Torrington, Connecticut, having read *Mars and Its Canals* "with great interest," asked for lantern slides of its illustrations and explained that "in my teaching of astronomy, I use a lantern, and slides of your work would not only be helpful, but very interesting."[31] One William M. Orr of Lincoln, Nebraska, thanked Lowell "for this most excellent work [*Mars and Its Canals*], declaring "it gives me an insight into the secrets of that sister world! as has not been had since time began."[32]

The publication of *Mars as the Abode of Life* brought similar expressions of admiration and appreciation for his work. A woman in Milwaukee, Wisconsin, for example, wrote that "I am only a layman but read and study astronomical literature almost exclusively. We have a copy of your book, *Mars as the Abode of Life*, and I have read it with great interest. I want to say that it has been the dream of my life for years to look through the lens of your big telescope where the air is so rare and unobstructed by atmospheric obscurations."[33] From a young man in Anthony, Kansas, came this missive:

I have had my mother read to me your book entitled *Mars as the Abode of Life*, and found it extremely interesting for one who enjoys science as I do. I am now having read your book *Evolution of Worlds*. ... I intend to take a course in science as soon as I finish the State School for the Blind in Kans. where there we read the N.Y. point system, and there isent [sic] many scientific books in this system.[34]

Such letters were quite typical. There are many others in a similar vein in the Lowell archives which, taken together, show clearly the impact of the Mars furor and of Lowell's theory on the minds of those of his generation, both in its intellectual scope and geographical extent. The volume of this correspondence increases beginning in 1905, reaches a peak at the 1907 martian opposition, and then continues at a declining rate over the ensuing years until Lowell's death, with lesser peaks at the 1909 opposition, the 1910 apparition of Halley's Comet, and again in 1915–16 when the Mars controversy experienced a brief revival in the popular press. Not all these letters, of course, referred specifically to one or another of Lowell's books. More often than not, they contained general praise of his work, although on occasion this was tempered with criticisms. Some offered explanations, more or less plausible, of the various points in Lowell's theory that he himself had left unexplained. Others sought a sympathetic hearing for their own theories, some of which reflect an abysmal ignorance of fundamental astronomy. Most frequently, letter writers asked simple questions and small favors, or sought Lowell's advice on astronomical matters. A few brief examples must suffice here as illustrations.

A grocer in Marietta, Georgia, for instance, inquired where he could obtain a copy of Schiaparelli's memoirs,[35] and a man in Decatur, Illinois, asked what would be the most suitable kind of telescope for amateur observation, adding:

I am proud that we have a man with such remarkable understanding of the universe and his grand observatory located on one of the most interesting spots of the whole globe.[36]

A mining engineer from Jerome, Arizona, a few miles south of his Flagstaff observatory, wondered "if an outsider could take a squint through the big telescope without interfering with your work,"[37] a not infrequent query. H. G. Tompkins, a Fellow of the Royal Astronomical Society, suggested that the "white spots" seen near the Mars equator might be tracts of alkali on the surface, a suggestion that Lowell agreed was at least a possibility.[38] One S. Millett Thompson of Providence, Rhode Island, who identified himself as "of Lowell descent," thought the canals might be the martians' way of leaching salts from their soil.[39] From Wilmington, Delaware, a man wrote that he had read "with much interest everything you have written on Mars," and asked specific details of the martian irrigation system:

When you get so far along as to be able to give me particular information as to the pumping system employed there, I shall receive the information with great interest as I am especially interested in mechanical developments.[40]

A long letter from a Watsonville, California, housewife lauded Lowell's martian work and then touched on a subject very much on Lowell's mind, although only a few associates were aware of his interest in it at the time:

For surely one who has done so much for the world is a friend to all mankind, and Prof. Lowell is a name loved from one end of our land to the other. I am not even a student of astronomy but love the science in the poor way a very very busy housewife may. For years I have been interested in Mars and of course delighted in every article I found on the subject. . . .
There are two planets beyond Neptune. The outer one of the two, is not likely to be discovered by astronomers, moving as it does in an orbit removed by 10,000 million miles from the sun; the other planet will likely be discovered by science through noting the disturbances in the orbit of Neptune.[41]

Certainly Lowell read these last lines with unusual interest, for even then he was feverishly seeking accurate data on the orbital motion of Neptune to help him in his search for what to him was "Planet X."[42]

Lowell's stature as an astronomical authority, at least in the public mind, also comes through in some of these letters. "Prof. Newcomb was always kind to answer my questions regarding astronomy," a Winter Haven, Florida, minister wrote in 1909, for example. "I turn to you now that he is dead."[43]

Inevitably, Lowell received queries about various other astronomical phenomena, real or imagined. A familiar one to astronomers, perhaps, came from a woman in Leesburg, Florida, asking: "May I know please where, and at what time, to look for the 'Star of Bethlehem,' and what is the astronomical name and magnitude of the star?"[44] A Thomas Nance in Lexington, Oklahoma, wrote that he was "the author of an exact measurement of the distance to the sun," and wanted to know whether "the government has a correct measurement for the same" and "if not, would they pay me for my rule should I have it copyrighted?"[45] A man in Orange County, California, demanded to know why he should see human faces and figures in a photograph of the Orion Nebula that appeared in the magazine, *Technical World;* apparently after receiving an unsatisfactory answer from Lowell's secretary, Miss Leonard, he retorted:

If it is from an unretouched negative whence do all the faces and figures come from? I have found over sixty and my friends see them also which proves that it is not an hallucination. Now just study it [the photograph] for fifteen minutes yourself before deciding that I am "nutty" and possibly you will find something more preposterous than the canals of Mars.[46]

Even after the Mars furor as such had spent itself, and through the final years of Lowell's life, the lingering impact of his Mars theory, and the worldwide reputation it had brought him, are reflected in the continuing flow of such correspondence. From Wisconsin, a businessman asked for photographs and drawings of Mars for a "club paper" he was preparing,[47] and a Texan, confiding that he did not merely read Lowell's books in the library but bought them, declared that "no other books have given me quite so much pleasure. I am listening . . . to know that from the Lowell Observatory in Flagstaff, the very Martians themselves have been revealed bodily from their hidings."[48]

From Maine, a bulky letter contained an article on "the Tidal Power of Earth and Mars as a Theory of how the water flows in the Martian canals," and asked Lowell's comments.[49] Another Texan belatedly reported his "important Discovery" in 1909 that Mars is inhabited, explaining that with "mirrors, reflectors, dials, and powerful magnifying glass" he had by accident caught "signals" from the planet. "If you desire any information," he added magnanimously, "I would be glad to give it to you."[50]

Early in 1916, a Vancouver, British Columbia, writer offered Lowell a new theory of gravitation, based on the supposed ubiquitousness of ether in space, which held that "the sun, the planets, etc., have no attraction for each other whatsoever, but . . . are *pushed* together."[51] And an Australian "irrigationist," apologizing for being presumptuous, suggested that "the Mystery of the Canals of Mars" could be explained by "huge rivers of water turned over the level country with check banks or even banks erected parallel miles apart to convey it across the surface. . . ."[52]

And—one final example—an old sailor scrawled a barely legible report of a total eclipse he had observed sixty-six years earlier at "Mauie, one of the Sandwich Islands," and wondered whether astronomers had noted the event. "Excuse my writing," his postscript read, "as [I] am nearly 87 and cannot see very well."[53]

If Lowell's popular prestige as an astronomer was largely a result of the Mars furor, the furor itself, in turn, was largely of Lowell's own making. This is to say that his long vigorous championship of his theory of life on Mars, together with the often vehement reactions this theory provoked, was primarily responsible for converting what had been a formal controversy among astronomers into a brief but worldwide cause célèbre; that it was not Mars itself, but Lowell's ideas about Mars, that caused all the fuss. The cumulative effect of his persistent promotion of his Mars theory focused public attention on the planet and its problems to a degree unprecedented before or since. This burgeoning of lay interest as Mars approached and passed through its highly favorable 1907 opposition is also evident in the increase in the number of informational essays and of speculative or argumentative articles on Mars appearing during this period in newspapers and magazines not alone in Boston, London, or New York, but from Budapest, Hungary, to Sydney, Australia.[54] The extensive publicity that attended the launching and progress of the Lowell expedition to South America further swelled the flood of "Marsiana," so to speak, as did the activities of such eminent and articulate advocates of Lowell's martian theory as Edward S. Morse, George R. Agassiz, Lester Frank Ward and, in Europe, the irrepressible Flammarion. In all this, almost nothing was printed about Lowell without reference to Mars, or indeed about Mars without reference to Lowell.

Lowell himself contributed to this journalistic outpouring. Following his successful Lowell Institute series, he gave other lectures on Mars—at Boston's University Club in December, 1906, for instance, and at Brown University in January, 1907—and these drew some notice in the press. Later, his terse reports of observations at Flagstaff and in Chile usually were distributed by *The Associated Press*, and in Europe by other general news agencies, and appeared quite widely in print.

Perhaps Lowell's main propagandistic effort in the pre-opposition months of 1907 was a signed article in the *Outlook* magazine in April. Here, again, he summarized the data he had amassed and the conclusions he had reached about Mars, and berated his critics anew for their continued inability to see the planet as he saw it, and thus he was sure as it really was.[55]

This article, certainly, contained nothing that was new, or that Lowell had not publicly proclaimed many times in the past. Typically, he did not take particular notice of the rising tides of emotionalism, of superstitious fears, or utopian hopes as the case might be, that were now swirling around the issue of martian life and its far-reaching implications. *Outlook's* editors, however, in their introduction, pointed out one such aspect of the controversy, noting the deep-seated religious opposition that Lowell's unorthodox theory had stirred and dismissing it as a mere anachronism from a scientifically unsophisticated age:

We suspect that the real reason for discrediting the conclusions of such observers as Mr. Lowell is this native egoism which persists in regarding man as the "lord of creation." The Copernican theory of astronomy put an end to the geocentric conception of the universe. If the not improbable hypothesis that Mars is inhabited by intelligent beings should be established, it would put an end to some of the semi-theological theories which we have inherited from a time when this world was thought to be the center of creation, the sun and the stars a mechanism for supplying it with light, and man the supreme and only son of God, for whom the universe was all made.[56]

Anachronistic or not, the argument thus summarized represented the traditional position of revealed religion on the question of extraterrestrial life and inevitably was raised against Lowell's insistence on the existence of intelligent martians. The idea of an anthropocentric cosmos, of course, was not directly involved in the Copernican revolution, for even if the earth was no longer the physical center of things, it could still be argued that it was uniquely favored by God as the abode of His supreme creation, man, and thus the center of His attention. The idea has proven exceptionally durable over the centuries since Copernicus, despite steady advances of science and of scientific modes of thought. It survived the strongly mechanistic and theistic attitudes that prevailed during the post-Newtonian Englightenment of the late seventeenth and eighteenth centuries, and it thrived on the so-called "romantic" reaction to the materialistic bent of science that set in with the advent of the nineteenth century. It gained new strength from the work of such brilliant pre-Darwinian reconcilers of science and theology as William Paley, who argued in his *Natural Theology* that the design so apparent everywhere in nature must logically imply a Designer, and William Whewell, who contended that the discoveries of microscopy, geology, and astronomy precluded a plurality of habitable worlds.

Furthermore, it had avoided direct confrontation with the revolutionary doctrine of evolution, although the Darwinian dethronement of purpose and plan in the development of terrestrial life-forms, which proved so upsetting to late nineteenth-century theologians, had its implications for extraterrestrial life as well. It is strange, surely to find that the twentieth-century champion of man's divine cosmic uniqueness against the challenge of postulated martians was the seminal evolutionary thinker who with Darwin reduced man's origins to a matter of random chance.

Alfred Russel Wallace was eighty years old in 1903 when he published his *Man's Place in the Universe* to argue against the general proposition that life exists elsewhere in the universe. Late in 1907, at the height of the Mars furor, he published *Is Mars Habitable?* to argue against the specific proposition that life exists or could exist on Mars. The earlier work, essentially a modernization of the Whewellian view, was reasoned on broad lines and did not particularly concern itself with the martian controversy as such. The later book, however, was nothing less than a direct assault on Lowell's martian theory and was designed to counter the deepening interest that his activities and ideas had progressively stirred in minds and emotions of growing numbers of people throughout the literate world.

"Popular belief in the habitability of our neighboring planet," the London *Daily Chronicle* pointed out in its lengthy review of *Is Mars Habitable?* "has been vastly stimulated, not only by the romances of the Vernes and the Welles, but by the reasoned conclusions of cold, calculating scientific observers like Mr. Lowell." Of Wallace's book, the reviewer noted:

A little while ago, he set forth with much logical acumen reasons for assuming that the earth is, after all, at the center of the visible universe. Now here he is accumulating evidence to prove that Mars is not only uninhabited, but uninhabitable. . . .[57]

Very little of Wallace's accumulated evidence, as it turned out, was new, for his book was largely a synthesis of his metaphysical views with those of the most vociferous of Lowell's critics over the years. He argued once again, for example, that the "canals" were simply cracks, drawing analogies to geological faults that run for miles in straight lines across the north of England, the Upper Rhine Valley, and Palestine. These cracks, he believed, resulted from the primordial cooling and shrinking of the planet and the subsequent reheating of its crust through meteoric bombardment. He also reiterated the argument that martian temperatures were too low to sustain life, and here he specifically disputed Lowell's figure of 48° Fahrenheit for the martian mean, declaring that Lowell had overestimated the amount of solar radiation penetrating the thin martian atmosphere and had underestimated the amount reflected back into space. And again, he cited the lack of water on Mars, noting that the spectroscope had failed to find any unequivocal evidence of water vapor there. He also argued that the polar caps were composed mainly of carbon dioxide and that at the most they could yield only enough water to flood the allegedly irrigated areas of the planet to a depth of two inches, an estimate he attributed to Miss Agnes M. Clerke, the English historian of astronomy.[58]

At one point Wallace reasoned logically and with telling effect that even if martians existed they could not have the high intelligence with which Lowell credited them. For the "canals" they were supposed to have built in many instances ran for thousands of miles across arid deserts and beneath clear cloudless skies, thus "losing enormously from evaporation, if we assume them to contain water. The mere attempt to use open canals for irrigation purposes would argue ignorance and stupidity. Long before half of them were completed, their failure to be of any use would have led any rational being to cease constructing them."[59]

Lowell, of course, had encountered most of these arguments before and had answered them many times in more or less detail. Now, his first reaction was to wonder how a biologist, no matter how eminent, could speak with any authority on matters of astronomy in general, and Mars in particular. Wallace's book, he confided to Mark Wicks, a young English amateur astronomer who later wrote a book about Mars and dedicated it to Lowell, [60] "is such a mass of misstatements and inadequate knowledge on the whole subject that it hardly seems worth replying to—nice man as Dr. Wallace is." [61]

But the book, in light of Wallace's great prestige in science, did require some answer, and this Lowell provided in the form of a letter to the editor of the British journal *Nature* in March 1908. In this he pointed out that Wallace was obviously not familiar with modern methods for determining temperatures and added that his own application of these methods to Mars had "expressly taken into account the blanketing effect of the air." Of Miss Clerke's estimate of a two-inch deep supply of water in the polar caps, he declared that his own estimate, as a specialist on the subject, was more nearly two and one-half feet and "so it is the argument of Dr. Wallace, and not the cap, which fails to hold water."

Again, his theory ... that the interior of Mars can have completely lost its heat in the very process of contraction, and yet later have suffered a meteoric bombardment sufficient to give it a heated outer layer, is mechanically whimsical, not to say impossible. [62]

"Misstatements," he concluded, "cannot be too carefully avoided in science, especially when a man, however eminent in one branch, is wandering into another not his own." [63]

Lowell did not specifically answer Wallace's evaporation argument, although he had long contended that the canals were mere threads in relation to the width of the vegetation bands along their banks and thus exposed a minimum of water surface to the arid martian atmosphere. Although he stressed this point anew in his subsequent writings about Mars, even suggesting that the canals might be covered to prevent evaporation, [64] Wallace's criticism persisted long after Lowell's death. [65]

Nor did Lowell concern himself with the theological aspects of Wallace's thesis, although these were discussed at some length by the press. For as Edmund Noble pointed out to the readers of the Boston *Sunday Herald* in mid-1908, Wallace's "account of man ... as the unique and supreme product of this vast universe is irreconcilable with Prof. Lowell's claim for the presence of life on the planet Mars." Wallace, Noble noted, contended that in the solar system, "with the exception of earth, there is not a planet in it, from Mercury to Neptune, which is able to support vitality or form the abode of any grade of intelligent life." For Wallace, the earth "has been especially favored by cosmic arrangements," he wrote, adding:

The whole of the available past life of the sun, Dr. Wallace claims, has been utilized for life development on the earth ... and the future will not be much more than may be needed for the completion of the grand drama of human history, and the development of the full possibilities of the mental and moral nature of man.[66]

Of Wallace's stand on the broader question of extraterrestrial life beyond the solar system, Noble commented:

We also get the anthropocentric theory elaborately modernized. ... The earth is central and not by accident, but because an original plan required its centrality, and man is the central and only intelligence in the cosmos, not by mere incident, but by virtue of the purpose which contemplated his appearance from the very beginning.[67]

While the *Herald's* full-page illustrated piece primarily dealt with Wallace's and Lowell's contrasting views, it also included this typically grandiloquent pronouncement by Flammarion on the ubiquitousness of life:

"Life," says Flammarion, "... is universal and eternal, for time is one of its factors. Yesterday the moon, today the earth, tomorrow Jupiter. In space there are both cradles and tombs."[68]

If the Mars that Lowell described so dramatically was alarming to Wallace and some others in post-Victorian England, it awakened vastly different emotions for large numbers of people elsewhere, particularly in America where the prevailing early twentieth-century mood was progressive and optimistic, with strong trends toward political and social reform. For indeed, a planet laced from pole to pole with an intricate geometric system of irrigation canals and irrigated oases implied a degree of political, social, and economic organization on the part of the martians that for many was more to be envied than feared.

"Mars may be the planet of the millenium," feature writer Lillian Whiting speculated for the St. Louis *Post-Dispatch* as the 1907 opposition neared.[69] And Flammarion exuberantly proclaimed Mars "the Monte Carlo of planets," adding that "the Martians must be less worried, less agitated, less nervous, more thoughtful, wiser and more prudent than we."[70]

Such a Mars seemed to have something to teach the people of a troubled earth, as Lester Frank Ward pointed out early in 1907 in a paper, entitled "Mars and Its Lesson," published in the *Brown* [Uni-

versity] *Alumni Monthly.* For if Mars was the "elderly planet" that
Schiaparelli and Lowell said it was, then life there should be more
advanced in its evolution, not only biologically but sociologically,
than on the earth. Ward argued the long tenure of terrestrial life and
pointed out that the human race had reached what he called "the
psychozoic age," or the "period of the living soul," only within the
past few thousands of years. "It is likely," the Philadelphia *Record*
summarized Ward's thesis, "that the people on Mars have trans-
formed and retransformed themselves, continuously improving and
developing higher and higher social efficiency."

> In the same way it is certain that the inhabitants of this earth will go
> from lower to higher stages of civilization in accordance with the
> principles of sociology now becoming understood and some of
> which Dr. Ward has been a pioneer in stating.[71]

Ward's paper proved to be exceptionally popular and, as Ward
himself wrote to Lowell in a long, rambling letter, "it has been copied
almost literally by scores, perhaps hundreds, of papers all over the
country. ... I have received many letters ... commenting on the
moral effect of the article. I only wish its readers would all read your
books instead."[72]

In this same letter Ward recalled his 1901 visit to the Flagstaff
observatory when he viewed Mars briefly through the 24-inch telescope, and quoted from his own notes made at the time to describe
the occasion to Lowell:

> "I was surprised at what I could see. The polar cap of snow was
> perfectly clear, and I could see great canals or long cavities in
> various directions. They made me draw what I saw and sign my
> name to it."[73]

The sociologist also confided to Lowell that he had been
speculating about martians for many years:

> The subject has always interested me greatly, and I remember dis-
> cussing with Professor Morse at least twenty years ago the probable
> nature of the life on Mars, should there be such. I also find myself
> "harping" on it in papers of mine as early as 1895. ... the aspects
> which most interest me and upon which I shall dwell at length are 1,
> *The Flora of Mars* (as I am a botanist), and 2, *The Sociology of Mars.*
> This last may make you smile, but you yourself have already referred
> to it.[74]

Lowell had indeed alluded briefly to what Ward and some others
considered the sociological and perhaps even socialistic implica-

tions of his theory, particularly in his *Mars and Its Canals*. Specifically, he pointed out that the ages-old human institution of war, symbolized for so many on earth by Mars, must have been outlawed and abolished long ago by martians in favor of concerted action to meet the challenge of the progressive desiccation of their planet. The sheer magnitude of the martian canal works alone logically demanded that the martians be not only intelligent but peaceful as well. "The first thing that is forced on us," he wrote in *Mars and Its Canals*, "is the necessarily intelligent and non-bellicose character of the community which could act as a unit throughout its globe."

War is a survival among us from savage times and affects now chiefly the boyish and unthinking element of the nation. The wisest realize that there are better ways for practicing heroism and other and more certain ends of insuring the survival of the fittest. It is something people outgrow. But whether they consciously practice peace or not, nature in its evolution eventually practices it for them, and after enough of the inhabitants of a globe have killed each other off, the remainder must find it advantageous to work together for the common good. Whether increasing common sense or increasing necessity was the spur that drove the martians to this eminently sagacious state we cannot say, but it is certain that it reached it they have, and equally certain that if they had not they must all die. . . . Difference of policy on the question of the all-important water supply means nothing short of death. Isolated communities cannot there be sufficient unto themselves; they must combine to solidarity or perish.[75]

Lowell's attitude toward war, as reflected in this passage, was quite unusual in an era more notable for its saber-rattling imperialism than its pacifism. but in this he was as consistent as in everything else. Hear him expound on the subject, for instance, in a 1903 "Commemoration Day" address to the citizens of Flagstaff:

War in itself is an unmitigated evil and it possesses glamour in exact proportion to the lack of development, the boyishness of the beglamoured. It is only commendable for the good things it brings about. Like physic, it is sometimes necessary and only incidentally enjoyable. To run risks is part of the joy of life, but to do so without sufficient cause marks the fool.[76]

Lowell's use of the word "boyish" in this context, surely, was aimed at the popular and largely self-created image of an exuberantly jingoistic and bellicose Theodore Roosevelt charging gloriously up Cuba's San Juan Hill with his Rough Riders during the Spanish-American War of 1898. Certainly "Teddy's" bluster and boisterousness offended Lowell's fine sense of propriety as much as his political progressivism offended Lowell's solid conservatism, their mutual Republicanism notwithstanding.

Lowell's purely speculative suggestion in *Mars and Its Canals* that martians must necessarily be peaceful folk stirred considerable interest. "Is it not a curious thing," Garrett Fisher wondered in his London *Tribune* review, "that the Red Planet, which has always been held sacred to the God of War, should thus prove to afford a most striking object-lesson to peace?"[77] The *Saturday Evening Mail's* reviewer proclaimed that "war has a new enemy and peace a new advocate in the person of Percival Lowell. And Prof. Lowell's argument against the horrors of armed conflict is as novel as Gen. Sherman's was terse and expressive."

Now what's all this got to do with war, you ask? Why just this, says Prof. Lowell: The Martians could never have built that immense system of canals if they had had any time to spend fighting. Or, having built it, they could never have kept it in working order if it were subject to the periodic danger of destruction by war.[78]

Even those whose fears of martian aggression might have been aroused by such fictional fantasies as *The War of the Worlds* could find comfort in the Lowellian view of Mars. "The evidence of peaceful conditions on our neighbor," the *Japan Daily Mail* informed its readers, "should diminish the fear of disaster that H. J. [sic] Wells' novel tended to create."[79]

But far and away the most singular comment on the sociological aspects of a populated Mars came in the form of a long poem, entitled "The Gospel From Mars," written by Edward Henry Clement, the erudite editor of the Boston *Transcript*. In four hundred sonorous lines of pure iambic pentameter, Clement summarized Lowell's theory and sermonized eloquently on its implications. "My main object," he explained mischievously to Lowell in April of 1907, "is to show that with such a system as has developed on Mars, the people of all nations must cease regarding boundary lines and have become one brotherhood, in fact and deed—so likewise the social classes. In short I am going to show why Mars is carrying through our Heavens the heart-red flag of socialism!"[80]

Clement had written the poem specifically for the annual meeting of the Delta Chapter of Phi Beta Kappa at Tufts College in Massachusetts that June. But a few days after that event, the New York *Evening Post* published it in its entirety—a tribute, perhaps, to the poetic talent of a fellow journalist but more likely an indication of the deep public interest in his theme. A few final lines of the poem must serve to illustrate its epic quality:

. . . . Suez! . . . Reds of France
Today halt chancellors who counsel war.
"All peoples," is their cry, "are of one blood
And one flag is enough for all the world;"
Again at Panama, a zone of Peace.
The conquering heroes of the world today
No more are butchers, but the engineers;
Construction, not destruction, is the word.
The era of the Canals begins on Earth.

The thousand years of peace on ancient Mars,
To millions lengthen while they hold at bay
The common enemy, the searing drouth,
The creeping desert; adding trench to trench,
Then great canals, until a hundred, broad
as lakes, and like meridians on our maps,
Score our neighbor straight from pole to pole,
Plain to our eyes at two-score million miles,
And thousands more belike we cannot see.
Thus only Mars survives: The dwindling store
Of precious moisture, at the poles locked up
In snow, is led alternately, south, north,
As roll the seasons, and a solemn march
Of vegetal awakening, cadence grand,
In measured tread advancing, marks in blue
The resurrection of the fields and fruits,
Of Corn and Vine; Processional of Life!
For death is conquered thereby Life and Love.
So come new hopes and faith fresh from the stars,
The Gospel sent us from the living Mars. [81]

Solving the mysteries of Mars, it was now evident to almost everyone as the planet came to its opposition in 1907, was no longer an exclusive concern of astronomy, as the Chicago *Record*, for example, pointed out editorially in July:

If the present favorable "opposition" should fail to advance our knowledge of Mars, the disappointment would be felt not only by astronomers and readers of books on popular astronomy, but by bold speculators who expect Mars to throw plenty of light on the "psychic" age of the earth, with its problems of race, national organization, international relations, war and peace. For Mars is supposed to have abolished war and established one grand brotherly republic, and what she can do, we can do. [82]

In Kansas City, the Reverend Charles Ferguson took the gospel of Mars according to Lowell, as interpreted by Ward, Clement, and others, as his text for a Sunday sermon in the All Souls' Church. "The investigations of Percival Lowell and his assistants," the preacher began, "afford a sufficiently satisfactory demonstration of the proposition that there are surveyors, mathematicians, and engineers at work on the planet Mars."

What on earth can rival the importance of this news? This is not a guess. It is not a Jules Verne dream. . . . Consider the implications of it, and how they bear upon our world problems. We see off there in space . . . a neighbor world, which shows at a glance by impressive signs that it is morally unified and civilized beyond the dreams of our reformers! . . . How would the jangling, class-riven nations of the earth fare under such natural conditions?
    I am inclined to believe that we should manage to make a living after a clean sweep of our inherited political, legal and economical superstitions. We should have to efface our international frontiers, our patriotic wars, our perpetual wasting class struggles, the incalculable frictional loss of our lawless commercialization—these are among the luxuries that an arid planet could by no means afford. [83]

Clearly, the problems posed by Mars had now acquired both a social and moral significance. Yet, as today there are those who question the value of spending huge sums to send men to the moon or for the spacecraft exploration of the planets, even then there were some who wondered if the fuss over Mars was worthwhile. Enough, at least, to bring the press to the defense of the intensive efforts expended on the planet in 1907. "Yes," the Chicago *Examiner* flatly declared, "it is worth while."

Every earned, energetic effort for the extension of human knowledge is worth while, even though it may have no value in dollars and cents. . . . The man who reaches the north pole or who proves the existence of the canals of Mars will have the supreme glory of doing something nobody else has done. That is worth while because such an achievement is the secret or expressed desire of all mankind. To be first, to be the only one who has done some notable thing, is the chief end of man when he evolves from barbarism. [84]

The New York *Christian Advocate*, too, conceded that studies of Mars would not reap any material rewards, but argued:

There is much advantage to these investigations. This is much better than spiritualism for the occupation of scientists and pseudo-scientists. There is one disadvantage: There is no probability of emigration to the planet; no mining schemes can be gotten up; no communities established; no wars of conquest to be fought; no martial strains fall on the human ear. But to the naked eye the star will lose none of its beauty. [85]

Lowell himself, incidentally, felt that some benefits to be derived from the study of Mars and other solar system objects would accrue to other branches of science, particularly geology, and this view, which certainly has been confirmed by the Apollo and Mariner missions to the moon and Mars, was shared by some geologists of his day. Professor William N. Niles was one, and Lowell warmly encouraged his interest in Mars. "That a geologist of your standing should realize the importance of the astronomical side of the subject is both gratifying and most pregnant with promise for the two sciences. The far view of the one is indeed part and parcel of the near view of the other and the greatest advance for both must be made by mutual help in the field common to each," he wrote.[86] Paleontologist G. R. Wieland of Yale University also assured Lowell of his "keenest interest" in the red planet. "Since reading your book on the canals," he declared, "I have scarcely been able to think of anything else than Mars and comparative planetary geology."[87]

As Mars approached and passed its 1907 opposition, however, Lowell focused his attention on the planet itself. His telegraphic reports of his own observations and those of his South American expedition were dramatically brief. On July 2, for instance, he announced in a single paragraph that the younger Slipher had succeeded in photographing the martian canals from Chile,[88] and on July 22, he wired tersely that both he and Lampland had photographed the double canal Gihon from Flagstaff.[89]

Early in August, as Mars began to fade in the night sky, he prepared a preliminary report of these observations for publication later that month in *Nature*. In this, while noting that he had "already counted fifty-six canals on my plates, including the Gihon double," he dwelt primarily on his visual observations of seasonal changes, claiming that these again followed the pattern predicted by his theory. The planet's aspect, he pointed out at the outset, was on "an even keel," with both northern and southern polar caps visible and the southern cap "dwindling" in the heat of a martian summer. Of the progressive changes observed in the appearance of the canal network, he wrote:

The process of evolution was in keeping with the method of development found here for the northern canals in 1903. The detection thus strikingly corroborated in the planet's other hemisphere, the southern one, the action then observed in the northern. The fact is of the nature of a prophesy fulfilled and not only supports the previous observations, but proves the theory deduced from them to have been correct. It is a direct *sequitur* from this that the planet is at present the abode of intelligent, constructive life.[90]

And to this, he added:

I may say in this connection that the theory of such life upon Mars was in no way an *a priori* hypothesis on my part but the deduced outcome of observation and that my observations since have fully confirmed it. No other hypothesis is consonant with all the facts observed here.[91]

The ghost of the accusation of preconception raised more than a dozen years before by Lick astronomers Edwin S. Holden and W. W. Campbell had not yet been exorcised, it would seem.

Lowell's report in *Nature* was widely quoted and commented upon in the popular press, and while his observations were generally praised, there was some tergiversation about the deductions he drew from them. "The essential fact of Professor Lowell's communication is the objectivity of the canals as shown by his photographs," the London *Daily Telegraph* opined. "Whether they are evidence of constructive intelligence is another question. Astronomers will hesitate to accept Professor Lowell's conclusions while admitting their deep interest." But, the paper also noted:

Whatever view may ultimately prevail, it is certain that nothing of such profound human concern has ever before been presented on equal authority with so much definiteness and assurance with regard to life on "other worlds than ours," as these results of the distinguished American astronomer's latest researches.[92]

The New York Times sought out the reaction of prominent astronomers both in England and the United States to Lowell's *Nature* article. "I can't see that Prof. Lowell has proven his case," the *Times* quoted Arthur Stanley Eddington, then England's chief assistant astronomer royal. "He has taken some wonderful photographs and they show the evolution he predicted in his work on Mars and its canals."

But we are more inclined here to accept the theory of Prof. [W. H.] Pickering of Harvard University, that these cracks are not artificial at all, but are the result of the planet's natural shrinkage. This view is generally held in England, and we are inclined to believe that Mars is played out, that its career is finished. . . . I can't quite follow Lowell when he says that because the canals evolved along predicted lines, it is a direct *sequitur* that the planet is at present the abode of constructive life.[93]

"As a matter of fact," Eddington added, "we have no right to discuss the question here, as practically all our information regarding Mars must come to us from America, as we are not situated where good observations are possible. For us, Mars barely rises above the tree tops, so we are wholly dependent on the observations of our friends in America."[94]

The *Times* also queried Princeton University's C. A. Lovett who "did not want to talk about" Mars at all. "Every two years, Mars is where it can be observed," the astronomer said bluntly, "and the public mind will be aroused."[95]

A few days later, the *Times* quoted Lowell in reply to Eddington's remarks: "It were wise to realize that the subject is not one of arm chair philosophy but of telescopic observation and deduction and that the interviewed opinions of investigators not engaged in the research are not of scientific value." The paper then went on to editorialize:

Meanwhile the layman may accept or reject what the different scientists say. But surely it is a fascinating theme. If Mars and all other members of the innumerable starry hosts are actually inhabited, the census of the heavens make the billion and a half or so of our earthly population seem a discouragingly small proportion of the prodigious total.[96]

The press, at least, had indeed found Mars to be a "fascinating theme" during that summer of opposition and had managed to keep the red planet and its enigmas before the public despite the fact that there was nothing really new to write about. Science writer Hector McPherson Jr., summarizing the old familiar arguments in a widely quoted article published in the London *Chronicle* on the eve of the opposition, could only add Italian astronomer C. Vincenzo Cerulli's espousal of the optical illusion theory of the canals and Lowell's wry remark that "Schiaparelli had the misfortune to be ahead of his time, and the yet greater misfortune to remain so."[97] In August, Vincent Heward simply rehashed the pros and cons of the controversy in an article for the *Fortnightly Review* in which he conceded some value to Lowell's observations but opted for Pickering's "cracks" explanation of the canals and concluded that as Lowell's theory "rests too implicitly on the assumption of terrestrial conditions," Mars was not habitable.[98]

In September, with Lowell's article in *Nature* and the appearance of the first direct reports emanating from his South American expedition, the pace of martian publicity picked up. From Chile's Tarapaca Desert, Mrs. Todd, wife of the expedition's director, wrote a long article about the summer's activities there for the *Nation*,[99] and the Los Angeles *Times* printed a full-page spread on the work of the expedition by a "special contributor."[100] In London, the *Sphere* published three drawings of Mars somewhat ambiguously captioned: "Canals as seen in direct photographs,"[101] and in Indiana, the Indianapolis *Star* capitalized on some local angles in the martian controversy with a full-page feature on Lowell's assistants, headlined "Solving the Riddle of Mars—Young Hoosier Scientists Doing Much to Add to World's Knowledge of Ruddy Planet."[102]

Some of this publicity was quite favorable to Lowell and his theory, but not all of it. The loyal Miss Leonard felt constrained to take strong issue with the New York *Tribune* for the skeptical tone of one of its articles. "Whether his chain of evidence that some form of life exists on Mars be such as 'would convince a court of law' is a matter of guess; that it convinces Professor Lowell and is influencing Schiaparelli is scientifically more to the point," she scolded the *Tribune's* editor.[103]

No one on the Lowell side of the Mars furor, however, complained about a *Scientific American* editorial later that month:

With the aid of photography, he has established beyond doubt the existence of a delicate tracery of lines on the sphere. ... There has been considerable criticism by prominent astronomers of the work done by Prof. Lowell as given out in his preliminary report. Prof. Lowell states that he is a specialist in the study of Mars, and he is better fitted than others of his own profession to judge the conditions on that planet. The idea of specialization in astronomy may appear to be somewhat new, although it is not at all unreasonable. No other branch of science presents so large a field of investigation, particularly in these days of the spectroscope.[104]

Certainly a *New York Times* editorial a few days later was a Lowell triumph. "While Lowell concentrates all his fine intellect and splendid energy on his investigations of the planet Mars," the *Times* declared, "a large number of other astronomers and scientific writers seem to be devoting too much of their time to disproving his theories. This ... is a great waste of intellect."

Prof. Lowell has the best of all the arguments about Mars, thus far, and he is in a fair way to learn a great deal more about the planet. He is no "sensationalist" or "yellow scientist" though he is certainly an enthusiast and probably proud of his enthusiasm.[105]

Scrawled on the margin of the clipping of this editorial in the Lowell archives, incidentally, are the penciled words: "Good for you!"[106]

The New York Times, indeed, although it dutifully published criticisms of Lowell and his ideas, became for a while at least one of the principal champions of life on Mars. In November, for example, the *Times* reported that Dr. Harold Jacoby, Rutherfurd professor of astronomy at Columbia University, had characterized Lowell's theory as "an example of inartistic research," and his sightings of martian canals as "seeing the unseen."[107] But in the same edition, the paper defended Lowell in an editorial, noting pointedly that he could "take comfort from the fact that Prof. Jacoby, in the same lecture, attacked Newton."

Lowell may find this consolation, if he bothers at all about adverse opinions of his work, which is doubtful. He is too deeply interested in his examination of the canals of Mars and his theories to bother about what folks are saying. .... Jacoby, and others like him, do not believe there is life on Mars, but they cannot prove what they say. Their arguments are purely negative. ... Prof. Lowell has the best of them. ... We not only sympathize with Prof. Lowell, but we envy him.[108]

The *Times* here was quite wrong on one point, for Lowell did indeed bother about adverse opinions and about what folks were saying about his martian theory. And occasionally he publicly answered criticisms, although in a general way and usually through the media of popular scientific journals rather than through the newspapers. The skepticism in the press in September, for example, he answered in October in an article in *Scientific American,* written under the pseudonym "A Mathematician." In this, he reiterated his definition of proof as a "preponderance of probability," and declared that "the mass of evidence in favor of the habitation of Mars is so strong that emotional prejudice can only attack it by denying the facts." The odds, he pointed out, were "millions to one" that such straight lines as observed on Mars could be the result of natural forces, and "one to an infinitely great number" that more than two such lines would cross or meet at the same point.[109]

Far more often, however, Lowell sought to answer criticisms of his theory quietly and out of the public eye through direct correspondence with the particular critic. In May 1907 when Newcomb again raised the optical illusion argument against the martian canals in a paper for the American Philosophical Society, Lowell obtained a copy of the paper and then wrote Newcomb a long letter replying point-by-point to his contention "that the seemingly straight lines on Mars are a psychic synthesis of other markings."[110] And when the short-lived journal the *American Astronomer* published an article by France's Professor L'Abbe Th. Moreaux complaining that the martian disk seemed "constantly befogged," Lowell wrote Moreaux to explain that "the apparent obliteration is not due to Martian but Earthly conditions, being a question of our own air-waves at the moment which are sometimes such as to leave the main contours perfectly distinct while confusing to obliteration faint details. . . . No such fog as you mention was seen here on those days."[111]

Perhaps the one criticism that bothered Lowell more than any other, however, was the inference that his martian theory was not original or was, at best, only a compendium of ideas previously advanced by others. He repeatedly denied such inferences publicly, of course, contending as he did in his *Nature* article that his conclusions were based on observations and not *a priori* reasoning. But he denied them far more emphatically and in greater specificity in personal letters to those who, in innocence or with malice aforethought, intimated that such might be the actual case. When Dr. H. H. Turner suggested in his London *Times* review of *Mars and Its Canals* that the original theory of martian irrigation had been W. H. Pickering's, Lowell reacted with a lengthy letter designed to set the Oxford astronomer straight on this particular point, as well as on Pickering's martian ideas in general:

I must correct one error, to wit: That my theory is not mine. This is a misapprehension due probably to the fact that in 1888 W. H. Pickering threw out the suggestion that the lines on Mars—a suggestion he has since abandoned for one supposing them to be natural cracks in the surface—might be irrigating canals. Now a mere suggestion like that, unsupported by any reasons, still less any proofs, can hardly be called a theory and doubtless has occurred to many people. I am certainly not indebted to it as is indicated by the fact that I called the junction spots oases while Pickering called them lakes proving that any irrigating idea of his must have been radically different from mine in its scope and purpose. Secondly, it was here at Flagstaff that it was first shown that there were no bodies of water on the planet—Pickering at the time believing that there were two of considerable size. The entire absence of bodies of water is the basis of my conception of the planet which is quite unlike his.[112]

And in a more general vein, Lowell added:

Pickering is all the time making suggestions of one kind or another, and upon their value light is thrown by an episode that happened the other day in my hearing at a scientific meeting in Boston at which in combatting the irrigation theory he delivered himself of the following startling statement: "That whether the canals were ditches dug or natural cracks, in either case vegetation would ensue and he didn't care which was correct as he had suggested both"!! From such a statement you can judge for yourself that his suggestions are not worth much even to his own eyes; for a man can hardly father two totally opposite ideas.[113]

The issue of the originality of Lowell's Mars theory also came up later in 1907 in his dealings with the *Century* magazine which was then preparing to serialize his October 1906 Lowell Institute lectures and to publish his special article on his South American expedition. And as in the case of Todd's uncorrected article for the *Cosmopolitan,* Lowell once again was forced to threaten a court injunction before this particular imbroglio was settled to his satisfaction.

Lowell perhaps should have been forewarned about the *Century's* attitude, for even before winning the bidding for the first publication of the South American photographs, its editors had objected to some of the illustrations, taken largely from such texts as Dana's *Manual of Geology,* that he had provided for the lecture articles. "You are under a slight misapprehension of the use of what you call 'text book illustrations,' " Lowell advised the *Century* in the same letter in which he accepted its offer for the South American photographs. "My thesis is quite new and I merely have to make employ in some of my steps of information from other sciences which without intention on their part lead up to my Martian conclusions. One might as well object to a picture of a pouted [sic] pigeon in Darwin's *Origin of Species.*"[114]

The first hint that Lowell received of the *Century's* apostasy had already come early in October in the prospectus for its November issue. "By the by," he wrote to associate editor R. U. Johnson, "I see in your prospectus for November that you 'hedge' a little as to the truth of the theory. Don't do it—it has come to stay. I have assumed nothing in the matter, the truth rests on massed evidence."[115] But when an advance copy of the November issue arrived in Flagstaff, the full extent of the magazine's "hedge" became painfully clear. Lowell reacted swiftly. "The advanced copy of the November *Century* contains a most regrettable editorial in which I feel sure you could have had no intentional hand, as it does me serious injury," he wrote Johnson:

I have no doubt that it was meant in good part but it so vitally affects my property that it will have to be corrected. I accordingly wired you this morning the following:

"Unless December editorial gives me sole credit for my theory and changes apologetic tone in November one shall regretfully be obliged to enjoin publication photographs. Answer immediately."[116]

Lowell then spelled out what the required correction would entail:

The December editorial should state:
1. No divided paternity for my child, please. If I did not speak with authority on this matter, I should not speak at all.
2. No apologies for presenting it. Unprofessional apologies for professional work are not only worthless and gratuitous but does it not strike you as most unwise in the magazine's own interest which is acting as godfather to the child?
As to Professor W. H. Pickering's idea that the canals are cracks, I have not only thoroughly disproved it but the very photographs you are about to publish do the like.[117]

The *Century's* editor, R. W. Gilder, now took a hand, pouring oil on the troubled waters, and Lowell replied in kind, taking the occasion to expound at length on the origins of his theory. "But though it meant no harm the editorial was not only interpreted by me as apologetic but elsewhere," he wrote, "and the harm such a misinterpretation can do is immense, for people are such sheep that if once given a lead one way or another they follow it blindly over everything."[118]

The theory is my own as are all the proofs of it which I have been patiently accumulating and thinking out for thirteen years. As it has been violently combatted even as Darwin's was and at last is now on the threshhold of general acceptance, it is not right to speak of it as a mere theory in the air at which all were working. It has overcome every objection brought against it. ... Even Schiaparelli himself, who started with quite other ideas, speaks now in no uncertain terms about it ... and his opinion outweighs that of all other critics put together.
The theory is not, as the editorial would imply, a conglomerate of reading joined to one's own observations. It is indebted to general scientific principles only; for in a peculiar manner it has had to depend for its data upon what has been accumulated here.[119]

No sooner was the *Century* straightened out on the point, than Lowell was complaining to publisher William Randolph Hearst about a clipping "from your Los Angeles *Examiner* via your Boston *American* which is such a fake that I know you will want to correct it. ... The theory brought out in the clipping has been cribbed directly from *Mars and Its Canals* published by me last year without the slightest reference to its source."[120]

In the final months of 1907, and despite persistent skepticism among scientists, public interest in the idea of intelligent life on

Mars ran higher than at any other time in history. Lowell, indeed, seemed to have good reason to believe that his theory was "at last now on the threshhold of general acceptance." In an era when the detective mystery was at a peak of popularity, the Boston *Transcript* could remark that "one of the New York literary critics pays Mr. Lowell's Mars hypothesis the lofty tribute of declaring that 'no detective story of fiction excels it' in interest."[121] In mid-October, the return of Lowell's South American expedition brought Lowell what today would be called a "good press," although the querulous Todd was reluctant to expand on his cautious statement that "it is reasonably certain that Mars has been inhabited in the past, and it is reasonably certain that it is inhabited now."[122]

Mars and Lowell's theory received more publicity in November when Lowell placed some of his photographs and drawings of the planet on exhibit at the American Museum of Natural History in New York, through the courtesy of its director, paleontologist Henry Fairfield Osborn. *The New York Times* reported the exhibit under the headline: "Theory of Martian Life Corroborated."[123] Later that month, returning to Boston after eight months in Flagstaff, Lowell staged a similar display at Massachusetts Institute of Technology, the Boston *Herald* commenting that "Prof. Lowell's pictures of Mars make a mighty interesting exhibit. The ordinary art shows are nothing to it."[124]

Early in December, *The New York Times* published a full-page illustrated article by Lillian Whiting entitled "There Is Life on the Planet Mars," which reported that Lowell, "recognized as the greatest authority on the subject, declares that there can be no doubt that living beings inhabit our neighbor. ... The discovery," Miss Whiting added, "is due to the brilliant genius, the persistent energy, and the marvelous power in research of Percival Lowell."[125]

Not to be outdone, the Boston *Herald* two weeks later printed a long, highly flattering, article about Lowell's life and work. "Endowed with plenty of scientific imagination, but also possessing the cool temper of the expert mathematician, Mr. Lowell confines himself wholly to the observed facts," the paper declared:

He comes back from Flagstaff, Ari., [sic] with nothing as to what "The Martians" are like, and does not trouble himself as to near or remote possibilities of communicating with them.
    The belief that life may exist on other planets is not a new acquisition of human thought, but Prof. Lowell was the first astronomer to devote himself to the task of converting that belief into certainty.[126]

So matters stood as the favorable opposition year of 1907 came to an end. Lowell would have a few more triumphs in 1908 before Mars retreated out of sight, and thus largely out of mind, for the better part of the next two years. In January he spoke to the National Geographic Society in Washington, D.C., on "The Geography of Mars," at the invitation of inventor Alexander Graham Bell, using the occasion to do some missionary work on behalf of his observatory with Capt. William J. Barnette, the U.S. Naval Observatory's superintendent.[127] In February, of course, he proclaimed Slipher's martian water vapor spectra to the world. His 1906 lectures appeared in the *Century* through May, and in Europe, Flammarion kept the Mars issue alive with articles in French and German journals.[128]

But his critics also remained active, although now he paid them less attention, explaining to Lester Frank Ward that "I do not think that to enter into controversy with one who has no first hand knowledge of the subject is ever worth while."[129] At any rate, he apparently did not deign to reply to a particularly virulent assault on his personal credibility made that March by one "Dr. Leffmann" before the Philadelphia Natural History Society. "When we consider how he has sought to influence the public and how well he must know the insufficiency of his data," the Pittsburgh *Press* quoted Leffmann on the subject of Lowell, "his actions are unworthy of a great scientist and are much more like those of a charlatan."

"No amount of scientific work or ability will excuse such methods. There is at present no proof that Mars is not inhabited, but it is to be regretted that any one should attempt to suggest solutions to the riddles of the universe without any more information than we have."[130]

When others at the Philadelphia meeting defended Lowell, the *Press* article reported, Leffmann replied that "because rash statements are made by an eminent astronomer does not make the statements any less rash. I repeat that Professor Lowell's ideas are out of the realm of real scientific thought."[131]

In 1908 too, there were non-astronomical things on Lowell's mind. In January, Flagstaff filed suit against "its famous scientist" to regain seventy-five acres of his observatory's land that its city council felt was superfluous for astronomical purposes, an action that was "deplored" by the Flagstaff Board of Trade.[132] And in April he arranged a demonstration by a Japanese Shinto priest, which involved walking on hot coals and climbing a ladder of sword blades, that delighted two hundred spectators and the newspapers in Boston.[133] Through that winter and into spring, Lowell also conducted an extensive correspondence with President W. L. Bryant and Professor W. A. Cogshall of Indiana University to obtain a doctoral degree for the elder Slipher. At first, he urged the university to award an honorary Doctor of Science degree, both in recognition of the achievements of Indiana graduates at the Lowell Observatory in general, and of Slipher's pioneering work in spectrography in particular. Indiana, however, insisted that Slipher take the "earned" degree of Doctor of Philosophy, agreeing that he could submit his spectrographic studies of Mars and the other planets in fulfillment of the requirement for a dissertation. After some argument over whether he should appear personally to be examined for the degree, he received his Ph.D. in June 1909.[134]

On June 10, 1908, aged fifty-three, Lowell suddenly turned from bachelor to benedict by marrying his long-time Boston neighbor, Constance Savage Keith, in New York's fashionable St. Bartholomew's Church. The couple spent a summer-long honeymoon in Europe which Lowell climaxed in September by ascending 5,500 feet above London by balloon to photograph the paths in Hyde Park and thus to determine the visibility of such linear features from high altitudes. His bride and his cousin, meteorologist A. Lawrence Rotch, made the spectacular ascent with him.[135]

As the year 1908 ended, Lowell's book, *Mars as the Abode of Life* appeared, advertised by its publishers, incidentally, as "science that reads like romance."[136] In December the Boston *Herald* took note of its imminent publication, declaring editorially that "nothing more audacious or captivating to the poetic and moral imagination is now underway among men" than Lowell's astronomical investigations:

On any discriminating list of contemporary Americans who have won international fame, Prof. Percival Lowell would be found. His forthcoming book ... may or may not convince astronomers and other scientists through its new evidence, but whether it does or not, there everywhere must be admiration for a pioneer in what Prof. Lowell calls "the science of planetology," and for a man whose field of investigation is so much nearer the cosmic than those of most men.[137]

And in December too, Lowell was one of four men including George Ellery Hale to be awarded a medal by the Astronomical Society of Mexico for their achievements in astronomy.[138]

Mars returned to another favorable opposition on September 24, 1909, but the event, in sharp contrast to 1907, went relatively unheralded by the press and unheeded by the public. It was not wholly without incident, however.

Mr. and Mrs. Percival Lowell

Six days after opposition, Lowell sighted two prominent canals near the Syrtis Major that he had never seen before and which he could not find on previous maps and drawings he had made of the planet. He concluded that they were what he called *"canali novae"*— "not simply new to us but new to Mars," he later wrote. "Owing to the long continued records at the Lowell Observatory, it has been possible to prove that these canals originated on Mars within the last few months," he proclaimed. [139]

But his announcement stirred little interest in the popular press, and where it was noted in scientific circles it was met with hard overt skepticism. In his observatory's name he felt obliged to protest vigorously to the *Scientific American* that it had "misrepresented" his discovery insisting that the canals in question were indeed *"novae* in fact and function and as such are the most important contribution to our knowledge of the planet in recent years."[140]

Of a more sensational character were press reports that British astronomers had sighted a "tremendous south polar cleavage" on Mars that subsequently spread "a gloomy yellow veil" over large areas of the planet. The phenomenon, *The New York Times* wired Lowell, suggested "a catastrophe transcending any ever known on earth" and the "total destruction of any life that may have existed."[141]

This clipping appeared in an unidentified Boston newspaper, probably in September 1908. A. Lawrence Rotch was Lowell's meteorologist cousin.

The "gloomy yellow veil" may well have been an unrecognized martian dust storm, such as those now known to occur quite regularly when the planet is near a perihelic opposition.[142] If so, Lowell missed a golden opportunity to make a new discovery. For he reported "no unusual changes, only seasonal ones" on the planet, adding that "canal development is progressing according to theory."[143]

Mars still looked the same to Lowell at least, but perhaps his attention had been diverted from the red planet by other astronomical matters that were very much on his mind.

CHAPTER

13

Planetology

"W E HAVE BEEN GETTING a little more astronomy than usual this autumn," *The New York Times* editorialized in November of 1907 while the Mars controversy still raged in full fury, "and some of it has been very poor astronomy. The inhabitants of Mars and their wonderful canals would have sufficed for one lesson. . . . But Prof. Lowell flew suddenly all the way from Mars to Saturn. . . . "

To make astronomical matters worse, Prof. Alfani of Florence, doubtless in envy of the new fame of Lowell and Flammarion, announces that huge new spots he has found on the sun will cause storms, floods, volcanic eruptions and earthquakes on earth this very week. Prof. Alfani has chosen a hard job. He must "make good" immediately, while Profs. Lowell and Flammarion have life commissions, which they may bequeath in good condition to their heirs.[1]

Lowell did not bother himself about Professor Alfani, his sun spots, or his dire predictions. But the problems of Saturn were very much his concern and, although he was busily engaged in both promoting and defending his embattled martian theory, this concern now plunged him into a brief brouhaha that resulted from a strange observation of the ringed planet.

This began on October 28 when W. W. Campbell of the Lick Observatory reported that "prominent bright knots" had been seen on the outer edges of Saturn's otherwise flat smooth rings which, in 1907, were oriented edge-on to the earth.[2] The knots were "symmetrically placed," he said, "two being to the east and two to the west" of the planet. Lowell, in Flagstaff, confirmed the observation for *The New York Times,* declaring that the knots, which he labeled "tores" because they gave the rings somewhat of an anchor-shaped appearance, were apparently condensations of material formed as the result of collisions among the billions of orbiting particles making up the rings. Because such collisions must involve a loss of orbital momentum by at least some of the colliding particles, he concluded that the "knots" proved that "the ring system is in the process of falling in upon the planet" in accordance with the long-established physical laws. [3]

"Rings of Saturn are Falling In," the unfortunate headline over this report proclaimed, evoking visions for some readers of a great calamity then in progress on the planet, and drawing hasty and thus perhaps ill-considered denials from some astronomers, most notably that crusty old canal critic and celestial mechanic, Simon Newcomb. Two days after the provocatively headlined report appeared, Newcomb, again in the *Times,* scornfully berated Lowell for his statements, citing a long list of astronomical authorities, and particularly Clerk-Maxwell's classic mathematical study of Saturn's rings in 1857, to deny that the rings were "falling in" at all and to defend their stability. [4]

But Newcomb apparently had not read much beyond the *Times* headline before publicly refuting Lowell's remarks. For he assumed that Lowell had claimed that the rings were being suddenly precipitated onto the planet all of a piece, in a sort of Saturnian catastrophe, while Lowell, who knew his Clerk-Maxwell as well as anyone else, had described the "falling in" as a gradual, continuing, particle-by-particle process. "Newcomb has been talking celestial mechanics rot to the papers to the effect that Saturn's rings are stable!!" Lowell wrote incredulously to Lampland in Flagstaff. "Think of it in the presence of collisions. Clerk-Maxwell knew better. I have settled him, however, with the *Times.*" [5]

In this reply, Lowell noted that the criticisms of his remarks regarding Saturn's rings had been "founded on a misapprehension of what has been observed at Flagstaff." The loss of energy resulting from collisions of particles in the rings, he explained in some detail, "must reduce the orbit of the particle, bringing it down at last upon the body of the planet."

No demonstration of celestial mechanics is more certain than this. Indeed, Clerk-Maxwell himself foresaw it. There is no catastrophe involved, as Prof. Newcomb supposes. The falling in upon the planet is a very slow process, but has been going on since the ring system formed and will go on for a long time to come. [6]

Newcomb had nothing more to say, in public at least, about Saturn and its rings, but Lowell did. At the end of November, he forwarded the text of an observatory *Bulletin* on Saturn's "tores" to Lampland for immediate publication along with instructions to carry out the observations and measurements suggested therein.[7] And in December, Lowell published a signed article in the *Scientific American* in which he not only described the "tores" and their significance but reported observations of a dark core in the shadow that the rings cast on Saturn itself, still another indication of the nonuniform distribution of their component materials. And typically, as no one else had observed this core, he could not resist some propagandizing on the point on behalf of his observatory:

The detection of the phenomenon speaks for the definition at Flagstaff, thus supporting the space penetration there shown for stars; for at Yerkes Observatory, Prof. Barnard had not caught it, as he told the writer a few days ago, and it has not been reported from the Lick, though it was visible to all observers at Flagstaff who examined the planet critically.[8]

Lowell had the best of it in Boston, too, where this incidental imbroglio over Saturn's rings was reported at length, and in terms largely favorable to Lowell, by the knowledgeable John Ritchie, Jr. in the *Transcript*.[9] But his triumph was not to be a lasting one. For when the rings again appeared edge-on to terrestrial observers a few years after Lowell's death, no "knots," "tores" or "cores" were seen and, as his brother and biographer A. Lawrence Lowell would later note, "it is needless to dwell more upon them."[10]

Lowell's seemingly impulsive flight to Saturn in the midst of the martian controversy was not either impetuous or frivolous. He had long considered the problem of the ringed planet a part of a grand pattern of planetary evolution that had been forming in his mind even before he founded his observatory at Flagstaff. This ingenious if somewhat ingenuous synthesis of his own observations and conclusions, with a *potpourri* of not always original ideas and theories propounded by others, concerned what he first defined as "the science of planetology"—a new and wholly separate branch of astronomy, which indeed it was to become. As the furor over Mars passed its peak and progressively began to fade, "planetology," along with his quiet, continuing search for a trans-Neptunian "Planet X," increasingly preoccupied his attention and his energies up to the day of his death. The massive stroke that ended his life on November 12, 1916, in his sixty-first year came after a night spent at the telescope observing not Mars but the inner satellites of Jupiter to confirm his belief that anomalies in their orbits were related to differential rotation of the giant planet. This shift of emphasis away from Mars and toward more general planetary studies is reflected in the fact that of the seventy-five observatory *Bulletins* published while Lowell lived, only twenty-eight do not concern his martian work in some way, and twenty-three of these appeared after 1907, the turning point of the Mars furor. Notably, too, the two volumes of his observatory's *Memoirs,* both published in 1915, concern non-martian matters, the first a trans-Neptunian planet and the second the rings of Saturn. His only other memoir for this post-1907 period was entitled "Origin of the Planets" and was published by the American Academy of Arts and Science in April, 1913.[11]

Lowell's ideas on planetary evolution, which appear in varying degrees of completeness and complexity throughout his astronomical writings, did not have the public appeal and impact of his martian theory. They were, perhaps, often expounded too technically for most laymen, and not technically enough for most astronomers. There was, moreover, nothing particularly new about most of them except for the deft synthesis he fashioned from them.

Lowell's overall concept of planetary evolution as it emerges from his various works included elements of the Kant-Laplace nebular hypothesis of solar system origins as well as the so-called "planetesimal" theory advanced in 1900–05 by Thomas C. Chamberlin and Forest R. Moulton of the University of Chicago to eliminate difficulties raised by the earlier hypothesis. It also embraced much of Sir George Darwin's tidal friction theory regarding the mutual effects of gravitational forces on rotating and revolving bodies and delved mathematically into the idea of commensurate periods which Daniel Kirkwood had used a half century before to explain the gaps in the belt of asteroids orbiting the sun between Jupiter and Mars. Further, he found analogies for the evolutionary course of planets in the epochal time scale postulated for the earth by historical geology and paleontology and in the evidence for and implications of Darwinian biology.

The general trend of his thinking along these lines occasionally surfaced in some of his earlier writings. But he first presented his evolutionary ideas in organized if somewhat sketchy form in his December 1902 lectures for the Massachusetts Institute of Technology, published the following spring in book form as *The Solar System.* Here he used the observed physical characteristics of Jupiter, Mars, and Mercury to assign the planets to progressive evolutionary stages which he referred to respectively in anthropomorphic terms as "youthful," "middle-aged," and "old."

Huge massive Jupiter, he said, was still hot enough internally to radiate its own heat, while small cool Mars had lost its seas and most of its moisture and was past its prime, and smaller sun-broiled Mercury was a long dead planet which, like earth's moon, had run its evolutionary course.[12]

In *Mars and Its Canals,* published four years later, he restricted his discussion of planetary evolution largely to the red planet, noting that its relatively small size and mass should accelerate its evolutionary progress and declaring that the desiccation, the "desertism," that plagued the planet had not only motivated its inhabitants to build a complex planet-wide irrigation system but marked Mars as being in the penultimate "terrestrial" stage of planetary development, its evolution far advanced physically and by inference biologically over the still "terracqueous," ocean-washed earth.[13] Such ideas, however, appear here mainly to supplement his martian thesis.

Planetology, particularly in its comparative geological ramifications, again becomes dominant in his 1906 lectures and 1908 book entitled *Mars as the Abode of Life,* where he devotes more than half of his exposition and the first half at that, to the subject. Finally in 1909 in his lectures and book on *The Evolution of Worlds,* Mars is relegated to the limbo of a passing paragraph and his new science of planetology, in all its mathematical and mechanical aspects, comes into its own.

"Planetology" is both the title of Part I of *Mars as the Abode of Life* and the subject of its introduction wherein it is defined. This "science of the making of worlds," he declares at the outset, "concerns itself with the life-history of planetary bodies from their chemically inert beginning to their final inert end" and constitutes "the connecting link in the long chain of evolution from nebular hypotheses to the Darwinian theory. It is in itself neither the one nor the other, but takes up the tale where the one leaves off, and leaves it for the other to continue." Furthermore, he contends, this is quite new:

Up to the middle of the nineteenth century, astronomy was busied with motions. The wanderings of the planets in their courses attracted attention, and held thought to the practical exclusion of all else concerning them. It was to problems of this character that the great names of the past—Newton, Huygens, Laplace—were linked. But when the century that is gone was halfway through its course; with the advance in physics celestial searchers began to concern themselves with matter, too. Gravitational astronomy had regarded the planets from the point of view of how they act; physical astronomy is intent upon what they are.[14]

This "new study," he adds, involves "the evolution of planets regarded as worlds" and has to do "not merely with the aggregation of material but with its subsequent metamorphoses after it has come together."[15]

Nevertheless in both *Mars as the Abode of Life* and *The Evolution of Worlds* the aggregation of material, and more particularly where the material came from in the first place, provided a dramatic prolegomenon for his discussion. Each lecture series opened and each book begins with his graphic description of the origin of the solar system in a near-collision of the sun and a "dark star," a description he embellished with a vivid verbal picture of what might occur in the unlikely event that the sun again encountered such a "tramp" in its flight through space.

This idea, of course, is basic to the Chamberlin-Moulton "planetesimal" theory, which held that the planets were formed by accretions of meteoric material torn from the sun by tidal forces raised by just such a cataclysmic near-miss and which Lowell thus accepted in principle if not in all its particulars.[16] The idea was first advanced by James Croll who in 1889 had proposed a collision of dark stars to account for the vast and then-unexplained store of energy in the sun.[17] It would become far more popular within astronomy, as expanded and expounded by Sir James H. Jeans and Harold Jeffreys in the years following Lowell's death and before developments in the field of astrophysics made it largely untenable.[18] It is inherently a sensational idea, and thus it is not surprising that it stirred a brief sensation when Lowell, in his *Evolution of Worlds* lectures at Massachusetts Institute of Technology early in 1909, speculated that the solar system not only began but might conceivably end with such a catastrophic event. He emphasized that such a calamity was quite improbable, but the press, seizing only on the remote possibility, reported his remarks as a flat prediction, unencumbered by any qualifications.

"The End of the World!" was the headline that many newspapers across the country put on this report.[19] The Omaha *World-Herald* proclaimed a "Danger to the Sun" from what the Minneapolis *Journal* called "Professor Lowell's Dark Star."[20] The Denver *Republican's* skeptical editor, in mock alarm, pleaded with "soothsayer-astronomer" Lowell "to name the day... The cruel part of his forecast to Ma Earth is that he leaves so much unsaid in the matter of when...."[21] The Salt Lake City News found a biblical parallel in Lowell's purported prediction of the ultimate destruction of the earth in a fiery encounter between the sun and another star, solemnly reminding its readers that:

"The day of the Lord shall come as a thief in the night; in which the heavens shall pass away with a great noise, and the elements shall melt with fervent heat: The earth also, and the works that are therein, shall be burned up (2 Peter iii, 10)."[22]

Lowell mildly chastised the press for its sensationalization of his opening remarks in his very next lecture, but his disclaimers did not reach readers much beyond the confines of Boston. "In his introduction," Ritchie wrote in the *Transcript,* "the speaker commented on the rather sensational reports that appeared in the morning papers following his last lecture, regretting that the interest should attach to the husk rather than the kernal of his remarks." And he quoted Lowell directly:

"So here the public as represented by the press evinces a natural if somewhat naive concern in the heavens which better understanding would regulate but not remove. To help impart a little of that comprehension, which far from detracting, only adds to its impressiveness and turns journalistic sensation into awe-compelling regard, is a hoped-for incidental to these very lectures."[23]

To the respected A. C. D. Crommelin of the Royal Astronomical Society, Lowell explained:

I see the English papers are crediting me with predicting a *rencontre* between the earth and a dark star. This was in consequence of some remarks in lectures here in which I carefully explained to them its extreme unlikelihood. But they ran away of course with the first bit in their teeth.[24]

W. H. Pickering tried to counter criticism that Lowell's misconstrued remarks had stirred, pointing out in a letter to the Boston *Transcript* that Lowell had properly emphasized "the remoteness of the ultimate end of the planet in time." Perhaps typically, Pickering added some dire predictions of his own about the earth's future. A more immediate concern of mankind, he declared, was the quality of the earth's atmosphere and the possible "exhaustion of oxygen through combustion, as suggested by Lord Kelvin, or of nitrogen, on which we must soon begin to rely as a source of fertilizers for our vegetation." And, he added, "what is closer at hand as a limit to the spread of our species is the inevitable rise in the cost of living."[25]

Lowell based his version of the collision hypothesis largely on what little was then known about novae and nebulae. The flaring of a nova, such as the well-known one in Perseus that he had observed fleetingly in 1901 with Sir Robert Ball, marked just such a stellar cataclysm and thus was an analog to the origin of the solar system. "Not otherwise was our own birth heralded in heaven," he declared.[26]

The debris swirled into orbit around a star by such a violent encounter, he argued, explained the nebulosities that had been observed around Nova Persei as well as some other historic nova after their first explosive brilliance had subsided and provided the material from which planets, satellites, and other bodies eventually formed through gradual gravitational accretion. The spiral nebulae, he suggested, best represented such incipient solar systems, revolving and evolving around a shattered star and were "exemplars of the way in which our own system came into being." In the case of Nova Persei, indeed, astronomers had "actually witnessed a spiral nebula evolved from a disrupted star."[27]

Meteorites were the "Rosetta stones" of the solar system, providing the key to the mysteries of its origins. As "pieces of the disrupted sun," they were "the oldest bits of matter we may ever touch, the material from which our whole solar system was fashioned."[28] Only in the nineteenth century, he pointed out, was the astronomical nature of "shooting stars" finally recognized, and only in recent years had they been shown to be a part of the solar system, "tiny planets" orbiting the sun "in the same orderly sense as ourselves."[29] The fact that they were composed of elements common to the earth argued for a common origin, and the occluded gases they had been found to contain testified to "great pressure, such as would exist in the interior of a giant sun." Thus, Lowell concluded, they were "fragments of a greater body" and "tell of an earlier stage before even the nebula arose to which we owe our birth."[30]

The evidence of meteorites also indicated that the cataclysmic event that led to the formation of the solar system was not an actual collision but a close encounter between the sun and another star. The only signs of external heating found on meteoritic specimens, he declared, came from atmospheric friction during the final plunge to earth, and thus "the parent body seems to have been torn apart without much development of heat." The circumstance "seems to hint not at a crash ... but at disruptive tidal strains."[31]

Meteors, then, were the stuff of which planetary systems were made, "the little bricks out of which the whole structure of our solar system was built up" by agglomeration within a "nebula of meteors" born from "the womb of the sun." He argued here that the so-called "white" nebulae, which had resisted resolution into stars by the telescopes of his day, were composed of meteors.[32] Sir Norman

Lockyer had proposed this idea as part of his "meteoric" hypothesis of cosmic evolution in 1887,[33] and, of course, it was part of the Chamberlin-Moulton hypothesis which had postulated that planets condensed out of a post-encounter nebula of meteor-like "planetesimals."[34] As an idea, it was as valid as any other at the time, for it had not yet been established that "white" nebulae, which included the spirals and were distinguished from "green" or gaseous nebulae, were even aggregations of stars, much less vast external galaxies of the order of the Milky Way itself.[35]

But all this was simply preliminary to planetology which concerned itself with what happened to the planets after they had been formed from the "nebula of meteors." In both *Mars as the Abode of Life* and *The Evolution of Worlds,* Lowell discussed the evolutionary processes of planets in terms of progressive stages.

The first of these was the "Sun Stage," when the newborn planet was still hot enough from heat generated by the gravitational infall of meteoritic materials to emit light. This was followed as the infant planet began to cool off by the "Molten Stage," represented by the four outer planets—Jupiter, Saturn, Uranus and Neptune. Both Jupiter and Saturn, at least, were apparently still "red hot," he declared, arguing from observational data relating to their albedo, or relative reflectivity.[36]

The third stage was the "Solidifying Stage," when a solid crust of lighter material began to form over the still molten body of the planet as it continued to cool. This was the stage of volcanism and metamorphic rocks when a planet "acquires a physiognomy of its own," when ocean basins were formed and mountain ranges raised by the "crinkling" of the crust as the cooling planet contracted within its new skin. He illustrated this with a photograph of the wizened visage of a dried-up apple. In this stage, he thought, the planets were stamped with the major crustal features "they are in fundamentals ever afterward to keep. Its face is then modelled once for all; and its face is the expression of its character."[37]

Earth itself was the only planet in the fourth, or "Terracqueous Stage," when seas filled the ocean basins and sedimentary rocks were deposited; while Mars was alone in the fifth, or "Terrestrial Stage," when seas disappear, the atmosphere becomes depleted, and the planet is slowly dying from progressive desiccation. The moon and largest satellites of the giant planets marked the sixth and last "Dead Stage" when even the atmosphere has departed.[38]

Although Lowell considered these stages to be both successive and progressive, the point at which the planets entered this evolutionary sequence, and their progress in it, was wholly dependent on their initial mass which, in turn, determined their temperature. "So soon as the meteorites started to fall together, they generated heat," he explained, "warming one another as the rubbing of two sticks together strikes fire. The amount of heat produced depended upon the number of particles concerned, or, in other words, upon the mass of the body the particles were busied to form."

Mass, then, is the fundamental factor in the whole evolutionary process, the determining departure-point, fixing what the subsequent development shall be. Though the bodies were in essence the same at the start, their initial quantity would change their very quality as time went on. What started like would become different; for the gathering together of the particles into a single body was the preface to that body's planetary career.

Not until the internal heat began to abate did what we call evolution set in. Up to then the growing temperature induced a devolution or separating into simplicity of what had been complex. The time taken by each planet to reach its maximum bodily heat differed as between one and another. The larger the body, the slower it attained the greatest temperature of which it was capable, both by reason directly of its mass, and indirectly of the pressure to which that mass gave rise.

At its heat-acme the picture each planet presented was all its own. Some may have been white-hot, some certainly were red-hot, some were merely darkly warm; for the one differed from another in self-endowment of warmth or light, each with a glory of its own.[39]

As the fledgling planet cools and contracts at a rate dependent on its mass and size, it sooner or later achieves the "Solidifying Stage" where "the crust finds itself too large for what it encloses" and, "to fit the shrunken kernel it must needs crumple into folds." This process, he reasoned, was "most pronounced where the heat to be got rid of is the greatest, and the surface to radiate it is relatively least," and therefore, "the larger the planet . . . the more mountainous its surface will be when it reaches the crumpling stage of its career." Furthermore, he added, volcanic action is similarly increased by contraction. Thus the earth, nine times the mass of Mars, should have a rougher surface than the red planet which, in turn, should be rougher than the moon.[40]

And indeed Mars "proves singularly devoid of irregularity" in contrast with the mountainous earth. From its smaller mass, and thus its low internal heat—which he calculated to be 2,000 degrees Fahrenheit or just under the melting point of iron—"we should have in consequence virtually no volcanic action," and "there could have been but little crinkling of the crust. . . . The planet should show . . . a remarkably smooth and level surface; and this is precisely what the telescope reveals."[41]

But the moon's surface, of course, was singularly irregular. "The first thought that strikes us," Lowell conceded, "is the glaring exception seemingly made by it to the theoretic order of smoothness."

The lunar surface is conspicuously rough, pitted with what are evidently volcanic cones of enormous girth and of great height, and seamed by ridges more than equal of the Earth's in elevation. . . . On the principle that the internal heat to cause contraction was as the body's mass,—and no physical deduction is sounder—the state of things on the surface of our satellite is unaccountable. The Moon should have a surface like a frozen sea, and it shows one that surpasses the Earth's in shagginess.[42]

How, then, did the moon get enough initial internal heat to account for its external ruggedness? Why from the earth, Lowell answered with a bow to Sir George Darwin and his tidal friction theory of the origins of the earth-moon system. If the gradual slowing of the earth's rotation and the corresponding expansion of the moon's orbit postulated by Darwin's theory were traced back in time, he explained, there came a point "when the Moon might have formed a part of the Earth's mass, the two rotating together as a single pear-shaped body in about five hours."[43]

Now the pregnant point in our present heat inquiry is that the face of our satellite indicates that the might-have-been actually was. The erupted state of the Moon's surface speaks of such a genesis. For in that event the internal heat which the moon carried away with it must have been that of the parent body—the amount the Earth-Moon had been able to amass. Thus the Moon was endowed from the start of its separate existence with an amount of heat the falling together of its own mass could never have generated. Thus its great craters and huge volcanic cones stand explained. It did not originate as a separate body, but had its birth in a rib of the Earth.[44]

Far from disproving the law and his conclusions "the seeming lunar exception, therefore, really upholds it," he added.[45]

In discussing the formation of topographical features in the "Solidifying Stage," Lowell made a curious, parenthetical point, if indeed it was meant to be a point at all. The continents of the earth, he noted, all appear to be roughly triangular with their apexes to the south, while the so-called "seas" of Mars, the dark areas, are also generally triangular, but with their apexes to the north. Thus Mars "stands as the negative aspect of the positive picture presented by the Earth," and if the martian "continents" and "seas" were reversed "to get their earthly proportions, with oceans preponderant, as on Earth, the two distributions are seen to typify the same action." What this action was, however, he did not say. [46]

The extent to which oceans would cover the surface of a planet was also directly related to its size because the "water each possessed would be as its mass, and when it collected into seas, these, if equally deep, would cover more of the surface on the larger planet, since it has less cuticle for its contents." Further, a larger planet is "better able to hold on to its gaseous elements," and thus retain more of what was to condense to water when the time arrived. The earth, Mars, and the moon "have, or have had, in all probability, judging from their present look, oceans of this order of size." [47]

The appearance of water in liquid form on the suface of a planet was a major milestone in its evolution, "for with the formation of water, protoplasm first became possible, what may be called the life molecule then coming into existence."

Of all the conditions preparatory to life, the presence of water, composed of oxygen and hydrogen, is at once the most essential and the most worldwide. For if water be present, the presence of other necessary elements is probably assured because of its relative lightness as a gas. Furthermore, if water exist, that fact goes bail for the necessary temperature, the gamut of life being co-extensive with the existence of water as such. It is so consequentially, life being impossible without water. Whatever the planet, this is of necessity true. [48]

In defining the vital midstages of his evolutionary sequence, Lowell relied mainly on the evidence provided by the earth and Mars. The other planets, he pointed out, contributed nothing to the knowledge of Stages III, IV, and V "because, like Mercury and Venus, they are too advanced, or because, like the major planets . . . they are not advanced enough." There were oceans, of course, on the earth, and Lowell as well as others considered the dark areas on Mars and the *maria* on the moon to be evidence of vanished seas. [49]

The contemporaneity of the appearance of oceans and life he argued largely on geological and paleontological grounds. "The geologic record proves that life originated in the oceans," he declared, " and lived there a long time before it so much as crawled out upon land." The *deus ex machina* in both origins "was the same—a gradual lowering of the temperature." [50]

The earliest remains of life found in sedimentary rocks of the paleozoic age represented what were already "somewhat advanced types," such as trilobites, he argued, and "the most remarkable characteristic" of "these fossils of the past" was their "world-wide uniformity." Evidence of this "latitudinarianism" was found not only in the Cambrian but in the subsequent Silurian, Devonian, and early Carboniferous eras of geologic time. The life forms thus so "coevally widespread" were "warmth-loving" as "betrayed by the fact that their nearest relatives now extant live wholly within the tropics."[51]

After life emerged from the tepid Paleozoic seas, its explosive profusion in the Carboniferous era testified to a vastly different environmental state than prevailed in later times. The earth then, he suggested, was first "warm everywhere with a warmth surpassing that of the tropics today; and second, the light was tempered to a half-light known now only under heavy clouds." And both these conditions "were virtually general in locality and continuous in time." The earth's hot and humid atmosphere was blanketed by "steamy clouds" and its surface covered by warm seas and "dank and dark" forests of lush shade-seeking cryptogams whose continuous growth, as indicated by the absence of annual rings in their fossilized stems and trunks, also indicated the absence of any seasons.[52] Indeed, it would seem that the earth then must have looked much like cloud-shrouded Venus does now, but of course Lowell did not draw this comparison for he had long insisted that the Cytherean atmosphere was clear and cloudless.

A number of hypotheses had been put forward both in geology and in astronomy to account for this seasonless "warm dawn of the early geologic ages," he noted, "but of the two kinds all alike fail."

Thus, merely a different distribution of land and sea will not explain it, because it was general, not local; and secondly, because this leaves untouched the problem of less light. Equally important is a change of position in the axis of the Earth; for were the axis so far changed as to point directly toward the sun, this would not do away with the seasons, but would accentuate them. Nor will an altered eccentricity of the Earth's orbit, which has also been suggested, prove more effective.[53]

Not less impossible, he added, was the suggestion that the sun was then so large "as to be able to look down on both poles of the Earth at once, and so to give our globe equal day and night everywhere and ... a substantially even temperature in consequence throughout." There was, he averred, "no place in modern cosmologies" for a sun "of such stupendous size," and further-more, such a bloated sun would necessarily have an "incredible tenuity."[54]

Planetology, however, provided an explanation for "this be-clouded hothouse state of things," for the earth's heat itself, sus-taining the warmth of the recently-precipitated oceans, "might well be responsible for Paleozoic conditions."

Simultaneously, a vast steaming must have gone up from the still warm waters, resulting in a welkin of great density. This would act in two ways to explain the phenomena. First, the welkin would keep the Earth's own heat in; and secondly, it would keep the sun's heat and light out. We should have a sort of perpetual tropical summer in a twilight of cloud; a climate superior to seasons because screened from direct dependence on the elevation of the sun. This is perfectly in accord with the half-light the vegetation vouches for, while the luxuriance of that vegetation testifies to the warmth and even suggests a further, though not the chief, reason for it in the great amount of carbonic dioxide its existence establishes as then present in the air.[55]

Later he suggested that fossil fauna might reflect these early environments physiologically, wondering, for instance, if Paleozoic trilobites had been blind.[56] After reading *Animals Before Man in North America,* by Frederick A. Lucas of the Brooklyn Institute, he wrote its author that "your mention of the small eyes of one of the great reptiles of Mesozoic times interested me and I believe it will be found to be a general characteristic of the period due to cosmic causes."[57]

In this "Solidifying," or "Self Sustained Stage" as he referred to it in *The Evolution of Worlds* to indicate a planet's relative indepen-dence from solar heat and light,[58] life "spontaneously evolved" through chemical processes. "So far as we have evidence, life is the inevitable outcome of the cooling of a globe, provided that the globe be sufficiently large," and he here delivered a passing thrust at the ancient concept of "panspermia," which had been modernized and revived by the Swedish scientist Svante Arrhenius in 1908:[59]

For life did not reach this earth from without. No fanciful meteorite bore it the seeds which have since sprouted and overrun its surface. Meteorites gave life, indeed, but in the more fundamental way in which all nature's processes are done, by supplying it with matter only from which by evolution life arose.[60]

While first this life was uniform and ubiquitous, it became progressively differentiated and localized as environmental condi-tions changed and grew more diverse with the continuing cooling of the earth. This "coincidence" he saw as a "vital parallelism" which "seems to be a general principle of evolution."

Thus all along the line we perceive that life and its domicile arose together. The second is necessary to the first and the first is always sufficient to the occasion. . . . Endless variation takes advantage of any opportunity so soon as it occurs. Life but waits in the wings of existence for its cue, to enter the scene the moment the stage is set.[61]

This process was accelerated as the planet passed gradually from the "Self-Sustained" into the "Sun-Sustained" stage in which the sun increasingly became the planet's dominant source of heat. "On Earth, the transition from self-support to solar dependence began with the first symptoms of atmospheric clearing in the time of the great reptiles."

The clouds that had veiled the whole Earth in the paleozoic period then began to dissipate; though it was probably not until much later that the sky approached the pellucid character we know. The Earth's own cooling thus first let in the sun.[62]

The earth was now entering the Mesozoic age of the geologists. In the fossil record of the Triassic era, gymnosperms, cycads, and conifers replaced the cryptogams of the Paleozoic, plants that "demanded more sunshine than their predecessors, and thus testify to the purifying air caused by the gradual cooling of the surface and the consequent less abundant generation of cloud."

That the Sun had not grown more insistent, but the Earth more open-eyed, the latitudinal character of the cooling shows. For it was not the absolute lowering in warmth, but the zonal differentiation of temperature that then set in, which is the noticeable thing. The tropics were as before; the climate was changing slowly toward the poles. Climatic zones began to belt the Earth.[63]

That this differentiation of the tropic, temperate, and arctic zones was progressively accentuated through the Mesozoic age Lowell argued from the decreasing latitudes which marked the range of corals during the Jurassic era, and from the appearance and spread of deciduous trees, as shown by fossil evidence first in the Cretaceous and then in the Eocene, Miocene, and Pliocene periods of the Cenozoic, or Tertiary Age, which culminated in the great ice ages of the Pleistocene era ending only some 10,000 years ago.[64] He attributed the earth's ice ages to a combination of increased precipitation and localized elevation of the land, rejecting Croll's then popular idea that they were caused by an increase in the eccentricity of the earth's orbit. Mars, he pointed out, had almost the same axial tilt but more than five times the orbital eccentricity of the earth and yet without sufficient precipitation, "the true *primum mobile* in the matter," there was no glaciation on the red planet. Thus, he added, did the science of planetology provide clues to the mysteries of the earth's past.[65]

Although planetology by his own definition ended where Darwinian evolution began, Lowell nonetheless traced briefly the ascent of man through the development of the brain to make the point that only intelligent beings were capable of marking the face of their planet in a way which would reveal their existence to neighbors in space. "Brain, indeed, now became the chief concern of nature," he declared. "The character of the habitat undoubtedly brought this about through the prizes it offered the clever, and the snuffing out to which it consigned the crass." Brain finally distanced brawn, and even in his savage state man became a being that other beings feared. Thus, man slowly "began to take possession of the earth."

For some centuries now this has been his goal, unconscious or confessed. The true history of man has consisted not in his squabbles with his kind, but in his steady conquest of all earth's animals except himself. He has enslaved all that he could; he is busy in exterminating the rest. From this he has gone on to turn the very forces of nature to his own ends. This task is recent and is yet in its infancy, but it is destined to great things. As brain develops, it must take possession of its world.[66]

This subjugation "carries its telltale in its train," he noted, "for it alters the face of its habitat to its own ends."

Already man has begun to leave his mark on this his globe in deforestation, in canalization, in communication. So far his towns and his tillage are more partial than complete. But the time is coming when the earth will bear his imprint, and his alone. What he chooses, will survive; what he pleases, will lapse, and the landscape become the carved object of his handiwork.[67]

This deduction, Lowell contended, appeared to be "equally applicable" to Mars and to martians. "Too small in body himself to show, it would only be when his doings had stamped themselves there that his existence could with certainty be known. Then and only then would he stand disclosed."

It would not be by what he was, but through what he had brought about. His mind would reveal him by his works—the signs left upon the world he had fashioned to his will. And this is what I mean by saying that through mind and mind alone we on earth should first be cognizant of beings on Mars.[68]

Indeed Mars proves "to occupy earthwise in some sort the post of prophet," he concluded. "For in addition to the side-lights it throws upon our past, it is by way of foretelling our future, ... from a scrutiny of Mars coming events cast not their shadows, but their light, before."

It is the planet's size that fits it thus for the role of seer. Its smaller bulk has caused it to age quicker than our earth, and in consequence, it has long since passed through that stage of its planetary career which the earth at present is experiencing, and has advanced to a further one to which in time the earth itself must come. ... [69]

This grim planetological prophesy of the earth's eventual death by progressive desiccation, however, would not be fulfilled in all particulars, he added. For "in detail no two planets of different initial mass repeat each other's evolutionary history; but in a general way they severally follow something of the same road." Still Mars had lost its oceans and most of its atmosphere through absorption into its crust or through gradual molecular escape into space, and there was abundant evidence available to show that these processes were already well under way on earth. Water was the key to this stage of a planet's evolution, and here again Lowell expounded and expanded his "desertism" concept to argue that terrestrial oceans are disappearing, and thus that "the earth is going the way of Mars."[70]

The outcome is doubtless yet far off, but it is as fatalistically sure as that tomorrow's sun will rise, unless some other catastrophe anticipate the end.[71]

Planetology also foresaw "the relative drying up of our abode" in *The Evolution of Worlds,* where his speculations on a planet's demise provide a dramatic finale to his exposition. Indeed, in this, his last astronomical book written and published for public consumption, he devotes his final chapter to the "Death of a World," noting that "heart-failure" through loss of water and air was only one of four possible ways in which the earth or any other planet might reach "that stage when all change on its surface, save disintegration, ceases." The least likely form of planetary death, Lowell opined, was "by accident," such as the cataclysm that had originally conceived them. But planets might also die from natural causes, through "paralysis" brought about by the relentless forces of tidal friction or, far more remote in the future, by extinction of the sun. [72]

In *The Evolution of Worlds,* however, Lowell took a wholly different approach, concentrating on physical rather than geological and biological ramifications of planetary evolution. That his thesis was of major import he left little room for doubt in the minds of his readers, at least. For there was an "infinite gradation" in the size and orbital shape of solar system bodies, from tiny meteors to giant planets, that was "kin to that specific generalization by which Darwin revolutionized zoology a generation ago," and was "as fundamental to planets as to plants." This "shows that the whole solar system is evolutionarily one." [73]

Again he began with a graphic description of the birth of the solar system by collision or near-collision of the sun and a "tramp" star, pointing out that both the decreasing density of the planets with distance from the sun and the progressively greater outward occurrence of lighter gases in the atmospheres of the major planets, as indicated by the elder Slipher's spectrograms, were "suggestive" of such an event. [74]

But now he set the stage for a new synthesis by briefly describing the physical and mechanical characteristics of the planets one by one, starting with Mercury and working through to Neptune, although giving only a passing mention to the earth and Mars. Rather than with planetary environments, he was now more concerned with their astronomical parameters—the inclination of their orbits and poles to the ecliptic, their periods and directions of rotation, their distribution in relation to distance from the sun, and the number and nature of their satellites.

From these data, as then known to him, he claimed a systematic if generalized unity for the solar system, finding congruities in its arrangements where others saw only incongruities and where, indeed, there were and are still only incongruities. On the hypothesis that a star passing near the sun would tear away some of the sun's substance and set it swirling in solar orbit in space, Lowell noted that the planets and lesser bodies forming from such meteoric material would "lie in the plane of the tramp's approach," and that the eight known planets did indeed lie within a few degrees of the ecliptic. [75] Under the nebular or any other hypothesis, the direction of their revolutions and rotations should be the same, he noted, yet incongruously the satellites of Uranus and Neptune's satellite Triton showed a retrograde, or backward, motion, implying to Lowell as well as to other astronomers of his day that in the absence of evidence to the contrary, the rotations of these two planets might also be nondirect.[76]

To resolve this seeming incongruity, Lowell carried the earlier applications of Darwin's tidal friction theory somewhat further than Sir George had cared to venture. For tidal forces, he suggested, not only gradually slowed the rotations of the planets and satellites to an eventual synchronous lock with their periods of revolution, but in so doing flipped them over a full 180 degrees on their axes. The original rotational direction of all these bodies, then, had been retrograde, and what the observed rotations and axial tilts thus revealed was "the systematic righting of planetary axes" by tidal friction, and the consequent reversal of the direction of their rotation. This process, Lowell proposed, had long been completed where tidal forces were strongest, as evidenced by the isochronic motion of Venus and Mercury, the earth's moon, and the inner satellites of Jupiter and Saturn. But it was still in progress where the tidal effects were proportionately weak, as shown by the extreme axial tilts of Uranus and Neptune, whose poles are tipped at angles of 82 and 35 degrees respectively to the plane of the ecliptic, and by the retrograde motion of their satellites and, presumably, the planets themselves. This explanation, he declared, was in accord with the principle of the conservation of momentum and, as tidal friction "tends to turn over the axis of a retrograde body," the nondirect motions in the Uranian and Neptunian systems, as well as some of the outer satellites of Jupiter and Saturn, were "survivals from an earlier state of things."[77]

Other "congruities," Lowell claimed, included the progressions of the axial tilts from Neptune to Jupiter and from Mars to Mercury; the synchronous periods of the innermost planets and the innermost satellites; a progressive increase outward in the orbital tilts of the satellites of Jupiter and Saturn; and the "remarkable congruity" of the "peculiar arrangement of the masses of the solar system" of which "little can as yet be said in interpretation." But the satellite systems of Saturn, Jupiter, and Uranus appeared to be "strangely reproductive" of this arrangement, and this "analogy with the solar system argues a common principle" in the origins of all.[78]

Lowell, be it noted, was somewhat diffident in advancing his planetary evolution hypotheses and certainly not as confident about them as he was about his martian theory. For he concluded with the qualification that he was submitting these ideas "with that distrust of which Laplace has so eminently spoken."[79]

Lowell's ideas on "planetology" were primarily syntheses of selected fact and extant hypotheses, tied together by more or less plausible if sometimes dubious assumption. The readability of the books was widely acknowledged, even by critics who deplored their slipshod science. Astronomer Charles Lane Poor, in his review of *The Evolution of Worlds* in *Science,* opined:

> The book contains many loose statements of scientific facts and principles and conclusions are drawn by special pleadings and by apt illustrations rather than any course of logical reasoning. Yet with all this, and in spite of exaggertions and obvious attempts to create popular excitement, the book gives the general reader, in attractive form, a more or less accurate conception of the latest ideas in regard to the evolution of our world. [80]

The congruities that Lowell found in the solar system were indeed strained and superficial, even by the relatively fragmentary body of data available in his day, and did indeed require "special pleadings" and "apt illustrations" to defend. The sweeping generalizations he proclaimed on the basis of a few specific geological, biological, or astronomical facts simply were not always valid. Even at the time he wrote, they had been made largely untenable by other data he conveniently ignored, loftily dismissed, or failed to take into account for one reason or another. Other critics, far less kind than Poor, took pains to point out these omissions and failures in glaring detail.

Ironically perhaps and certainly unintentionally Lowell, by going outside astronomy and eclectically drawing material from other disciplines to develop his ideas on planetary evolution, had both laid himself open to attack on his flanks and had swelled the ranks of his attackers. A geologist, in fact, mounted the most virulent assault on his concept of "planetology," touching off a bitter debate which erupted in April 1909 in a blistering review of *Mars as the Abode of Life* in *Science* and raged through most of the year in the pages of that journal.

"Although it is improbable that these words will be read by more than a small proportion of those who have seen or heard of Mr. Percival Lowell's *Mars as the Abode of Life,*" the University of Wisconsin's Eliot Blackwelder prefaced his vitriolic review, "it seems worthwhile to point out to the scientific workers of the country the gross errors which this book is propagating."

It is not surprising that Mr. Lowell, an astronomer, should have only a layman's knowledge of geology, but that he should attempt to discuss critically the more difficult problems of that science, without, as his words show, any understanding of the great recent progress in geology, is astonishing and disastrous. One can not but recall the old adage that "fools rush in where angels fear to tread."[81]

Blackwelder then proceeded to detail how Lowell had "twisted and squeezed" the geologic history of the earth to fit his "preconceived opinion of Mars" through "fallacious" statements of facts and "fanciful" interpretations of dubious and often obsolete theory.

Not the least of Lowell's geological sins, he began, was a seeming ignorance of such fundamentals as the actual nature of metamorphic rocks. Then he marshaled a mass of evidence to contradict Lowell's postulate of a warm, damp, clouded, seasonless climate for earth during the Carboniferous era, pointing out that the lush flora of the time was localized rather than general in distribution and that "the Carboniferous plants show by their structures an adaptation to severe rather than genial climatic conditions." The "shade-loving" ferns that Lowell insisted proved "a shady half-light" in Mesozoic times, he noted, "now stand out isolated on the brushy hills of equatorial Africa under the blazing tropical sun," a fact apparently unbeknownst to Lowell. "Under the circumstances," the geologist added with dry sarcasm, "he would have found the services of a botanist advantageous."[82]

The Wisconsin scientist also effectively demolished Lowell's cherished "desertism" hypothesis by pointing out that the late James Dwight Dana, whose textbook on geology was frequently cited by Lowell, had "many years ago propounded the opinion that the lands had grown steadily larger from small beginnings" but that "if Dana were alive to-day, he would doubtless repudiate the idea, for it is wholly contrary to the mass of facts more recently made known." Indeed, these facts show "that there have been fluctuations of land and sea throughout recorded geologic history, and these changes show no general tendency."

Having assured his readers that the earth is drying up and that it will sooner or later "roll a parched orb through space," he cited as proof the alleged fact that deserts are increasing in size. This is the beginning of the dreadful end which "is as fatalistically sure as that to-morrow's sun will rise, unless some other catastrophe anticipate the end." Here again the proverb applies, "a little knowledge is a dangerous thing." ... Had he inquired into the recorded facts of geologic history, he would have learned that deserts have existed in many parts of the world ever since the earliest times, wherever topographic and atmospheric conditions were favorable.[83]

These examples alone, Blackwelder declared, were sufficient to "show what kind of pseudo-science is here being foisted upon a trusting public." Lowell's book, he noted, "is avowedly a popular exposition of science, not a fantasy," written by "a highly educated man of distinguished connections and some personal fame . . . in a vivid, convincing style with the air of authority."

The average reader believes him, since he can not, without special knowledge of geology and kindred sciences, discern the fallacies. He has a right to think that things asserted as established facts are true, and that things other than facts will be stated with appropriate reservation. . . . The misbranding of intellectual products is just as immoral as the misbranding of products of manufacture. Mr. Lowell can not be censured for advancing avowed theories, however fanciful they are. . . . But I feel sure that the majority of scientific men will feel just indignation toward one who stamps his theories as facts; says they are proven, when they have almost no supporting data; and declares that certain things are well known which are not even admitted to consideration by those best qualified to judge.[84]

"Censure," Blackwelder concluded his devastating critique, "can hardly be too severe upon a man who so unscrupulously deceives the educated public, merely in order to gain a certain notoriety and a brief, but undeserved credence for his pet theories."[85]

Strangely perhaps, Lowell did not immediately answer this searing indictment and indeed never replied to it in specific detail except to state that he had taken the geologic facts in *Mars as the Abode of Life* "from recognized sources, chiefly Dana, [Sir Archibald] Geikie, Dr. Lapparent and recent research."[86] His initial silence, in fact, might have forestalled further criticism had it not been for the egomaniacal Dr. Thomas Jefferson Jackson See who, since his shadowy dismissal from the Lowell Observatory staff in 1898, had been employed as a mathematician at the U.S. Naval Observatory at Mare Island, California.

See now penned a letter to the editor of *Science,* entitled "Fair Play and Toleration in Science" and published late in May, in which he not only defended Lowell fervently if somewhat irrelevantly, but boastfully proclaimed his own unacknowledged preeminence as a cosmogonist.[87] In this grandiloquent self-promotion, See blithely dismissed the recently promulgated Chamberlin-Moulton "planetesimal" hypothesis as merely an "inconsistent and destructive" criticism of the Laplace nebular hypothesis of solar system origins and breezily concluded that except for his own exhaustive studies of the subject over the previous twenty-five years "most of the recent speculations on cosmogony are not worth the paper they are written on."[88]

See's exaggerated pretentions, in turn, resulted in new attacks, although these were now directed largely at Lowell's erstwhile champion rather than at Lowell himself. In a letter to *Science* early in July, Yale University geologist Joseph Barrell, while expressing his complete agreement with Blackwelder's bill of particulars, offered a paragraph of qualified praise for Lowell along with several pages of unqualified damning of See and all his works:

As an offset, however, to the necessarily severe criticism of "Mars as the Abode of Life," cordial recognition may well be given at the same time to that great enthusiasm manifest in all of Lowell's work, which has led to the founding of a magnificent observatory and has contributed to astronomy much of real value. A coming generation of scientists will find much to regard highly in Lowell and will see in his work a stimulus to further knowledge, but will hold it as unfortunate that the same temperament which led to those results should have given rise to writings which called forth such severe criticisms as have appeared from his contemporaries in order to separate errors of promise and conclusion from that which is of real value.[89]

Three weeks later, again in *Science,* astronomer Forest Ray Moulton of the University of Chicago, coauthor of the planetesimal concept, did not even mention Lowell's name in his scornful denunciation of See's scientific braggadocio. Moulton carefully reviewed cosmogony from Laplace's time and then demonstrated that See's "alleged" twenty-five years of study had produced only two papers—one on spiral nebulae printed in *Popular Astronomy* in 1906, and another on planetary orbits published by *Astronomische Nachrichten* and *Popular Astronomy* and summarized by See in the Chicago *Record-Herald* early in 1909. These papers, he pointed out from liberal quotations, were neither original nor consistent and the cosmogony in them seemed strangely reminiscent of his and Chamberlin's own ideas promulgated in 1900 and thereafter in professional journals:

I wish to point out that notwithstanding the evidence furnished by the 1906 paper of his familiarity with our work, and in spite of the fact that at his request I furnished him reprints of my papers several months in advance of his recent publication, there is in it no direct or indirect reference to Professor Chamberlin or myself. Ordinarily such conduct justifies the use of strong terms in characterizing it, but in the present case I believe astronomers and others who are familiar with the situation will fully agree with me that these aberrations are more deserving of pity than of censure.[90]

See, of course, replied to these charges, declaring that because Barrell had "evaded the issue... his long-drawn-out discussion requires no further notice," and dismissing Chamberlin and Moulton's results as "negative," while insisting on the priority of his own cosmogonic work.[91] But a more important effect of Moulton's remarks was Lowell's direct if belated entry into the running controversy.

In September, again in *Science,* Lowell attempted to discredit the Chamberlin-Moulton hypothesis, citing mathematical errors he claimed to have found in Moulton's 1906 book, *Introduction to Astronomy,* in which the theory was described at length. After explaining these with formulae and calculations, Lowell confidently concluded that "from what we have said it will be seen that the hypothesis will not work."[92] He also made this claim in letters to such eminent scientific acquaintances as A. A. Michelson, Lester Frank Ward, and England's John Ellard Gore.[93]

The errors Lowell cited were quite trivial, as Moulton noted in November in a bristling retort in which he ridiculed what he labeled Lowell's "new logic."

Quite apart from the validity of the allegations it is, to me, a novel idea in logic that errors made in trying to support a proposition become thereby "disproof of it." One might infer by this sort of reasoning that the errors of the class-room have long since destroyed all of the principles of mathematics. The logic of the present case is all the more remarkable in that two of the four alleged mistakes do not occur in my discussion of the planetesimal hypothesis at all, while the two that do relate to it are really one, and it is not shown that even this one has any critical relation to the hypothesis.[94]

Moulton did, however, confess to one of the errors cited by Lowell, calling it "inexcusable" in light of the fact that he had made the point correctly in earlier publications, but pointing out that this error occurred in his book eighty-three pages before his exposition of the planetesimal hypothesis began. "This seems to fix a new principle in logic with a quantative function: a mistake in expounding one proposition, if made within 83 pages of the discussion of another proposition, throws discredit on the latter.[95]

And finally, with biting sarcasm, Moulton quoted passages from *Mars as the Abode of Life* and from Lowell's article on "The Revelation of Evolution" in the August 1909 *Atlantic Monthly* to demonstrate that "Dr. Lowell has a real affection for the main features of the planetesimal hypothesis."

If I had not been so unfortunate as to have utterly destroyed it (according to the new logic) by the blunder in my book 83 pages before I took the hypothesis up, he might have almost reconstructed it from his own recent writings. I am wondering whether in his forthcoming book on "The Evolution of Worlds" he will not give additional proof of his affection for the planetesimal theory, though perhaps under some other name, or in some nameless form more congenial to that mysterious "watcher of the stars" whose scientific theories, like Poe's visions of the raven, have taken shape at midnight.[96]

This ended the debate, with Lowell clearly a loser. He was not a man to concede defeat, graciously or even grudgingly, yet he was practical enough to recognize it and shrewd enough to ignore it when it happened to him. Henceforth, he would no longer dwell publicly on the subject of "planetology."

Planet X

P ERCIVAL LOWELL touched only briefly on the concept of commensurability of periods in his books on planetology, but in the final years of his life the idea became more and more central in his work. He eventually developed it into what he considered to be a universal "law" of planetary origins, and used the concept in his long mathematical search for a trans-Neptunian "Planet X."

Commensurability involves simple ratios such as 1:2, 2:3, or 3:5 between the orbital periods of bodies revolving around the same primary. When such simple ratios exist, maximum gravitational forces are exerted on the bodies recurrently at the same point in their orbits and the resulting resonances create unstable zones centered on the commensurable points. The concept was used by Danial Kirkwood in 1866 and again in 1876 to point out that gaps in the belt of asteroids beyond Mars, which are still called "Kirkwood gaps," seemed to occur at points commensurable with the orbit of massive Jupiter. [1] Lowell too, in his studies of planetary evolution, used the idea to contend that the divisions in Saturn's rings represented orbits that were commensurate with the orbits of its two inner satellites, Mimas and Enceladus. [2]

But more importantly, Lowell used commensurability to construct a "general planetary law" from which he made a prediction regarding a trans-Neptunian planet. The law he stated in a short paper on "The Origin of the Planets" presented to the American Academy of Arts and Sciences in April 1913 and published in its *Memoirs*. It decreed that:

Each planet has formed the next in the series at one of the adjacent commensurable-period points, corresponding to 1:2, 2:5, 1:3, and in one instance 3:5, of its mean motion, each of them displacing the other slightly sunward, thus making of the solar system an articulated whole, an inorganic organism which not only evolved but evolved in a definite order, the steps of which celestial mechanics enables us to retrace.[3]

Lowell likened this "law" to Mendeliev's Periodic Law of the Elements in chemistry, noting that "it, too, admits of prediction."

Thus in conclusion I venture to forecast that when the nearest trans-Neptunian planet is detected it will be found to have a [semi-] major axis of very approximately 47.5 astronomical units, and from its position a mass comparable with that of Neptune, though possibly less; while, if it follows a feature of the satellite systems which I have pointed out elsewhere, its eccentricity should be considerable, with an inclination to match.[4]

The 47.5 astronomical units* that Lowell here gives for the distance of his Planet X from the sun is, as it so happens, "slightly sunward" of a commensurable period at 48.4 a.u. which yields simple ratios with the orbits of Neptune of 2:1 and of Uranus of 4:1.[5] This circumstance, as will be seen, is significant in Lowell's mathematical attempts to predict the position of a new planet beyond Neptune. For a planet's mean distance from the sun, that is to say the radius vector of its elliptical orbit, is a key factor in calculating the orbital elements, or the "theory" of a planet. When the planet is known its distance can be determined with great accuracy. But in the difficult problem of predicting the location and elements of a planet whose presence is merely suspected because of its apparent perturbing effects on a neighboring body, this factor—usually designated a' in the pertinent equations—must of necessity be assumed. In the history of planetary discovery, there have been two main grounds on which this assumption has been based.

France's Urbain J. J. Leverrier and England's John Couch Adams, in the computations that led to the discovery of Neptune in 1846, relied on a curious but spurious mathematical progression known as the Titius-Bode law to assign values for a'. This "law," stated in a footnote by Johann Daniel Titius in 1772 and quickly seized upon, restated, and promulgated by Johann Elert Bode, roughly relates distances of the planets to a progression expressed

---

*An astronomical unit is the mean distance from the sun to the earth.

by the formula: $a + 2^n b$, in which $a = 0.4$, $b = 0.075$, and n = the planet's number, counting from the sun. For the fourth planet Mars, for instance, the formula gives $0.4 + 2^4 \times 0.075 = 1.6$, which is fairly close to the planet's actual distance of 1.52 a.u. The progression in fact worked quite well for all the planets out to Saturn, and William Herschel's discovery of Uranus in 1781, which turned out to be 19.2 a.u. from the sun, was widely hailed as validation of the law, which had indicated a distance of 19.6 a.u. for a trans-Saturnian planet. In 1801 and thereafter, a gap in the progression between Mars and Jupiter at $n = 5$ (2.8 a.u.) was also filled in apparent conformity with Bode's law with the first discoveries of the asteroids.[6]

The discovery of Neptune, however, demolished the law's predictive pretentions, for the new planet swung in an orbit 30.2 a.u. from the sun rather than the expected 38.8 a.u. Leverrier and Adams, with the Bodean progression in mind, had initially assigned values of 38 and 38.4 respectively for the term a' in their equations although, as their laborious calculations progressed, they revised these values downward somewhat, giving 36.15 and 37.25 a.u. in their final prediscovery elements—still quite wide of the mark.

But Leverrier, interestingly enough, also had set a lower limit of 35 a.u. for the distance of his proposed trans-Uranian planet and thereby inadvertently introduced the idea of commensurability into the many-faceted controversy that followed Neptune's dramatic discovery. The American mathematician Benjamin Pierce quickly pointed out that Leverrier's inferred limits included a commensurable point at 35.3 where the ratio of the orbital period to that of Uranus would be 5:2. The gravitational influence of a planet in this orbit would cause "remarkable irregularities" in the motion of Uranus, he contended, and thus introduce a discontinuity into the computations that would constitute "a complete barrier to any logical deduction." Changes in the character of the perturbations at 35.3 a.u. were such, Peirce claimed, "that the continuous law by which such inferences are justified is abruptly broken at this point, and hence it was an oversight by Mr. Leverrier to extend his inner limit to the distance 35." In passing through this commensurable point, in short, Leverrier had failed to take into account the "peculiar and complicated" effects of commensurability on the perturbations of Uranus, and accounting for these perturbations was, after all, what the problem was all about in the first place. Peirce concluded "that the planet Neptune is not the planet to which geometrical analysis has directed the telescope; that its orbit is not contained within the limits of space which has been explored by geometers searching for the source of the disturbances of Uranus; and that its discovery by [J. G.] Galle must be regarded as a happy accident."[7]

The details of Peirce's argument and the objections raised to it by Sir John Herschel, Adams, and others, as well as other aspects of the controversy over Neptune's discovery, are not germane here.[8] But it may be noted that Peirce also pointed out that Neptune's actual distance of about 30 a.u. fell just short—"slightly sunward," Lowell might have said—of a commensurable point at 30.4 a.u. where a planet would revolve around the sun in 168 years, twice the orbital period of Uranus. And it may also be noted that Lowell, who later studied under Peirce for two years at Harvard, not only relied on commensurability in subsequently making his assumptions about the distance of a trans-Neptunian planet but claimed to have refuted his mathematical mentor in the process.[9]

Lowell, of course, was not the first to propose the existence of another planet beyond Neptune. Leverrier, in fact, advanced the possibility shortly after Neptune's discovery.[10] In 1877 David P. Todd, the Amherst astronomer later associated with Lowell in his north African eclipse observations in 1900 and as leader of Lowell's expedition to South America in 1907, predicted a planet at a distance of 52 astronomical units with a period of 375 years, and searched for it briefly but unsuccessfully. In 1879 France's Flammarion suggested that such a planet was indicated by the movements of comets, specifically by the apparent clustering of the outermost points in the orbits of some periodic comets at or near the orbits of the giant planets. And in 1880, George Forbes of Edinburgh used this idea to predict not one but two trans-Neptunian bodies with periods of 1,000 and 5,000 years respectively orbiting at 100 and 500 astronomical units from the sun.[11] Lowell, incidentally, relied on the idea to proclaim the certainty of a planet beyond Neptune in his lectures and book on *The Solar System* in 1902–03, declaring that comets thereby were "finger-posts to planets not yet known."[12]

While Todd, Flammarion, and Forbes, and later W. H. Pickering used more or less empirical methods for predicting planets, H.-E. Lau in Copenhagen and M. A. Gaillot in Paris after 1899 applied the laborious analytical method that led Adams and Leverrier to Neptune in 1845–46 to the problem of a trans-Neptunian planet and came up with two each—Lau's at 46.5 and 71.8 a.u. and Gaillot's at 44 and 66 a.u.[13] This rigorous mathematical solution involves exhaustive computations to find the most probable orbital elements for the suspected exterior planet from the so-called "residuals" of Uranus, that is, the differences between its theoretical and actually observed positions in the sky over a period of years, and thus the amount of perturbations in its motion not accounted for by known gravitational influences. When such residuals are greater than the range of probable observational errors, a reasonable assumption is that the unexplained perturbations are being caused by some undetected body, or bodies.

When early in 1905 Lowell decided to launch a serious search for a trans-Neptunian planet, he elected to attack the problem both observationally, through the systematic comparative photography of likely regions of the sky, and mathematically, through the application of progressively sophisticated methods of celestial mechanics. Eventually, he brought the full rigor of the analytical method of Adams and Leverrier to bear, carrying his calculations well beyond the point to which these two geometers had been obliged to go. For the mathematical problem was now far more difficult than it was in Adams' and Leverrier's day. The inclusion of Neptune's perturbative effect in the computations, for one thing, sharply reduced the values of the Uranian residuals. Lowell himself later noted that in 1845, "the outstanding irregularities in Uranus' motion had reached the relatively large sum of 133."0 [arc seconds]. Today, the residuals do not exceed 4."5 at any point of its path."[14] Furthermore, the residuals of Uranus differed considerably according to the theory of the planet that was used to calculate them. Lowell began his long computations with Leverrier's theory, published in 1876[15] and brought up to 1906 with subsequent observations made at Greenwich. But he later favored the theory published in 1910 by A. Gaillot[16] which, although differing in the values adopted for the masses of Jupiter, Saturn and Neptune, "otherwise perfected Leverrier's theory besides bringing it up to date."[17]

Neptune itself, moreover, was no help in the investigation, for as Lowell also noted the planet had completed only about a third of its 165-year orbit since its discovery and "a disturbed body must have pursued a fairly long path before the effects of perturbation detach themselves from what may be well represented by altering the elements of the disturbed." Nevertheless, he attempted an analysis of the residuals of Neptune and found that they "yielded no rational result."[18]

Once his decision was made to seek a trans-Neptunian planet, Lowell launched his new project with typical vigor. He began in February 1905 by requesting up-to-date charts and photographic surveys of the sky in the region of the ecliptic from his friend Flammarion and Harvard's E. C. Pickering, assuming in accord with his planetological views that an unknown planet, like the known ones, must lie on or very near the "invariable plane."[19] When John Duncan arrived in Flagstaff that summer to initiate the Lawrence Fellowship program, he was assigned the task of systemically photographing the sky along the ecliptic, and this task was continued the following year by the second Lawrence Fellow, young E. C. Slipher.[20] Such photographs were useful for comparison purposes, and of course there was always a chance, however remote, that the sought-after planet might show up on a plate and be recognized for what it was. The photographic survey ended in 1907 without success, however.

But the odds were somewhat better for a mathematical solution, Lowell believed, despite the major assumptions to be made and the uncertainties of both the Uranian residuals and the observational errors in the data. In March he asked Walter S. Harshman of the Naval Observatory's Nautical Almanac office in Washington, D.C., to recommend a capable computer who would be willing to take on the onerous task of calculating the residuals for both Uranus and Neptune. Harshman quickly answered that William T. Carrigan, "one of our most efficient men," had agreed to undertake this chore "outside of office hours."[21]

Over the next ten years, the search for Planet X runs like a seam through the fabric of Lowell's life and work, alternately appearing and disappearing in the warp and woof of his many other interests and activities. Despite repeated frustrations in the search, Lowell never abandoned either his conviction that a Planet X existed or his energetic efforts to find it.

Carrigan went to work immediately, beginning on Lowell's instructions on the residuals of Uranus for the years 1781, when the planet was discovered by William Herschel, through 1820, using observations of the planet made during the period at the Greenwich and Paris observatories.[22] But it was slow tedious work[23] and by January 1907 Lowell began to get impatient for results. "I have been looking over your computations," he wrote Carrigan, "and I find that I have the residuals for the years 1801 to 1810 inclusive and 1814 to 1818 inclusive. Is this all up to date?"[24] A week later, he wondered:

Would it not be possible to get a general preliminary view of the residuals through the century by taking up the data in a less exhaustive manner—the work of the last twenty years especially? I am anxious to have such a preliminary survey as soon as possible.[25]

Lowell was now actively working on the problem himself: "Can you tell me what mass you have been using for Neptune in the calculations for the perturbations of Uranus," he inquired some few days later, "is it 1/19380 of the sun? And do you know what Newcomb's recently deduced value is, and can you give me roughly how it would affect the residuals?"[26] After another week went by, Lowell assured Carrigan that he had "nothing to criticize about your method," but was "curious to see how your residuals for the end of the eighteenth century will come out after your revision. I have, of course, my own theory in the matter," he added.[27]

At this point, Lowell apparently put aside his work on Planet X temporarily to devote his attention to the forthcoming favorable opposition of Mars and the popular furor that accompanied it. But by March 1908 he again was querying Carrigan about the Uranian residuals.[28] In November of that year, his efforts abruptly took on a new urgency, for another hunter had joined the search—his one-time Flagstaff associate and the post-Flagstaff skeptic of martian canals, W. H. Pickering.

Pickering revealed his interest in the problem of a trans-Neptunian planet in a paper before the American Academy of Arts and Sciences, and in a letter to Lowell. "I am looking up the whole subject myself," Lowell replied tersely, "and have not yet got far enough along to undertake any visual search. When I get a position I will let you know."[29]

Pickering's interest not only stimulated Lowell's strong competitive spirit but apparently also provided inspiration in a more practical way. "I was led to take up the search for X again because of a very striking paper . . . by W. H. Pickering the other day before the American Academy," he confided to Carrigan. "The method is as old as Sir John Herschel but I had not supposed the residuals so salient as Pickering found."[30]

Lowell had, at this time, encountered problems with the inconsistency of his data that were destined to plague him for a long time to come. "Are there any tables of residuals geocentric or heliocentric of the planets except those in the Greenwich observations from 1836 to the present time?" he asked Carrigan:

In these, it is unfortunate for my purpose that the theories never continue long in one stay. . . . Now what I want are sets of residuals from one and the same theory for Uranus, Newcomb's computed and then Leverrier's ditto. Is it not possible to let me have these with sufficient accuracy and without a great deal of calculation, by suitable interpolation? The things such sets bring out with regard to perturbation, as Jupiter's and Saturn's show, is [sic] very remarkable. I shall greatly appreciate what you can do for me in these matters and I know you yourself will be interested when you learn the results.[31]

Through the final days of the year and into 1909, Lowell continued to fire questions at Carrigan concerning "discrepancies" in the various orbital theories of Uranus and Neptune and in the results he derived from them, and to request, insistently, more residuals. "I am as full of work as can be," he wrote Lampland in Flagstaff late in December 1908, "searching theoretically for a trans-Neptunian planet; talk and article on Halley's Comet to say nothing of Mars' work."[32] But the discrepancies persisted and in January 1909 he urged Carrigan to take a vacation from his official duties and come to Boston to go over the work. "There are points in the investigation which will interest you," he appealed. "See if you can manage it and I will see to the expenses."[33]

Carrigan's labors, however, soon came to an end, for early in May Lowell advised him that his most recent calculations were "not what I wanted."[34] One week later, he announced in a final letter to Carrigan that "I have completed my investigation (I regret without being able to use any of your computations) and I find evidence of an external planet at 47.5 astr. units from the sun, magnitude 13, heliocentric longitude 287±. My paper will soon be published."[35]

This pronouncement, of course, was premature, for no paper was forthcoming for nearly six years. But Pickering, whose approach to the problem involved a relatively simple graphical method combined with analyses of cometary orbits, published his prediction for a "Planet O" that same year in the Harvard Observatory Annals. "Planet O," the first of at least seven new planets that Pickering postulated over the next twenty years, was 51.9 a.u. from the sun, with a period of 373.5 years and a mass of twice the earth's, he found. Lowell, incidentally, penciled comments and corrections in his copy of Pickering's paper, noting at the end that "this planet is very properly designated O [and] is nothing at all."[36]

Lowell apparently was not fully satisfied with his own conclusions, and in July he sought new data on residuals of Uranus and Neptune for 1908–09 from England's A. C. D. Crommelin, explain-

Halley's Comet, Venus, and meteor trail just above head of comet. The photo was taken by Lampland in 1910.

ing that "I have been investigating the matter of a trans-Neptunian planet and have a solution by variation of the elements of Uranus arising from such perturbations—I assume the radius vector of X from other indications—which when compared with the outstanding residuals seems to imply such a planet and perhaps its place."[37]

The search, clearly, was far from over, but while Carrigan was no longer a participant, Lowell now enlisted an attractive young mathematician named Elizabeth Williams and a staff of computers which eventually reached four in number to work on the project in his Boston office. In a letter to Lampland early in December 1910, he reported on the status of his continuing search:

Miss Williams and I have been pegging away at it, have constructed the curve of perturbations due X if at 47.5 astr. units including all terms of the first power of both eccentricities and examining the most important terms in their squares. We find some interesting things ... but we do not find the planet for except in periods the theoretical curve swears as the residual one. It is of course possible that Leverrier's theory is not sufficiently exact.[38]

Not until March 13, 1911—on his 56th birthday—did Lowell give the order to begin a new photographic search for Planet X. "Please begin to photograph region ecliptic where south with 40-inch," he telegraphed Lampland that day. "Hope to wire position in a few days. Calculations tremendously long."[39] Miss Leonard added some details soon after in a chatty letter:

Dr. Lowell has no objection to your going in for the hunt there as long as he tells you where to look from this end. It has been a long computation and Dr. Lowell and Miss Williams have been faithful to it all winter. We feel rather heavy with it but *when* it is found it will brighten our hearts and make us glad. I am sure you appreciate what the work has meant, and now your part in it will soon follow.[40]

But no strange planet appeared on Lampland's first plates, and late in April, Lowell, still busy with his calculations, sent new instructions:

I telegraphed you last week to look in heliocentric longitude 235°. Our residuals, from necessities in the case very uncertain, seem to indicate for Planet X a heliocentric longitude of 239° for it. Please take plates in neighborhood for 2° or 3° on either side of the fundamental plane and devise best method of comparing them for the stranger.[41]

Still X remained an unknown, and by July Lowell was hard at work on more computations, using residuals from the newly published theory of Gaillot. Early that month he wired Lampland that the results now "indicate that search should be carried further east. ... Don't hesitate to wire anything by day letter."[42] Only days later, however, he corrected himself, telegraphing V. M. Slipher that "Lampland should photograph west not east with a probable longitude about two hundred ten."[43] Late in July Lowell discovered a "suspicious object" retrograding on a set of the latest plates and ordered a further observational check, but it apparently was only an asteroid.[44]

For the next eleven months, Planet X vanished temporarily from Lowell's correspondence to appear again early in June 1912 with a telegraphic query to Lampland concerning the limiting magnitude of his X plates and a request to "make the plates of the same density no matter what the exposure."[45] The search was on again: "Have not received X plates for long time," he wired the elder Slipher on June 12.[46] To Lampland, he wrote:

I have had no Planet X plates in a long time. ... It is important that they should be taken at once and received at once so that if anything is found it can be followed up. The plates of last year that I have examined do not need to be duplicated. I think I should go more south than north of the fundamental plane. I have found nothing so far which cannot be interpreted as variable and what strikes me as very odd: I have run across but one asteroid. In a short time we shall not be able to take these plates at all as the motion of X would not be apparent. The pair of plates should only be a day apart if possible.[47]

A steady stream of communications now began to flow from Boston to Flagstaff that was to continue for the next three years.

"Better take plates two or three degrees south of the ecliptic ... because of greater chance of X having there been overlooked," he wired on June 14. On June 20 he telegraphed a probable longitude between 239° and 243°, adding that he had found a variable double star on a set of the X plates and asking Lampland to notify him "if not already found." On July 10 he directed Lampland to "photograph east as far as longitude two-sixty" and advised him that Duncan, who had obtained his Ph.D. degree in 1909 and was now at Harvard, would return briefly to Flagstaff to help in the observational search.[48]

Lowell had also resumed his intensive calculations with Miss Williams and an augmented staff. "That you have not heard from me about Planet X is because we are hard at work on the latitude," he wrote in August. "We are also at work on a more precise longitude determination. Do the best you can … till further notice."[49] But some two weeks later, a note of discouragement began to appear in his communications:

I am very sorry—first that the planet should now lie so near the sun due to being near its aphelion*—this sounds contradictory but it is not—and secondly that the latitude determination is not yet finished. In consequence, I have delayed sending you any instructions as to where to look. For I feel convinced that the inclination is considerable, both because of the size of the residuals in latitude and because of the great eccentricity. … [50]

Early in September he telegraphed Lampland: "Solution Uranus residuals in latitude including X give considerably smaller residual squares than Gaillot's solution without it," but added: Inclination uncertain owing to necessary uncertainty of mass."[51] A few days later he seems to have decided to rework the whole problem:

We are revising and extending our work on Planet X, results of which we are, unfortunately, not yet in a position to communicate to you. Every new move takes weeks in the doing.[52]

Each move did, in fact, require long laborious computations, and now a new factor conspired to slow progress. For Lowell became ill again in October, apparently a recurrence of the neurasthenia which had incapacitated him from 1897 to 1900. This time, however, the attack, accompanied by painful shingles, was far less enervating to his mind at least, and he was able to continue some direction over the planet X work from his home. Indeed, activity at his Boston office now reached its peak. "We have four assistants here now plus the old standbys," Miss Leonard reported to the elder Slipher in Flagstaff early in November. "It is a very busy place!"[53]

Both Miss Leonard and Mrs. Lowell kept the observatory staff informed of Lowell's progress. "Mr. Lowell is improving," the latter wrote to Slipher on November 8, "and looks and acts much better."[54] But a month later, Miss Leonard noted that Lowell had been absent from his desk for seven weeks. "It is nervous exhaustion," she advised Slipher, "and he is *up* and *down*. Some days he cannot even telephone. He gets nervous about the work and impatient for things to come from Flagstaff."[55]

*The point in an orbit farthest from the sun.

Although Lowell did not completely recover from the effects of this illness for nearly a year,[56] he was back in his work routine within a few months, and once again telegrams began to flow to Flagstaff. On January 22, 1913, he wired Lampland two possible positions for X, urging him to rush his observations as the planet was "getting in bad position," and adding that because X's magnitude was then "very possibly 10 or 11," five minute exposures should be sufficient.[57] A week later, Lowell informed him that while latitude was still an uncertain quantity, a figure of 7° north of the ecliptic appeared "most probable."[58] On February 3 and again on the 5th he telegraphed new data, putting its inclination at 2°30′, again noting that the figures were uncertain.[59] Finally on February 8 he began a long letter to Lampland with the announcement: "Here are our new elements of a new planet." Then he listed the full results of his latest computations with corrections for precession and date, and added:

You will be interested to know that by solving for other distances as well, my original supposition of 47.5 [a.u.] turns out almost exactly right. ... Please note everything interesting as you may catch other fish, and notify *us* of the haul. Good luck![60]

But now exhaustion and illness overtook Lampland in turn and Lowell urged his reluctant assistant to stop all work and take a long rest. "Now you *must* do as I did: go off and get well. I cannot afford to loose [sic] you. The X work can be carried on by Dr. Slipher with Mr. Slipher till your return."[61] A few days later Lowell decided that a brief sojourn at his observatory might be in order, wiring his Flagstaff confidant Edward M. Doe to "expect the worst blizzard of the season: Me—by Limited March 8."[62]

Back in his Boston office in May, Lowell sought data on the most recent residuals of Uranus from Gaillot, and in July he wired Lampland: "Generally speaking, what fields have you taken? Is there nothing suspicious? Not yet here in position to give X latitude nor even longitude any better as we have been solving including the second power of the eccentricities."[63] Even while Lowell vacationed briefly at his sister Katherine's home in Maine that summer, new longitude estimates were flashed to Flagstaff—232° on July 24, for example, 245° on August 4, and 239° and 241° on August 21.[64] "Use these," he directed on this late date. "Suspect inclination large, probably south. Am presently still on the retired list. Await another excitement proving true. Any news grateful."[65]

Lowell made another visit to Flagstaff in December, remaining there until the spring of 1914 while his staff of computers continued their labors in Boston, forwarding results to him from time to time.[66] He was back east in April to leave on what would be his last European trip, although his departure was delayed until May by his wife's surgery for an ulcer.[67] Before sailing, he sent new instructions for Lampland: "Most probable place for X now seems to be: hel. long. 261° and perhaps a little north of ecliptic. . . . Don't hesitate to startle me with a telegram—'FOUND.' "[68]

He returned early in August, by way of Halifax incidentally, for his ship, the British liner *Mauretania,* had been ordered to run for that Nova Scotian port with the outbreak of World War I.[69] Once he reached home, however, the elusive Planet X again became the first order of business, and indeed the long wearisome search would soon come to a climax. In mid-August he asked Lampland to bring him up to date on the photographic surveys. "I should like some rough statement of the fields covered for X; merely to be sure that they connect with the old ones and also that they extend eastward well beyond the most probable 262°," he explained.[70] To the elder Slipher he confided that "I feel sadly, of course, that nothing has been reported about X, but I suppose the bad weather and Mrs. Lampland's condition may somewhat explain it."[71] In September he was back in Flagstaff to supervise the work from there, staying on Mars Hill well into November.

But X remained hidden, and by December Lowell decided that the time had come to announce at least the theoretical and mathematical aspects of his search. Thus, on December 21, in forwarding new longitudes for X of 84° and 88°, and a latitude of "?," he informed Lampland that "I am giving my work before the Academy on January 13," suggesting wryly that "it would be thoughtful of you to announce the actual discovery at the same time."[72]

The January 1915 meeting of the American Academy of Arts and Sciences was apparently a frustrating one for Lowell in more ways than one. For although Lampland was elected to membership,[73] he had no announcement to make, and Lowell's Planet X paper itself caused no great stir, the Academy not even accepting it for publication. Moreover Lowell embroiled himself in a petty controversy with the Academy's president John Trowbridge over the membership of a committee that was to award a gold medal which Lowell was providing in recognition of outstanding work in what he considered to be traditional astronomy.[74] Even a month after the meeting, Lowell was still furious. "The stagnation and old fogyism in consequence of which the Academy is run by a set of men certainly not broad [of] view or judgement," he fumed to an acquaintance at the Massachusetts Institute of Technology, "is greatly to be deplored and ought to be changed."[75]

Lowell published his Planet X paper as his observatory's first *Memoir* that spring.[76] In this handsome, 105-page volume he reviewed his method, displayed the results of his voluminous calculations in numerous tables, and finally summed up his conclusions. In only two of them, however, did he express any confidence, and both were incidental to the main problem. The first related to the fact that the only element of Leverrier's and Adams' proposed planet that turned out to agree at all with the actual elements of Neptune was the longitude at epoch, and thus its position in the sky:

A point in passing which these results serve to establish is the fallacy that though the elements of a disturbing unknown may not be found, the direction of that unknown can be discovered. The excuse for the statement seems to have been a mistake of Sir John Herschel's. ... Sir John asserted that *Neptune* was found at the best time for its detection because its residuals from having reached a maximum had begun to rapidly decrease and that this showed it, the then unknown, to have been in conjunction with *Uranus* on or about 1822, such being the date of the maximum disturbance. Because at conjunction the tangential component of the perturbing force changes sign, he argued that the excess of longitudes must there be greatest. Not so. The maximum deviation is not reached until the perturbing force or component after its reversal has had time to annul its previous effect on the velocity, a time long after its own reversal.[77]

The other point his investigation established, Lowell noted in confutation of his old math professor Peirce, "is the continuity of the solution in spite of the commensurability period positions through which it may pass. The fact was not understood at the time of the discovery of *Neptune* and the controversy to which that discovery gave rise."[78]

For in his computations, he worked out solutions for five different values for a', the mean distance from the sun, ranging from 51.25 a.u. sunward to 40.5 a.u., and passing through the commensurable point at 48.4 a.u.:

The outcome is conclusive. From charts constructed from the several values calculated, it is evident that there is no discontinuity in the results at the points of commensurable period, the function passing through them without a break. Thus the *a priori* opinion at the time of the *Neptune* controversy that such points barred transit from one region of space to another proves to be unfounded. .... [79]

His other conclusions, however, were preceded with strong qualifications:

Owing to the inexactitude of our data, we cannot regard our results with the complacency of completeness we should like. Just as Lagrange and Laplace believed they had proved the eternal stability of our system, and just as further study has shown this confidence to have been misplaced; so the fine definiteness of positioning an unknown by the bold analysis of Leverrier or Adams appears in the light of subsequent research to be only possible under certain circumstances. Analytics thought to promise the precision of a rifle and finds it must rely upon the promiscuity of a shot gun after all, though the fatal fault lies not more in the weapon than in the uncertain bases on which it rests. But to learn of the general solution and the limitations of a problem is really as instructive and important as if it permitted specifically of exact prediction.
For that, too, means advance. [80]

Lowell then noted that all his various solutions, reached "by the most rigorous method, that of least squares throughout," were substantially in agreement. With the admission of an unknown perturbing body into his equations, the outstanding squares of the residuals of Uranus were reduced between 90 to nearly 100 percent, depending on the number of observational equations used in the computation. But "though this would indicate an absolute solution to the problem," he cautioned, "it must be remembered that the actual against the probable errors of observation might decidedly alter the result. ...."

A slight increase of the actual errors over the most probable ones, such as it by no means strains human capacity for error to suppose, would suffice entirely to change the most probable distance of the disturber and its longitude at epoch. Indeed the imposing 'probable error' of a set of observations imposes on no one familiar with observation, the actual errors committed, due to systematic causes, always far exceeding it. [81]

Lowell then presented two sets of elements for his Planet X with longitudes around 0° and 180°, and declared these to be "on the whole the best solutions...." [82] It will be noted that his "original assumption" of 47.5 a.u. for a' did not turn out to be "exactly right" after all, and that he finally reduced X's distance because his least squares checks indicated that the lesser values were the most probable. His elements were:

For X$_1$		For X$_2$
$E' = 22°.1$	Longitude at epoch	$E' = 205°.0$
$a' = 43.0$	Mean distance from sun	$a' = 44.7$
$e' = 0.202$	Eccentricity of orbit	$e' = 0.195$
$\overline{w}' = 203°.8$	Longitude at perihelion*	$\overline{w}' = 19°.6$
$m' = 1.00$	Mass	$m' = 1.14$
	(Unit = 1/50,000 of sun)	

The heliocentric longitude of the unknown for July 0, 1914, that is, its direction from the sun on that date, he noted, was 84° for X$_1$ and 262°.8 for X$_2$. The result, he added, "indicates for the unknown a mass between *Neptune's* and the *Earth's;* a visibility of 12-13 magnitude according to albedo; and a disk of more than 1″ [arc second] in diameter." His investigations of the unknown's latitude had "yielded no trustworthy results," but by analogy with the known planets "in which eccentricity and inclination are usually correlated" he predicted its orbit would be inclined to the ecliptic by "about 10°," noting that "this renders it more difficult to find." And his final conclusion was "that when an unknown is so far removed relatively from the planet it perturbs, precise prediction of its place does not seem possible. A general direction alone is predicable."[83]

With the publication of his Planet X memoir, the subject virtually vanishes from the correspondence in the observatory archives. Only a single letter in mid-summer of 1915 inquires of Lampland: "No news from X?" and suggests a new longitude of 264° and a latitude of plus-or-minus 10° from the ecliptic.[84] After Lowell's death in November 1916 the search was entirely suspended for thirteen years.

But then on March 13, 1930, the seventy-fifth anniversary of Lowell's birth and fourteen years after his death, V. M. Slipher announced that an object, whose "position and distance appear to fit only those of an object beyond Neptune, and one apparently fulfilling Lowell's theoretical findings," had been discovered at Flagstaff seven weeks earlier. The new planet had been found by 23-year-old Clyde W. Tombaugh while scanning a series of photographic plates of the sky near the star Delta Geminorum with a blink microscope. The plates had been exposed on the nights of January

---

*Perihelion is the point in a planet's orbit that is nearest the sun.

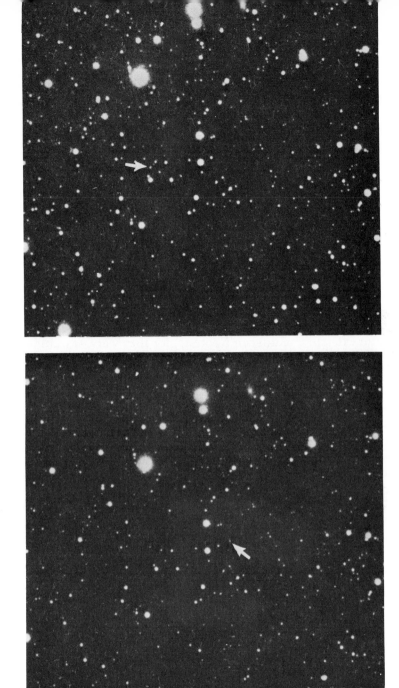

Lowell Observatory photographs (top: Jan. 23; bottom, Jan. 29, 1930) showing sections of the plates from which Tombaugh discovered the planet Pluto on February 18, 1930.

21, 23, and 29, 1930, with the observatory's new 13-inch Lawrence Lowell telescope, specially built and installed a year earlier for the reactivated search for Lowell's Planet X. [85]

The discovery was reported on the front pages of newspapers all over the world and moved *The New York Times* to quote lines from Keats' "On First Looking Into Chapman's Homer."

> Then I felt like some watcher of the skies
> When a new planet swims into his ken.

The English poet, the paper noted, "was doubtless thinking of Herschel. ... "

But the experience of these Arizona watchers who actually saw this "Planet X" swim into sight was rather more like that of Leverrier in France and Adams in England who predicted by their reckonings the existence of unseen Neptune, for they were searching for a wanderer which he who had founded the observatory knew to be there . . . . [86]

The *Times* also had a suggestion for naming the new planet:

One would like to have Lowell's name associated with this new-found planet, as Leverrier's was at first with Neptune. ... If another Olympian name is to be given, why not Minerva—the planet that sprang to human view full panoplied from the mind of man? [87]

The planet was named Pluto, of course, the first two letters of the name as well as its planetary symbol of the superimposed letters "P" and "L" standing for the initials of Percival Lowell's name.

Pluto had, in fact, been discovered only about 6° from the position given by Lowell's elements for Planet X_1—not as close as Neptune had been to Leverrier's and Adams' predicted locations, but nevertheless certainly indicating "a general direction." Moreover, taken all together, Lowell's X_1 elements were in far better agreement with Pluto's actually observed orbit than Leverrier's or Adams' elements were with those of Neptune. This circumstance quickly prompted leading mathematical astronomers of the day to make "a fresh examination" of Lowell's work, as Yale University's respected Ernest W. Brown put it. Brown himself reanalyzed Lowell's calculations and concluded that Pluto's discovery—shades of the Neptune controversy!—was "purely accidental." [88]

Brown first pointed out that oscillations in the Uranian residuals "seem to have periods too short for an explanation on the basis of

The A. Lawrence Lowell 13-inch photographic refractor with which Pluto was discovered.

an exterior planet, and neither of the two hypothesized planets of Lowell seem to account for them." It was also an "accidental circumstance" that because "the interval used is near the synodic periods of the two planets," there "must be approximate symmetry about the middle of the interval" which "gives at once a fictitious longitude at epoch and longitude of perihelion which are almost exactly those of Lowell's hypothetical planets." Again it appeared "that the actual values he obtains for distance, mass, and eccentricity substantially depend on three groups of observations made before 1783, having large probable errors." Furthermore, the residuals of Uranus since the last observation used by Lowell "bear no resemblence to those which either of his solutions require. . . . The information concerning the newly found planet at this moment is scanty, but it appears sufficient to prove that it could not have been predicted from its effect on Uranus."[89]

Brown, however, conceded that Lowell had carried out his computations "in great detail and apparently with high accuracy," and added:

Original dome of the 13-inch Lawrence Lowell refractor

It is unfortunate that, if my analysis is correct, so much careful and laborious work can lead to no result. However, in so far as it has stimulated a search for an outer planet which has proved success-ful, one cannot regret its completion and publication. Perhaps there are other planets beyond the orbit of Neptune, with masses similar to that of the prediction or smaller, awaiting discovery.[90]

To this, astronomer Henry Norris Russell, with specific refer-ence to the "large probable errors" in the pre-1783 Uranian observa-tions, added the comment:

The question arises, if Percival Lowell's results were vitiated in this way by errors made by others more than a century before his birth, why is there an actual planet moving in an orbit which is so uncan-nily like the one he predicted?
    There seems no escape from the conclusion that this is a matter of chance. That so close a set of chance coincidences should occur is almost incredible, but the evidence assembled by Brown permits no other conclusion.[91]

Lick Observatory astronomer Ernest Clare Bower also noted "the extraordinary agreement between the elements of X_1 and of Pluto," adding that the circumstance "leads a few astronomers to accept Lowell's work as a trustworthy prediction with merely an overestimate of the mass." And he presented the elements of Low-ell's X_1 and X_2 and of Pluto side-by-side to show the degree of their agreement:[92]

	X_1	X_2	Pluto
a' (Mean distance)	43.0	44.7	39.5
e' (Eccentricity)	0.202	0.195	0.248
i' (Inclination)	$\pm10°$	$\pm10°$	17°
$\bar{w}'$ (Perihelion)	203°.8	19°.6	222°.7
Longitude July 1914	84°	262°.8	90°.4
Magnitude	12-13	12-13	14-14.5
Apparent diameter	>1"	>1"	<4"
Mass (Earth = 1)	6.6	7.5	0.7

As can be readily seen from this, given the many assumptions and uncertainties the problem involves, Lowell's Planet X_1 ele-ments are indeed fairly close to the initial elements found for Pluto, deviating most seriously in the case of the mass. But as Bower pointed out, Pluto's mass had not yet been determined satisfacto-rily, although from its magnitude and the "best assumptions" for its albedo and density, he felt an upper limit of no more than 0.3 of the earth's mass was indicated. "Until a disk is actually seen," he added, "the most probable value of Pluto's mass may be taken to be about 0.1 of Earth's"—an estimate that is remarkably accurate in view of the most recent modern value of 0.11 of earth's mass, computed in 1971.[93]

Others, notably Crommelin in England and later V. Kourganoff in France, defended the validity of Lowell's Planet X prediction; even W. H. Pickering, one of whose Planets O was a brief rival to Lowell's X in the controversy, conceded that Lowell had been surprisingly successful. The details of this long-continued debate, however, are beyond the scope of this present study, and it must suffice to say that the question of whether Pluto really is Planet X has not yet been settled satisfactorily for many astronomers and, possibly, never will be.

The discovery of Pluto turned up a number of oddities and ironies. Lowell certainly had been on the right track in his oft-stated belief that both the eccentricity of X's orbit and its inclination to the ecliptic were considerable, although his values of 0.202 and "about 10°" for these elements still fell short of the actual values of 0.246 and 17°.1, as most recently determined in 1967.[94] Pluto's eccentricity, in fact, is such that at its perihelion passage— which Lowell, incidentally, predicted for 1991 and which is now expected in 1989—the planet actually swings inside of the orbit of Neptune, thus temporarily forfeiting its trans-Neptunian status. W. H. Pickering, interestingly enough, suggested as much in 1928 in another of his predictions relating to what he preferred to call "Planet O."[95]

The high inclination of Pluto's orbit, too, provided a small irony in retrospect, as Tombaugh later pointed out. For in Lowell's initial photographic survey of the invariable plane in 1905–07, while the three-hour exposures then used were sufficient to record the planet, its high inclination placed it just outside the camera's field. And during Lampland's later searches, when exposures as short as five minutes were sometimes used, with larger telescopes, Pluto was in a remote segment of its orbit and fainter by half a magnitude than at the time of its discovery. Thus his early attempts to find X had failed.[96]

But there were greater ironies than this. For Pluto, as it turns out, had not only been recorded photographically during Lowell's lifetime, but twice—on March 19 and again on April 7, 1915—had been recorded on plates exposed at the Lowell Observatory.[97]

Post-Pluto staff of the Lowell Observatory. From the left are V. M. Slipher, C. O. Lampland, C. W. Tombaugh, E. C. Slipher, and (seated) observatory assistants Allen Cree and Kenneth Neuman. Not shown is Henry L. Giclas, who took the photograph.

In fact, some sixteen prediscovery observations of the planet have since been found on photographic plates from various observatories, including three in 1914 for a total of five prior to Lowell's death, four in 1919, two in 1921, 1925 and 1927, and one in 1929.[98] The 1919 observations, incidentally, were made at the Mt. Wilson Observatory at the request of W. H. Pickering who that year had again predicted a position for a trans-Neptunian planet, this time on the basis of his study of the residuals of Neptune. The faint image of Pluto on the plates, however, was not recognized for what it was, and thus fame passed Pickering by.[99]

Still another irony perhaps, as Bower pointed out, was the fact that Pluto's orbit at 39.5 a.u. is nearly commensurate with that of Uranus at a ratio of 3:1 and with that of Neptune at a ratio of 3:2, and falls just outside the lower limit of 40.5 a.u. to which Lowell had carried his computations.[100]

But if the discovery of Pluto, like the discovery of Neptune, is to be considered an "accident," Lowell, like Leverrier and Adams, deserves and indeed has been given a share of the credit for finding the new planet. Tombaugh, the actual discoverer, has pointed out that:

Some doubt has arisen as to whether Pluto is really Dr. Lowell's predicted "Planet X" because of the apparent smallness of its mass. On the other hand, Pluto was found within 6° of his predicted place, and the predicted and actual elements of the planet's orbit are in remarkably good agreement.[101]

Russell perhaps has summed up as well as anyone the consensus of opinion concerning Lowell's role in planetary discovery:

In any event, the initial credit for the discovery of Pluto justly belongs to Percival Lowell. His analytical methods were sound; his profound enthusiasm stimulated the search; and, even after his death, was the inspiration of the campaign which resulted in its discovery at the Observatory which he had founded.[102]

The discovery of Pluto, then, represents a triumph, albeit a posthumous one, of Lowell's intellectual vision and tenacity of mind. But that "Planet X" was not found while he yet lived, his brother and biographer A. Lawrence Lowell has noted, "was the sharpest disappointment of his life."[103]

CHAPTER
15

Lowell and Mars

THE FAILURE of his Mars theory to win general acceptance in the scientific world of his own time was also a disappointment for Percival Lowell during the final years of his life—years that were marked with other disappointments, large and small.

During those years he fought his battles with no less gusto than before, and his fame as an astronomical authority, at least in the eyes of an uncritical public, remained largely intact. But times and tides were changing, and while he now increasingly berated his critics in science for what he considered to be their blind unreasoning refusal to go beyond traditional and orthodox modes of thought, he was no less tenacious in clinging to his own ideas and beliefs, although these now progressively came under an intensified attack on a broad front.

Even the political and social trends of the day were not all to his liking, and he watched the advances of socialism in Europe and of progressivism in the United States, even in his own cherished Republican Party, with a mounting concern. Early in 1911 the narrow margin of the reelection of his friend Henry Cabot Lodge to the U.S. Senate in Massachusetts prompted him to remark ruefully to his Flagstaff crony Edward M. Doe: "I cannot help thinking that the country is bound on a long, political debauch."[1] The schism in Republican ranks brought on by Teddy Roosevelt's progressivist "Bull Moose" defection, and the resulting election triumph of Democrat Woodrow Wilson and his "New Freedom" in 1912, would confirm this foreboding for Lowell and other Republican conservatives.

Lowell's lively interest in politics surfaces only occasionally in the Lowell Observatory Archives and then more often in the later years. For the most part, these references are brief passing comments occurring in correspondence dealing primarily with other matters, but they nonetheless reveal his attitudes on some of the main issues of his time. He did not, for instance, look with particular favor on women's suffrage;[2] he was against unrestricted immigration,[3] and he strongly opposed prohibition, considering it a "violation of the rights of citizens" and "a curtailment of our freedom."[4] He was, in turn, in favor of capital punishment, assuring the warden of the Arizona State Prison, who had asked for his views on the subject, that "it is both deterrent and prevents criminal propagation."[5]

Lowell had opposed statehood for the Arizona Territory. His activities in this connection show up in the Archives in late 1905 and early 1906 when he sent $200 to Doe "to be used for the Anti-Statehood League,"[6] conferred with Senator Lodge on the attitude of Congress on statehood, and sought to coerce Flagstaff's *Coconino Sun* to abandon its pro-statehood editorial position by putting economic pressure on editor C. M. Funston, who was under a financial obligation to Lowell.[7]

At this same time, parenthetically, Lowell briefly involved himself in a proposal to build a tourist hotel in Flagstaff and a tramway to the top of the 12,600-foot San Francisco Peaks north of the town, remarking in regard to this latter project that "I dislike, of course, to spoil the virgin simplicity of our beloved arid zone! But I think it will be a good thing for Flagstaff."[8]

Lowell maintained his political interests side by side with his astronomical ones and could, for instance, lecture on "Astronomy Today" at the Harvard Union and on "Aspects of Socialism" at Boston's Victorian Club within the same week with equal facility and to the apparent delight of both audiences.[9] Occasionally he mixed his politics directly with his astronomy. A particularly interesting example of this is his October 1911 lecture, "Two Stars," at Kingman, Arizona Territory, where his audience, rather than members of an affluent intelligentsia, was composed largely of hard-rock miners. Here his text was astronomical but his lesson was political, and here, incidentally, he drew sociological implications from his Mars theory that were quite different, although not less utopian, than those Lester Frank Ward and some others had found in it earlier.

Far from being democratic in any way, martian society as envisioned by Lowell was a benevolent oligarchy of the intellectual elite in which each individual martian was assigned his place and was content to remain in that place without questioning the judgment or wisdom of his superiors. On Mars, the dearth of water "necessitates a community of interest under penalty of death," he declared, and consequently the martians must have abolished such terrestrial institutions as nations and war. "Undoubtedly a uniform government controls the economic activities of the whole planet," and "a single interest sways the whole body politic." But:

If oneness of purpose is vital to Mars, it is no less imperative for us. In disunion lies inefficiency. We shall never pull our strongest until we pull together. At present we are engaged in exactly the opposite endeavor, ignorance being set in killing capital and self-crippling labor as if the two were not inseparably linked and to hurt both were to help either. Understanding is a plant of slow growth, and the average man is not far-sighted enough to foresee the end from the beginning. The dupe of catchwords, he calls progressive what Mars stamps exactly the reverse. . . . [10]

The study of Mars, he continued, "teaches us that well-being lies not in strife but in mutual interaction."

For rightly read its rede tells not only of a well-ordered community but of how that unity can alone be fashioned and guaranteed. Think of the intelligence and far-contrivance necessary to execute and maintain a system of irrigation worldwide in its scope and meticulously dependent upon the seasons for its functional activity. Every drop of water is precious. . . . The very ablest intelligences that Mars can produce alone are fitted to conceive and direct so universal and vital a matter. Such only can be in command. On the other hand, each member of such a community must equally carry out his part to the utmost of his capability. . . . From top to bottom each individual has his place and fills it. . . . To know one's place is the best qualification. . . . All are given a chance to rise, but only the worthy do. We may be very certain that in the Martian world-economy, the fittest only have survived. [11]

But while intelligence and ability are the prime leadership virtues for martians, ignorance and ambition dominate terrestrial polities:

Our earthly communities now suffer more or less from . . . a sort of political St. Vitus dance, presided over by persons who would do excellently perhaps in subordinate positions but who are capable only of comicality or worse when raised to supreme command, their aspirations swelling every day due to the semi-education to which they appeal. . . . If in masses we judged the demagogue as we do when we are alone, the demagogue would become the dodo, an extinct species.[12]

Lowell, at least, considered this lecture a small triumph. "Certainly I made some friends, even among the Socialist miners which was my aim," he wrote Miss Leonard some days later from Flagstaff:

One of them whose views were quite subversive, now loves me—to my immense surprise. ... I had shown how the solidarity of the Martian canal system points to an efficient government in which the best men are at the front and then I went on to show its applicability to us.[13]

Politics and other matters notwithstanding, Lowell's astronomical interests, and the problems that increasingly beset them in the years after the martian furor reached its peak, continued to hold top priority. The year 1907 not only marks the turning point of the controversy over Mars but of Lowell's astronomical career. For where before he had aggressively proclaimed his theories and confidently brushed aside the skepticism they stirred, he was now everywhere on the defensive, reacting petulantly and even picayunishly to a steadily increasing number of specific criticisms which, moreover, now came not only from within astronomy but from other scientific disciplines as well. At the same time he found it more and more difficult to get his ideas or even the results of his continuing observations before the public or to other astronomers. Several incidents are both symptomatic and illustrative of the problems he now began to encounter.

The first of these had its beginnings in 1907 when E. C. Pickering of Harvard College Observatory, a clearing house for astronomical information, had refused to distribute some of Lowell's telegraphic bulletins about his martian observations. Early in 1909, as another quite favorable opposition of Mars loomed, Lowell moved to set up his own distribution system for planetary news, arranging to receive and disseminate such data in America through his Flagstaff observatory and in Europe through Dr. H. Kobold, editor of *Astronomische Nachrichten,* in Kiel, Germany.[14] When Pickering ignored the new service, however, a furious Lowell called on his brother, newly installed as Harvard's president, to get Harvard's observatory director out of the planetary news business. "E. C. Pickering," he wrote A. Lawrence peevishly, "got himself made the distributing center for astronomic telegrams in America for subscribing observatories ... in a not very creditable manner (ask John Ritchie, Jr.). I was one of the subscribers."

Two years ago he wrote me a letter refusing to transmit any of my telegrams about Mars on the ground that the other subscribers were not interested—Today I receive notice of such a telegram on Mars from [W. W.] Campbell of the Lick and in regular course to all the subscribers. Meanwhile, I have been obliged to start such an inter-communication on my own account with Kiel, mine covering the planets. This observatory was thus discriminated against two years ago and now his institution is poaching on its preserves.

Now I want Pickering to tell Campbell that such telegrams should be sent here.[15]

Director Pickering apparently got the message from President Lowell for in November astronomer Lowell wrote Kobold that their distribution system had been a great success:

To my surprise the greatest possible interest has been aroused by it in the press and elsewhere. After the arrangement had been con-cluded with you, Pickering wrote to me asking permission to take on the distribution in America—which of course I refused. You may like to know what a sensation this much-needed change has caused and how much the general interest in Mars is increasing.[16]

But Lowell himself, as it turned out, was not immune from the charge of discrimination in the dissemination of astronomical news. As Halley's Comet passed through perihelion in May 1910, Ed-mund Stover of the *Associated Press* in Los Angeles complained to Lowell that his observatory was discriminating against the big news-gathering agency:

F. S. Breen has resigned as AP correspondent at Flagstaff because of seeming failure to obtain the usual press courtesies at the Lowell Observatory.... Enclosed find clipping from a Los Ange-les paper, indicating that the embargo upon Lowell Observatory news is laid against the Associated Press only.[17]

Lowell was more than willing to provide news of his observa-tory's work to the *Associated Press*, but he was insistent that such news should appear precisely as given, without relevant back-ground material or comment to put it in perspective. Conse-quently he advised Stover:

This observatory is willing to furnish such news provided that it ... is printed verbatim which has not been the case in the past, and that you will understand that there are times when its whole staff is too busy to give attention to newspaper requests. And also that what it sends must be given a place to itself and not be associated with or made a part of similar news from other sources.[18]

Stover replied with some terse advice of his own:

I hope that you will understand that the Associated Press can only deliver the news to the daily press of the country, and cannot exercise control of their handling of it thereafter. Permit me, sir, to suggest that brevity safeguards a dispatch against blue-pencil mutilation in the newspaper offices.[19]

The advent of Halley's Comet marked one of the relatively rare instances in the post-Mars furor years when the press took the initiative in seeking out news of Lowell's astronomical activities, in sharp contrast to the active media interest up through 1907. The comet's return after its 76-year journey to the outer reaches of the solar system was greeted by sensational publicity, reminiscent of that given Mars only a few years before. As the earth was due to pass through a portion of the comet's tenuous, millions-of-miles long tail, believed to contain poisonous gases, there were some fearsome forecasts concerning the possible mass asphyxiation of all terrestrial life to assure public excitement.[20] Lowell did not contribute to such speculations, however, and in his articles and his lectures on Halley's Comet in particular and on comets in general, he confined himself to reviewing known facts, and avoided sensation. His reputation as an astronomical authority, however, inevitably brought a flurry of letters to his observatory that reflect the popular curiosity and concern over the event. "I am totally deaf and interested in the comet," a Norward, Colorado, man wrote, for example. "Please inform me of how I can get photographs, information and Data of Halley's Comet. ... and you will oblige an old soldier."[21] A Baptist minister in Woodford, Oklahoma, had more urgent inquiries:

Will you please inform me as to what time it will be nearest to the earth? If it strikes the earth will it be destructive to this earth? On what day will it strike the earth? How far does it travel per hour? How large is the comet? How long is the tail of it? Is it solid or is it just a shell? What city is it nearest now, I mean over what part of the country is it now moving?[22]

One such correspondent, an erstwhile commentator on the international political scene, garrulously admitted that he was "sceert."

I saw in the paper that 7 comets will come on this year, and I am getting sceert. I have heard you have good glasses, a clear sky and a good man at the head, and if you will be so good as to mail me the address to where I can get to find out what to expect from these comets and not be fooled with bad reports, as we do in the newspapers.... One paper said the comet we saw this winter was "coming round the sun"—I hope the Old Sun will make it hot for some of the 7 wise men who surely know what's coming and won't tell the American people, so's to save them some unnecessary work. I've been sceert ever since I am [sic] took that Panama job; good job for John Bull, just like the Suez Canal. Hans and John Krapo are getting in shape to knock him out, if those seven visitors don't do us all up.[23]

Lowell and his observatory staff probably chuckled over such letters, but their reaction was somber two months later when the mail brought news of the death of Schiaparelli in Milan. Lowell, who had not only lost an ally in his battle to establish the reality of the martian canals but a friend, sadly agreed to write a memoir on the Italian astronomer's life and work for *Popular Astronomy*.[24]

Lowell's comparatively subdued activities during the brief public frenzy over Halley's Comet may have been a result of the sharp new attacks on his martian theory that followed the favorable opposition of Mars in September 1909. As has been noted, during this opposition E. M. Antoniadi at Meudon and George Ellery Hale and his associates on Mt. Wilson, among others, had turned larger and more powerful telescopes than were available to Lowell toward the red planet and had unanimously reported that the "canals" so vital to Lowell's logical scheme had been dissolved into a complex of spots and blotches.

To counter these well-publicized observations, Lowell could only reiterate what he had been proclaiming for the past fifteen years, for his new 40-inch reflector was not ready in time for the opposition[25] and his 24-inch refractor again revealed to his eye the same familiar Schiaparellian network of *canali* that he had so often pictured in all its gossamer glory. Even his "discovery" of two "new" canals and his claim that they proved that martians were still hard at work constructing their vast irrigation project stirred hardly a ripple on the swelling sea of skepticism over the existence of any "canals" at all on Mars.

After 1909 Mars entered a cycle of progressively less favorable oppositions through 1916, and as no new or significant discoveries were forthcoming or indeed even expected, the press and the public and to a lesser extent astronomers turned to other interests or concerns. Lowell himself, as has been seen, devoted more time to other problems, although he and his staff at Flagstaff continued to make routine observations of Mars which refined some of their earlier data but revealed nothing new or sensational.[26]

After 1909 Lowell's defense of his Mars theory in particular took on a somewhat different tone, at once more general and philosophical and now often tinged with bitterness as it became increasingly obvious that his Mars work would not be accorded the recognition he was still confident it deserved. In these later years he seldom replied specifically to specific criticisms that appeared sporadically but persistently in both the popular and professional media. Rather, he answered with sweeping statements on behalf of his conclusions and against his critics' credibility, referring those who sought further details to his earlier books on Mars.

"You are quite right in your idea about the canals of Mars being skilful [sic] engineering processes—which they undoubtedly are," he wrote a Baltimore engineer who asked about particular points in his theory in 1912:

As to carbonic acid: you will find this point fully explained in my "Mars and Its Canals." Large telescopes are not so good as small ones; for effective observations one must diaphragm down as no air is so perfect as to admit their full aperture. And with regard to newspapers: what you glean from them is not trustworthy—the supposed authorities are not such in fact.[27]

And in May 1913, to another engineer, this time in London, he wrote in a similar vein:

It is interesting that your engineering work should have lead you to explain mechanicaly [sic] the canals of Mars, for of the fact that they are artificial there is no manner of doubt. The criticism of their artificiality comes only from those who are non-experts in the matter. All the best observers agree in seeing them as geometric as they have been depicted. The opposition to accepting them is of the same kind that led to the violent diatribes against Darwin.[28]

At this time, Lowell declined to comment in print on a critical article that appeared in the English journal *Knowledge* while defending the originality of his theory to its editor:

A copy of your journal for May, 1913, has been sent to me, apparently to call my attention to and perhaps secure my comment upon an article denying the reality of the Canals of Mars. Comment on the argument is unnecessary but the statement on the first page that Schiaparelli is responsible for the theory of their artificiality should be corrected. The blame for the discovery rests wholly with me. His theory was that they were natural channels, though with the magnanimity and openmindedness of genius he wrote to me before his death: "Votre theorie devient de plus en plus probable."[29]

Lowell's problems in keeping his own oft-proclaimed martian theory before the public, and the predominantly critical tone of the articles about it that did appear on occasion, are perhaps best reflected in the fact that in the late summer of 1913, he actually hired a "press agent" to write favorable articles about his Mars work and to get them published in the popular press. "Accept suggestion of permanent retainer at one hundred a month," Lowell wired one Victor Emanuel of Brooklyn, New York, in September. "You to write anything necessary and see to its publication. How soon can you possibly begin?"[30] A letter written the same day provided Emanuel with supplementary details of his new job:

In further explanation . . . I wish to add that you would not be officially connected with the observatory as that would defeat the whole purpose of having you at all. Whatever the observatory wishes to publish itself it already does and is not of a popular character. You would of course write over your own signature only as an outside writer upon the subject.[31]

Emanuel apparently did submit an article which Miss Leonard returned to him with a long list of corrections and a request: "Could you temper your style a little, making it more dignified?"[32]

Lowell also tried to keep his professional fences in repair. In November 1913 he wrote to James R. Worthington, a Fellow of the Royal Astronomical Society who was amenable to Lowell's ideas,[33] to note that a French astronomer, M. Jarry-Desloges, had recently published maps of Mars "that are almost an exact replica of our own."

As they were made without my knowing M. Jarry-Desloges or having had any correspondence with him they afford a complete refutation to the whole skeptical school. The mathematical probabilities of such coincidence in the case being so enormous as to afford absolute certainty not only that the markings exist but that they exist in the fineness and geometrical manner that we see them in Flagstaff.[34]

A few days later he wrote a note to W. H. Pickering. After briefly discussing the telescopic appearance of the "wisps" of Jupiter, Lowell advised him:

As to the canals of Mars: they are not difficult to a good observer with the fineness and directness and the geometric characteristics we have drawn. To see them without these characteristics is not to see them at their best and is either the fault of the observer or his air. Independent observations of different observers with us have shown this conclusively. In bad air only are they broadened and a blur and it should not be forgotten that certain eyes are incapable of seeing them any other way though this, again, is complicated by psychological reasons. When the "Mars canals" come to be generally recognized it will be surprising to see how many people will see them as they are in their fineness.[35]

Again, as the "least favorable" opposition of 1916 approached, the controversy over Mars was revived briefly in the press with a spate of critical articles based largely on Antoniadi's 1909 observations of the so-called "canals." As early as June 1915, Waldemar Kaempffert of the *Scientific American* sent Lowell the text of such an article for his comments. Lowell replied:

The enclosed kind of article . . . is written by a man with no first hand knowledge of the subject and is entirely incorrect in the statements it makes. It is merely repeating objections met and disproved long ago as the files of our publications will show. . . . He is, of course, naturally anxious that you should publish, and thus endorse, his *a priori* ideas which are those of the unprogressive scientific British public represented by [E. W.] Maunder.[36]

By November, Kaempffert was writing an article of his own about Mars, and Lowell attempted to convince him that there was nothing really controversial at all about the martian canals:

"The Controversy" is merely between observers of Mars and non-observers; between science and nescience. Those who object to what science has found either have not looked at the planet or have done so under unsuitable conditions of planet and instrument.[37]

Lowell read Kaempffert's article in proof in mid-December and sent it back to him with warm praise, noting: "As you see I have practically nothing to change except as to the scientific theory of intelligence on Mars: I did not 'champion it;' it may be poor, but it is my own."[38]

But as the February 1916 opposition neared, the explanation of the martian "canals" advocated by Maunder, Antoniadi, and others was clearly on the ascendant over the Lowellian view, and Kaempffert, at least, still expressed doubts which Lowell sought to down. Antoniadi, he argued to the veteran science writer, simply was not an expert in the observation of Mars:

The chief trouble with Antoniadi is that he is a man without knowledge of how to observe. I see that you speak of the canal question as unsettled. This is a mistake. All the experts are perfectly agreed on them. Indeed, we are now engaged in much further research in which they are merely the first stage. [39]

Some days later, he sent Kaempffert an early chart of Mars by Antoniadi to cinch his case against that astronomer's credibility as a martian observer:

The enclosed map of the canals of Mars will continue to open your eyes to Mr. Antoniadi's testimony when you see his signature in the corner as having seen and drawn all these canals. After he left M. Flammarion he experienced a change of heart! and now denies what he asserted before. He does not understand the use of a large telescope like the Meudon one, especially in bad air. [40]

Kaempffert, however, apparently remained unconvinced, and Lowell's last word in the matter came in the form of a letter to the *Scientific American's* editor:

My eye was caught by an editorial on Mars founded clearly on a misapprehension of the facts. To an expert, such a thing is rather startling as it awakens doubt about other subjects on which he has no direct knowledge. In the present case the facts are that among astronomers qualified to judge there is no difference of opinion as to the existence of the "canals" of Mars, opposition arising solely from those who without experience find it hard to believe or from lack of suitable conditions find it impossible to see. [41]

Such letters, of course, clearly show a speciosity in Lowell's arguments. In citing Jarry-Desloges' charts, for instance, he was well aware that his own maps of Mars were generally available and thus the "mathematical probabilities" for the "coincidence" between them were hardly "enormous." Moreover Antoniadi, Hale, and others were competent experienced observers and thus Lowell's statements to the contrary were not only self-serving but slanderous. And surely Antoniadi cannot be faulted as a scientist for changing his mind in the face of new and better data. Interestingly, while Lowell privately assailed Antoniadi for reversing his opinion, he publicly berated scientists in general for not reversing theirs.

In his public lectures and writings during these later years, Lowell frequently fulminated against science, and in fact against all mankind, for failing to recognize and acknowledge the cosmic importance of his Mars work. This failure, he insisted over and over, was rooted in human nature itself. Man, he argued, is essentially conservative and instinctively fears the new and the different, strangers and the strange, on earth or in the heavens. This, he explained repeatedly, was why his theory was scorned, why it had not been accepted as part of "the received scientific fashion of the time." The history of any great idea shows that it was rejected out of hand in its youth, vehemently and unreasonably. "A new departure inevitably shocks," he had written in 1905 before the Mars furor had run its course, "and conservatives are peculiarly affected by a kind of snow-blindness which dazzles their prejudices."

As an instructive example we commend to such the Owen-Huxley controversy of 1860–62 in which Owen would not have it that the ape's brain bore similarity to man's and when facts opposed him loftily ignored them. The case is pertinently parallel to the preconceived absurdity of an organically inhabited Mars, as if the Earth had a guarranteed [sic] monopoly of life in any form whatsoever. [42]

Now with the Mars furor over and Mars itself not readily available for confirmatory observations, he relied more and more on this theme to rationalize the deepening skepticism pressing in on his martian ideas from all sides. He used it in 1909, for example, in a lecture at Prescott, Arizona Territory, in discussing his observatory and its work:

Every new departure from a beaten path has its youth against it; and the more so in proportion as it is destined to revolutionize man's thought. ... Only the accustomed and the commonplace do men take kindly at once. The strange terrifies them. It is with ideas in men as with unfamiliar sights in beasts. Both shy at first at what they have never seen before. Scientist and layman alike are afraid to commit themselves to that upon which they have not been brought up. [43]

The theme also appears in another of its variations in his *The Evolution of Worlds* in a passage written about this same time:

The unpardonable impropriety of a new idea, I am aware, is as reprehensible as the atrocious crime of being a young man. ... Boasted conservatism is troglodytic, and usually proves a self-confessed euphemism for dull. For conservatism proceeds from slowness of apprehension. It may be necessary for certain minds to be in the rear of the procession, but it is of doubtful glory to find distinction in the fact. [44]

If it is strange, perhaps, to find a thorough-going conservative like Lowell flailing conservatives for their conservatism, it should be remembered that he was seldom conservative where Mars was concerned.

Lowell made the point, too, in his lecture series in February 1910 at Sigma Xi chapters at colleges and universities in Illinois, Iowa, Kansas, Missouri, and Nebraska.[45] And in propounding his politics at Kingman in 1911, he explained to the "Socialist miners" there why his martian results were "being scanned with a critical, and none too friendly eye by mankind at large."

For man is so profoundly jealous of peers that he looks askance at anything resembling himself, like a savage at his own reflection in a looking glass, viewing with anticipated displeasure all departures which threaten his own permanence.[46]

Lowell's most comprehensive exposition of this idea of an innate human hostility to change, however, came just weeks before his death and during his September-October 1916 lectures at college and university campuses in Idaho, Washington, Oregon, and California. One of the lectures he prepared for this tour is devoted almost entirely to the subject and will be quoted here at some length, for it clearly reveals his own confident assessment of the significance of his martian work. "The road to discovery is not an easy one to travel," he warned the young students in his audiences:

There is to add to its forbiddingness no warm compensating reception at its end, except in one's own glow of attainment. For progress is first obstructed by the reticence of nature and then opposed by the denunciation of man. Nature does not help and humanity hinders. If nature abhors a vacuum, mankind abhors filling it. A really new idea is a foundling without friends. Indeed a doorstep acquisition is welcome compared with the gift of a brand new upsetting thought. The undesired outsider is ignored, pooh-poohed, denounced, or all three according to circumstances. A generation or more is needed to secure it a hearing and more time still before its worth is recognized.[47]

Here, he cited the opposition to Darwin's theory of evolution by scientists as well as theologians:

This is not because either scientists or clergy of the time were dull. The antagonism they manifested to the view was normal human hostility to the unaccustomed. . . . Little discoveries are not received in this hostile manner; it is only the great ones that are. The reason is not far to seek. A new idea is, as has well been said, a menace to the community. It threatens the somnolent self-contentment of all conservatives who constitute always the majority of mankind. So that between private jealousy and public fear the new idea fares ill.[48]

"An idea," he declared, "is a mode of motion, not in any figurative sense but in an actual one, and as such possesses inertia."

That is, it takes force to start it going; and once set going, force to make it stop. A new idea entering the field encounters opposed to it the momentum of those already there. It has got to overcome this before it can prevail. Then it in turn grows hard to oust. The ripple of yesterday has become the roller of today. The despised is now the despotic. [49]

The textbooks of science "abet this delusion," he explained, "by presenting us the great ideas of the past in their last epiphany. They are silent on the struggles of their early youth, due to their initial crime of being a new idea, and give us merely their pictures in their habilitated states." The moral of this, he concluded, "is never to allow yourself to be disheartened or discouraged by what your contemporaries say."

Gauge your work by its truth to nature, not by the plaudits it receives from man. In the end the truth will prevail and though you may never live to see it, your work will be recognized after you are gone. [50]

That Lowell himself had not become disheartened or discouraged by what his contemporaries had been saying for so long comes through clearly in these lectures.

In one, "Far Horizons of Science," he referred only briefly to his Mars work, commenting: "I shall simply say that we have obtained positive proof of intelligence on Mars. . . ."[51] But in another, "Mars and the Earth," he argued his theory of martian life in considerable detail and confidently proclaimed:

Twenty-two years ago we had proof sufficient to warrant the publication of the fact. . . . In the time that has elapsed further and further proof has been accumulated from long continued systematic observation of the planet. . . . Visual observation has stated the facts; photography has given its imprimatur to what observation disclosed; spectroscopy has yielded confirmation; and mathematical physics from testimony of its own has shown that we ought to see just what we perceive. . . . To fully appreciate the combined weight of this evidence it is necessary to consider it all, singly and collectively. . . . It must be read in its entirety and then anyone able to judge of evidence will see that its testimony is overwhelming. [52]

Finally in a fourth lecture entitled "Mars: Forecasts and Fulfillments," he gave his theory a detailed review, climaxing his exposition with his "discovery" of the *"canali novae"* in 1909:

This last alone would have been conclusive of the presence of intelligent life, but here we have every link in the chain of reasoning leading to that end separately secured. Singly and together they make an impressive picture of forecasts fulfilled.[53]

These final lectures again attracted notice in the press, and his familiar martian conclusions now moved *The New York Times* to protest, with gentle satire, that perhaps Lowell had not gone far enough:

It would be grossly unfair to Professor Percival Lowell to assume from the latest reports of his discoveries that he has abandoned any part of his admirable theory relating to the hand-made canals of Mars. He has always contended that the lines he fancies he can see on the surface of that planet are zones of vegetation watered by canals, or channels, laboriously dug by the inhabitants. . . . What we object to is the assumption credited, fairly or unfairly, to Professor Lowell, that the creatures who, by the construction of these mighty works, have defied nature and preserved life on their planet, are necessarily human beings. If we must grant that the human being is the highest living type that we know anything about, it requires no stretch of the imagination, surely, to conceive of a higher type.[54]

"Humanity," the *Times'* writer philosophized, "has not, thus far, made such a shining success of its birthright that we need assume that no more rational and purposeful living things exist than men. To be sure, all the gods of mythology are but supernatural men, but they are symbols of man's egotism."

Professor Lowell has got as far away from the earth as Mars; he has made splendid discoveries about its seasons and its climate and the engineering genius of its inhabitants. He would not be going much further if he could admit the possibility of the existence of quite superior beings on his inhabited planet, and all the others, in our little solar system and throughout the universe. . . . There is no limit to the productiveness of nature. Why should there be to the imagination of Professor Lowell?[55]

Obviously the *Times* had not been paying strict attention to what Lowell had been saying, and he apparently straightened the paper out forthwith on the point of martian humanity, for it promptly published a second editorial:

Percival Lowell, as we inferred . . . does not entertain the theory that the beings which inhabit that planet Mars are human. He has no idea that they are anything like human beings. He does not speculate as to what they may be like in shape or size. . . . From his own

clearly defined point of view he does not speculate about them at all.
... He feels that he knows, and has proven by scientific processes,
that they exist and possess a high order of intelligence, that they
have performed on the surface of their planet an engineering feat
quite beyond the present powers of self-satisfied man, though comprehensible to his intelligence.[56]

Then the paper praised Lowell and his Mars work warmly:

The world cannot afford to withhold from Professor Lowell the
homage that is due to a tireless, expert, indomitable explorer of
astronomical space. He has reached his conclusions as to the existence of the canals of Mars and their artificial origin and the purpose
they serve after twenty-two years of persistent observations, relying
also upon the observations of the Italian astronomer Schiaparelli,
begun seventeen years before the discoveries began at Flagstaff. He
has answered in strictly scientific terms the objections of his critics.
He has triumphantly exploded their theory that the white caps at the
Martian poles are composed of carbon dioxide instead of snow and
ice, that the lines he sees so clearly and photographs so plainly and
draws so accurately are similar to those observed fitfully on other
planets, such as Mercury, Venus and Saturn. They are not the
same. He is convinced that his proof of the existence of intelligent
life on Mars is as clear as the proof that there is sodium in the sun.[57]

Two weeks after this editorial appeared, a massive stroke ended
perhaps the most colorful, and certainly the most controversial
career in the history of astronomy.[58] Lowell's ideas, the impact of
his mind, live on.

The full extent of this impact on subsequent generations is
seldom recognized. As far as the public is concerned, Lowell's
intelligent, canal-building martians have been largely relegated to
the realms of fantasy and science fiction. The process, indeed,
began as early as 1898 with H. G. Wells' classic *The War of the
Worlds*, and increasingly thereafter martians in one imaginative form
or another were celebrated in both song and story.[59] It was given
particular impetus beginning in 1912 when Edgar Rice Burroughs
published *A Princess of Mars*, the first of a popular series of eleven
novels detailing the adventures of an American named John Carter
and his martian princess, Dejah Thoris, on the canalized red planet
that Burroughs' green-skinned martians themselves called "Bar-
soom."[60] At least one modern student of Marsiana has pointed out
that Burroughs incorporated Lowell's ideas in fictional form in these
books, thus disseminating them to an even greater audience.[61]

Martians of various shapes, sizes, and degrees of malevolence were also familiar to readers of the daily press in the late 1920s and the 1930s during the heyday of the so-called "comic strips." An entire generation of Americans grew up following the adventures of a "Buck Rogers" or a "Flash Gordon" as, equipped with spaceships, "flying belts," and "disintegrator" guns they cavorted on Mars and other planets known or unknown to science. Some of these comic strip characters, in turn, became stars of serialized films and grade B motion pictures that were seen as "extra added attractions," or as the second half of a "double feature," by countless thousands in theaters throughout the United States and Europe. Such films, as well as latter-day counterparts inspired by the advent of the "space age," have been common fare for late evening television viewers since the late 1950s. Martians in particular and extraterrestrial beings in general have been the subjects of a number of popular, long-running, serial programs on American television.

Lowell's martians, perhaps, may not always be immediately recognizable in these modern media manifestations, the evolutionary line as it were having been modified considerably by the selective factors of imagination and a progressively favorable technological environment operating over several successive generations. But as Lowell so often insisted, their minds betray them. Since Lowell's time, fictionalized martians, whether belligerent or beneficent and however conceived in body, have possessed in common the identifying trait of superior intellectual development which Lowell assigned to them. Before Lowell, in short, martians were usually conceived to be simply human; after Lowell they have been almost universally imagined as superhuman, at least in so far as intelligence is concerned.

If the more or less credulous layman sitting before his television set or reading a science fiction thriller has casually accepted fanciful martians and other imaginary beings from outer space, a more or less incredulous science has consciously committed itself to the possibility, even the probability, that life in some form might exist on Mars or elsewhere in the universe. This is not to say, of course, that scientists have accepted Lowell's Mars theory in any of its particular details but that they recognize as they once did not that Lowell's fundamental premise of the probability of extraterrestrial life is a valid subject for scientific investigation. This shift in thinking, important both in the history of science and in intellectual history, can be attributed largely to the work and thought of Percival Lowell.

Forest Ray Moulton, surely no Lowellian, conceded as much some twenty years after Lowell's death. Writing in 1935 Moulton noted the long refusal of science to consider the question of extra-terrestrial life as one worthy of serious study, pointing out that "even in recent years it has been repeatedly argued, particularly by British writers, that the solar system is the only planetary system in the universe at all suited for the abode of life, that in our solar system the earth is the only inhabited planet, and that man is the climax, and hence the goal, of creation."[62] The reference here, of course, is to evolutionist Alfred Russel Wallace and his anti-Lowellian followers who advocated the doctrine of an exclusive creation of life according to a divine plan. And, Moulton continued:

A generation ago astronomers regarded it as unscientific or perhaps beneath their dignity to raise the question of the habitability of planets other than the earth. . . . Whatever the reason may have been for this aloofness of our predecessors, it is not shared by present-day astronomers. The change in attitude has undoubtedly been brought about in considerable part by Percival Lowell's persistent and able attempts to prove the existence of life on Mars.[63]

To this, perhaps, it is only necessary to add that the probability of the existence of extraterrestrial life is now widely accepted not only by astronomers but by scientists in general and the where-abouts and nature of this life, as well as the conditions under which it exists, are subjects of intensive study by such exotic new disciplines as exo-biology, astro-geology and cosmo-chemistry. Distinguished scientists and scholars have published hundreds of papers and a number of books on the overall problem, or on its specific ramifications, and have seriously discussed the question in not a few symposia.[64] Nor has the irony in all this been entirely lost. Educator and philsopher Jacques Barzun in commenting on a symposium on "Life in Other Worlds," held in New York City in 1961, noted not so long ago:

Percival Lowell, one of the first to present astronomical observations in support of the view that life exists on Mars, did not live to hear the five scientists, including the President's scientific advisor, speculate about the notion for which he himself was ridiculed.[65]

And—shades of the Mars furor—there have even been serious systematic efforts made to detect "signals" that an extraterrestrial civilization may be beaming toward earth from an unknown planet circling a far distant sun. The plain assumption underlying "Project Ozma," proposed in 1959 and subsequently implemented in the early 1960s by Frank D. Drake at the National Radio Astronomy Observatory at Green Bank, West Virginia, is that these out-of-this-world broadcasters are at least as intelligent as man.[66]

But if science has now embraced the basic premise of Lowell's Mars theory, it has continued to be skeptical of the theory itself. "After all," Moulton wondered back in 1935 in summarizing contemporary astronomical opinion, "is Lowell's dramatic story of life on Mars anything more than a ploy, representing something not obviously impossible and having, indeed, a degree of plausibility? Is it more than a fanciful vision emanating from a brilliant mind?"[67]

But nonetheless, the astronomer added, "even if the observational material Lowell used is doubtful and if all his principal conclusions are erroneous, still it is not a waste of time to follow his speculations. ... There is, of course," he noted, "nothing which compels us to either accept Lowell's hypothesis or to conclude that there is no life on Mars. As a matter of fact, it is not improbable that there is life on the planet."[68]

While the dogmatism and emotionalism that marked the Mars controversy during Lowell's life vanished after his death, his work and ideas provided the focal point for the more subdued debate that continued, largely within astronomy, through the intervening years. Since Lowell's death, a number of competent and respected observers, including Gerard de Vaucouleurs and Audouin Dollfus in Europe and Gerard P. Kuiper and E. C. Slipher in the United States, devoted a considerable portion of their working time to problems posed by the red planet. But the results of their work were often conflicting and, of necessity, tentative.

In the 1950s, for instance, Dollfus, following the pioneering work of Bernard Lyot in the 1920s, undertook a series of polarization studies of Mars that suggested the polar caps were composed primarily of hoarfrost, that ice might temporarily form at higher elevations, and that some clouds at least were composed of ice crystals as are cirrus clouds on earth.[69] About this same time Kuiper compared the reflectivity of the polar caps in infrared light and found a similar suggestion of hoarfrost.[70] The carbon dioxide hypothesis, however, was not thus disposed of, and indeed prevailed to receive new support from the 1971 Mariner 9 mission to Mars.[71]

Kuiper in the late 1940s and early 1950s also expanded on two inconclusive investigations undertaken earlier by V. M. Slipher. One of these concerned chlorophyll on Mars, with Kuiper concluding that the reflection spectrum of the dark areas more nearly resembled that of terrestrial lichens, although he quickly pointed out that he was not specifically postulating the existence of lichens as such on Mars but only the possibility of "some very hardy vegetation" there.[72] In 1947, Kuiper completed a task to which Lowell had assigned Slipher in 1916 by making the difficult spectrographic identification of the carbon dioxide molecule in the martian atmosphere, a result that has been amply sustained by Mariner 9's findings.[73]

The techniques of spectroscopy and polarimetry were also used during this period in attempts to determine the nature of the reddish-ochre martian "soil." Lyot in 1922 had tentatively identified this material with limonite, a hydrated iron oxide, but Kuiper later turned a spectrograph on the bright "desert" areas and suggested felsitic rhyolite.[74] Mariner 9's infrared interferometer spectrometer has now indicated that the spectrum of the material stirred up by the great martian dust storm of 1971 corresponds to that of terrestrial rocks containing between 55 and 65 percent silicon oxide (SiO_2).[75]

Compounding the problem of the nature of the martian surface material for telescopic observers was the fact that not only the color but the changes—Lowell's seasonal "wave of darkening"—had to be accounted for. And plausible and imaginative explanations for this phenomenon have not been wanting over the years. In Lowell's lifetime the Swedish scientist Svante Arrhenius contended that the soil of the dark areas at least was composed of hygroscopic salts that darkened as they absorbed water from moist air sweeping down from the poles in the martian spring.[76] More recently and more intriguingly it has been suggested that the martian surface is covered with a polymer of carbon suboxide (C_3O_2) and thus the observed changes represent differences in the degree of polymerization.[77] Perhaps the most fully developed of the pre-Mariner explanations was that advanced in the mid-1950s by University of Michigan astronomer Dean B. McLaughlin who theorized that the changes in the surface appearance of Mars could be accounted for by volcanism and aeolian deposition of dust.[78] Cornell University astronomer Carl Sagan, on the basis of Mariner 9's photographs, has proposed windblown dust as the principal agent for the differences and variability of martian surface features, noting that McLaughlin's hypothesis appears to have been "remarkably prescient."[79]

Both de Vaucouleurs and the younger Slipher have been among the most indefatigable and systematic observers of Mars in the post-Lowellian, pre-Mariner years, yet their work led them to somewhat different conclusions, the former remaining skeptical and the latter optimistic about the "canals" and any possible life on the planet.

Slipher continued the rigorous program of planetary photography that Lowell and Lampland initiated in 1903 and which he himself took up in 1906—a program which yielded more than 200,000 images of Mars over fifty-five years—and remained convinced of the reality, if not the artificiality of the canals. "While their existence as true markings on the planet has been clearly established," he wrote shortly before his death in 1964, "there is room for difference of opinion as to the interpretation of the canals," adding that "because of the artificial character of these markings, they have become the most mysterious and yet the most discussed and widely debated of all planetary features."[80] He also cautiously opted for some form of vegetable life on Mars, noting that "there is no evidence whatsoever that intelligent animal life exists" but arguing that "nothing we know of" except vegetation would explain the appearance and seasonal changes of the martian blue-green areas. "Since the theory of life on the planet was enunciated some fifty years ago, every new fact discovered has been found to be accordant with it," he concluded:

Not a single thing has been detected which it does not explain. Every year adds to the number of those who have seen the evidence for themselves. Thus theory and observations coincide.[81]

De Vaucouleurs on the other hand has relied primarily on visual observation and has long denied the reality of the canals, although conceding they presented an "irritating enigma."[82] Of the possibility of martian life, he has written:

If we were to sum up in a single sentence our present knowledge of the physical conditions of Mars, we might liken them to those which obtain on a terrestrial desert, shifted to the polar regions and lifted to stratospheric level. We leave it to the reader to decide whether under such circumstances Mars can be "the Abode of Life" or not.[83]

Plainly, the problems of Mars have been discussed within a Lowellian framework and in Lowellian terms since Lowell's death.

It must be noted that four brilliantly successful Mariner Mars probes have sharply modified nearly all the old ideas about the planet, and indeed astronomers have been forced to change many of their new ideas after each successive Mariner flight.

Mariner 4, which executed a historic first "fly-by" of Mars in

July 1965, obtained only twenty-two photographs covering a thin sliver of the martian surface, and of these only a half-dozen or so showed any appreciable detail. But these few revealed a crater-pocked lunar-like landscape that astonished most laymen and not a few astronomers. Mariners 6 and 7, in late July and August 1969, made similar fly-bys which produced some two hundred far- and near-encounter pictures of considerably better quality. These now showed the martian surface to be actually quite varied and not at all as moonlike as the Mariner 4 photographs had indicated.

Finally on November 13, 1971, Mariner 9 was placed in an elliptical orbit around Mars, which swung it to within a thousand miles of the surface. More than 7,200 superb photographs, covering almost all of the planet's surface, were obtained during the spacecraft's active "life" of nearly a year, and these, along with the mass of spectrometric and radiometric data telemetered back to earth, have given scientists a wholly new and exciting, if hardly less puzzling, view of Mars.

These productive Mariner missions, it is important to note, were neither intended nor expected to prove or disprove the presence of life in any form on Mars, nor have they incidentally provided any conclusive evidence to resolve this still intriguing question. But nevertheless, the Mariners, and particularly Mariner 9, have produced some surprises. For it now appears that Mars, long considered a geologically "dead" planet, is very much alive, that tectonic forces, volcanism, and other geologic processes are at work there as on earth. Mariner 9 cameras also have revealed a far more rugged and varied terrain than most astronomers, and most certainly Lowell, believed existed on the planet. Towering volcanic mountains, estimated to be up to 17 miles in height, and chasms far vaster and deeper than the earth's great Grand Canyon in Arizona, have been recorded, along with many other unexpected and even bizarre topographical features. [84]

But the most dramatic revelation of the Mariner 9 pictures is the evidence of erosion and sedimentation that has led some scientists to suggest, by analogy to similar geologic features on earth and in the absence of a better explanation, that water once flowed on the surface of Mars. "We are forced to no other conclusion but that we are seeing the effects of water on Mars," Dr. Hal Masursky, an astrogeologist and a principal investigator in the Mariner 9 mission, told a press conference on June 14, 1972. [85]

If water did indeed flow on Mars, it flowed in sinuous eroded channels rather than in straight narrow "canals," and in fact the

Mariners have finally disposed once and for all of any possibility that a planet-sized network of waterways of the scope and geometricity so often described by Lowell exists or ever existed on Mars. The canal controversy, after running for nearly a century, is thus ended. This is not to say, however, that it served no useful purpose, as English historian of astronomy R. L. Waterfield pointed out more than twenty years after Lowell's death:

Now the story of the "canals" is a long and sad one, fraught with backbitings and slanders; and many would have preferred that the whole theory of them had never been invented. Yet whatever harm was done was more than outweighed by the tremendous stimulous the theory gave to the study of Mars, and indirectly to the planets in general. Whether in a positive way to champion it, or in a negative way to oppose it, it attracted many able observers who otherwise might never have taken an interest in the planets. ... So the pistol which Schiaparelli had so unwittingly let off, though it shocked the finer feelings of many, had undoubtedly been the starting signal of that race for discovery which the planetary astronomers are still successfully pursuing.[86]

Otto Struve, who cited this same passage, specifically stressed the impact of Lowell's work and ideas, however imaginative, on the subsequent development of planetary astronomy, devoting a full third of his chapter on the planets in his 1962 review of twentieth-century astronomy to Lowell's activities. He prefaced the chapter by noting:

Up to the end of the [1800s,] planets were visually observed mainly by amateurs using relatively small instruments. At the present time, most large observatories devote a considerable portion of their ob-serving time to planetary studies, and some of the best-known astrophysicists are attempting to interpret these observations. This change in attitude can to a large degree be attributed to the efforts of one man, Percival Lowell, and to his part in one of the greatest astronomical controversies of this century: the existence of life on Mars.[87]

Lowell's death on November 12, 1916, at his observatory on Flagstaff's Mars Hill was reported in the press throughout the world.[88] Eulogies were uttered and memorials published in the proceedings and journals of many scientific societies where his astronomical work had been received, discussed, or debated over the final twenty-two years of his life. In this obituary outpouring—in itself evidence of his impact on the collective mind of his generation

Percival Lowell, about 1916

—his contributions were generally acknowledg
prominent figure in astronomy," the Council of the
nomical Society of Canada declared, to cite one examp.
many researches made at the Lowell Observatory will h
history of science. Indeed, it is not too much to say that our ᵢ
edge of the planets has been almost reorganized through the wo.
this institution."[89]

Yet these necrological tributes seldom glossed over the more
controversial aspects of his career which had, of course, made him
famous in the first place and for which he is usually remembered
today. Even his long-time friend and ally George Russell Agassiz felt
obliged to refer somewhat apologetically to the Lowellian furor over
Mars in his carefully written formal memorial for the American
Academy of Arts and Sciences:

His work and theories on Mars are most widely known. His observations on that planet have accumulated an amount of data greatly in
excess of the total results of all other observers combined. If the
theory, which he deduced from this data, that Mars is inhabited,
seems fanciful to many, it should at least be borne in mind that it is
deduced from observed facts by logical reasoning. Furthermore, no
other satisfactory explanation of the facts has ever been offered.
His theory would doubtless have made much more headway in the
scientific world had it been less dogmatically presented.[90]

The facts are different now of course. But certainly in going
beyond his own data for his more sensational conclusions, Lowell
ventured into the realm of the "fanciful" as his critics so often
charged. And most certainly he expressed these conclusions dogmatically, trusting in analogy too much perhaps, and being too
willing to accept the merely plausible for the actual in the absence of
any immediate alternatives. But it was the imaginative quality and
the dogmatism that stimulated the intense reaction to his ideas
among scientists and laymen alike. And whether this reaction was
critical or credulous, it served to focus unprecedented attention on
problems which Lowell insisted were not only basic to man's understanding of the universe, as indeed some of them are, but were also
properly subjects for scientific investigation.

In retrospect, while Lowell's grander theories have come to
naught, much of the work on which they were based was sound and
often innovatively productive. The advances in spectrography and in
planetary photography made by the Sliphers and Lampland under

Lowell's aegis, while perhaps crude by the standards of modern astronomical technologies, were nonetheless on the frontiers of those technologies in their time. Lowell's "velocity-shift" method for determining spectroscopically the composition of planetary atmospheres, probably his most original idea but one which could not be successfully applied in his day, has since proven to be a viable observational technique. More importantly the elder Slipher's pioneering spectrographic studies of the nebulae, initiated and encouraged by Lowell, opened up rich new vistas in astronomy which in more recent years have vastly expanded man's concept of the cosmos.

If Lowell's ideas on planetary evolution, or "planetology," were superficial and scientifically naive, his repeated insistence on the importance of the atmosphere in astronomical observations was well founded. If Lowell could see spurious markings on the cloud-shrouded surface of Venus, he also saw clearly that the progressive deterioration of the quality of the earth's atmosphere would become a major problem in the future, and not alone for astronomy. If Lowell offended astronomers of his day by declaring that the study of the planets was a specialized field, astronomy has since agreed, and indeed the American Astronomical Society now has a flourishing Division of Planetary Sciences with a membership in the hundreds. If Lowell argued speciously over supposed mathematical errors in the Chamberlin-Moulton "planetesimal" hypothesis, his own mathematical labors on the problem of a trans-Neptunian planet were rigorous enough, perhaps not unworthy of a Leverrier or an Adams, and led to the discovery of the ninth planet of the solar system.

If Lowell's canal-building martians are today in total disrepute, the possibility of extraterrestrial life which was the fundamental premise of his controversial Mars theory has become not only widely accepted in science but has provided a rationale for a number of highly imaginative investigations which, if far more sophisticated, are hardly less "fanciful" than the one that Lowell embarked on at the turn of the twentieth century.

Reference Material

Appendix A

A Standard Scale for Telescopic Observations

1. *Present State.* At present there exists no criterion among astronomers for the weight to be attached to any given observation due to the atmospheric conditions under which it is made. Yet these atmospheric conditions are among the most important factors entering into an astronomic observation. They are far more to the point than the size of the instrument. For our telescopes have long since outstripped the conditions under which they are put to work; the great bar to advance today, whether visually, photographically, or spectroscopically being not instrument but atmosphere. Each man realizes this but marks his own work on his own scale; as if he should take his own foot as the unit of length.

2. *Difficulties of This Condition.* In consequence no absolute value is assignable to any man's work and no comparison between different men's work is possible whether in accuracy or credibility. The practical outcome is that the only test is the test of time and while the world is waiting for confirmation of any new result just so many years are lost. As important is the incapacity to leave permanent records of observations, capable of being compared with newer ones as time rolls on.

3. *A Change is Necessary.* A change in this state of things is imperatively needed. It is time a standard scale for observations were introduced similar to what the metric system is that it may do what that does for physics generally.

4. *Possibility of a Criterion.* Until lately a scale has not been feasible owing to ignorance of the conditions upon which it must be based. Studies, however, directed to that end first at Arequipa and then at Flagstaff during the past few years have resulted in the knowledge of the conditions which constitute good or bad seeing and have thus enabled an absolute scale to be constructed.

Note: The typewritten text of this material, prepared probably in the Spring of 1902, is in the Lowell Observatory Archives.

[315]

5. *The Criterion.* The basis of the matter lies in the discovery that systems of waves traverse the air; several of these systems being present at once at various levels above the earth's surface. The waves composing any given system are constant in size and differ for the different currents all the way from a fraction of an inch to several feet in length. If the wave be less than the diameter of the object-glass from crest to crest the image is confused by the unequal refraction from different phases of the wave. If the wave be longer than this a bodily oscillation of the whole image results. The first is fatal to good definition, the second to accurate micrometric measurement. It is possible to see these waves by taking out the eye-piece and putting one's eye in the focus of the instrument when the tube is pointed at some sufficient bright light. It is further possible to measure their effect by careful noting of the character of the spurious disk and rings made by a star and the extent of the swing of the image in the field of view. By combining the amount of confusion with the degree of bodily motion of the resulting image, the definition at any time and place can be accurately and absolutely recorded. The increasing perfection of the optical image of a star testifies to the increasing lack of damaging currents with reference to the object-glass used. It records all the waves below a certain wave-length. Similarly the amount of bodily motion registers all those above that length. The two taken together give account of all the currents independent of the glass.

6. *The Scale.* It is, therefore, necessary only to agree upon some size of glass for making the fundamental tests and then to reduce the results to any aperture by relations which will be found set forth in a pamphlet by Mr. Douglass made at this observatory entitled: Scales Of Seeing.

The most feasible size for comparison purposes seems the six-inch aperture.

The scale it is proposed to adopt is, therefore, as follows:
With a Six-Inch Glass.

0 Disk and rings confused and enlarged
2 Disk and rings confused but not enlarged
4 Disk defined; no evidence of rings
6 Disk defined; rings broken but traceable
8 Disk defined; rings complete but moving
10 Disk defined; rings motionless

Synchronous determination of the amount of bodily motion of image in seconds of arc.

—The Lowell Observatory

Appendix B

Expedition for the Ascertaining of the Best
Location of Observatories

In order to discover the best place or places for the location of telescopes in the future it is proposed to send observers furnished with similar instruments and identical instructions to all promising parts of the earth's surface.

Two desert regions girdle the earth in the subtropical regions of Capricorn and Cancer and from the meteorological conditions there prevailing these belts offer the greatest promise to the astronomer. In the northern hemisphere the belt shows itself first in the Sahara of Africa, then in Arabia, then in the desert of Gobi, and crossing the Pacific crops out again in Arizona and Mexico. Of these, the two with the greatest height for their plateaux are Arizona and Mexico and the desert of Gobi. In the southern hemisphere we have the veldt of southern Africa, the western part of Australia, and finally the west coast of Peru and Bolivia. Of these, the last is the highest, and the Transvaal the next.

With regard to these places, we have the most systematic series of records from Arizona, the next so from Peru, some slight knowledge of the Sahara and next to none of any other locality.

Although the desert belts promise the best, other localities widely different should also be examined. Chief among these perhaps are the islands of the Pacific.

It is desirable, therefore, to send out observers somewhat as follows:

1. To the desert of Gobi
2. To the veldt of the Transvaal
3. To the Samoan Islands.

The observations made at these points could then be repeated elsewhere till the earth's surface should be known from an astronomical point of view.

Note: The typewritten text of this material, prepared probably in the Spring of 1902, is in the Lowell Observatory Archives.

[317]

Each observer is to be armed with a 6-inch glass, all of the glasses made by the same maker (for instance Alvan Clark & Sons Corporation) and to report according to the proposed standard scale of "seeing."

It is important that the said scale should be agreed to by astronomers generally before the various expeditions start.

—*The Lowell Observatory*

Appendix C

Publications in the Archives

The following is a partial list of the newspapers and other periodicals of a somewhat popular nature which are represented by identifiable clippings in the Lowell Observatory Archives:

Argonaut, The
Arizona Daily Gazette
Arizona Silver Belt
Athenaeum, The
Atlantic Monthly

Baltimore (Md.) *American*
Bangor (Me.) *Commercial*
Boston *Advertiser*
      Commonwealth
      Gazette
      Globe
      Herald
      Journal
      Post
      Times
      Transcript
Bloomington (Ind.) *Chronicle*
Brooklyn (N.Y.) *Eagle*
Bruxelles (Belgium) *Matin*
Budapest (Hungary) *Hislop*
      Peater Lloyd
Buffalo (N.Y.) *Times*
Burr-McIntosh Monthly

Catholic World
Century, The
Chicago *Examiner*
      Record-Herald
      Tribune
      World

Christian Science Monitor
    (Boston)
Cleveland *Plain Dealer*
Coconino Sun (Flagstaff)
Couriere Italia
Cosmopolitan

Die Wocke
Denver (Colo.) *Republican*

Fortnightly Review (London)
Frankfurter *Zeitung*

Hamburger *Nachrichten*
Harper's

Illustrated London News
Independent, The
Indianapolis *Star*
Irish Catholic

Japan Daily Mail
Jersey City (N.J.) *Journal*
Journal de Rouen

Kansas City *Times*
Knowledge and
    Scientific News
Knoxville (Tenn.) *Sentinel*

La Croix
Le Cosmos

Literary Digest, The
London *Chronicle*
     Graphic
     Mail
     Mirror
     Morning Post
     News
     Telegraph
     Times
     Tribune
     Standard
Los Angeles *Times*

Ma Revue
McClure's
Milan Coniere
Miliane Francais
Minneapolis *Journal*
     Tribune
Montreal (Canada) *Gazette*
Muncie (Ind.) *Star*

Nation
Nature
New Orleans *Times-Democrat*
New York *American*
     Christian Advocate
     Commercial
     Evening Post
     Evening Sun
     Herald
     Mail
     Morning Telegraph
     Staats-Zeitung
     Times
     Tribune
     World
Nineteenth Century, The
Norfolk (Va.) *Pilot*

Omaha (Neb.) *World-Herald*
Outlook

Pall Mall Gazette (London)
Petique Republique
Philadelphia (Pa.) *Inquirer*
     Press
     Record
Pittsburgh (Pa.) *Dispatch*
     Press
Popular Astronomy
Popular Science Monthly
Portland (Me.) *Press*
Prescott (Ariz.) *Journal-Miner*
     Courier
Providence (R.I.) *Journal*
Putnam's Magazine

St. Louis *Globe-Democrat*
     Post-Dispatch
     Times
Salt Lake City *News*
San Francisco *Call*
     Chronicle
     Examiner
     Weekly Post
Saturday Evening Mail (London)
Science
Scientia (Italy)
Scientific American
Scientific News
Scottish Review
Spectator, The (London)
Sphere, The (London)
Springfield (Mass.) *Republican*
Sydney (Aus.) *Daily Telegraph*

Tombstone (Ariz.) *Prospector*

Wall Street Journal (N.Y.)
Washington (D.C.) *Star*
Westminster Gazette (London)

Chapter Notes

With the exception of direct correspondence, sources such as manuscripts, typewritten texts or drafts, newspaper and periodical clippings and other materials in the Lowell Observatory Archives are noted by the abbreviation, *L.O.A.*

Notes to Introduction

1. Agnes M. Clerke, *A Popular History of Astronomy During the Nineteenth Century*, London, 1902, 280.
2. Otto Struve and Velta Zebergs, *Astronomy of the 20th Century*, New York, 1962, 26.
3. V. M. Slipher to W. A. Cogshall, Dec. 21, 1916.

Notes to Chapter 1

1. Samuel Glasstone, *Book of Mars*, Washington, D.C., 1968, 2. While Mariner spacecraft photographs of Mars from as close as 900 miles show no evidence of life, the question is still an open one. Glasstone (220) among others has noted that photographs of the earth from satellites only a few hundred miles above its surface show no conclusive evidence for terrestrial life.
2. The summary of the early observational history of Mars which follows here is drawn largely from Agnes M. Clerke, *A Popular History of Astronomy During the Nineteenth Century*, London, 1902, 274 ff., and Glasstone, *op. cit.*, 15 ff.
3. W. Herschel, "On the remarkable Appearances at the Polar Regions of Mars," (1784), *The Scientific Papers of Sir William Herschel*, London, 1912, Vol. I, 156.
4. W. Herschel, "Letter to Nevil Maskelyne, Astronomer Royal," (1780), ibid., Vol. 1, xc–xci; and "On the Nature and Construction of the Sun and Fixed Stars," (1795), ibid., Vol. 1, 470. Herschel believed that the spots on the sun indicated that "under the outward flaming surface is most probably contained a solid globe of unignited Matter." See Herschel's "Short Account of some experiments upon light that have been made by Zanottus, with a few remarks on them," ibid., Vol. 1, xcvi.
5. Patrick Moore, *Guide to Mars*, London, 1956, 112–13.
6. Lowell, *Annals of the Lowell Observatory*, III:xiv (1905).
7. Garrett P. Service, *Other Worlds*, New York, 1901, 93 ff. Service is translating G. V. Schiaparelli from *L'Astronomie*, I:217 (1882).
8. Ibid.
9. Antonie Pannekoek, *A History of Astronomy*, New York, 1961, 378–79, quoting C. Flammarion, *La Planète Mars et ses conditions d'habitabilité*, Paris, 1892, 591.
10. The program was based on H. G. Wells' *The War of the Worlds* and starred actor Orson Welles. The reaction was extensively reported in the press; see *The New York Times*, Oct. 31, 1938, and Howard Koch, *The Panic Broadcast*, New York, 1970.

11. Rudolph Thiel, *And Then There Was Light,* New York, 1957, 243.
12. Walter Sullivan, *We Are Not Alone,* New York, 1964, 178–81. Also re: Edison and Galton, unidentified newspaper, January 1897; re: Tesla, unidentified newspaper, January 1901, and New York *Herald,* June 27, 1907; and re: Marconi, *The New York Times,* April 1, 1907; *L.O.A.*
13. "The Signals From Mars," *Popular Astronomy,* 3:47 (1895).
14. Clerke, *op. cit.,* 281.
15. W. W. Campbell, "An Explanation of the Bright Projections Observed on the Terminator of Mars," *Publications of the Astronomical Society of the Pacific,* 35:103 (1894).

Notes to Chapter 2

1. W. E. Garrett Fisher, London *Times,* June 1906; *L.O.A.*
2. "The Riddle of Mars," *Scientific American,* July 13, 1907; *L.O.A.*
3. *Wall Street Journal,* Dec. 28, 1907; *L.O.A.*
4. Samuel Glasstone, *Book of Mars,* Washington, D.C., 1968, 24.
5. N. J. Berrill, *Worlds Without End,* New York, 1964, 52.
6. "The Man Who Explored Mars," *Literary Digest,* Dec. 2, 1916; *L.O.A.*
7. P. Lowell, "Comets," text dated March 10, 1910, of an article for the Boston Society of Arts; *L.O.A..*
8. A. M. Clerke, *A Popular History of Astronomy During the Nineteenth Century,* London, 1902, 323 ff.
9. The biographical summary that follows here is drawn largely from A. Lawrence Lowell, *Biography of Percival Lowell,* New York, 1935, and from Ferris Greenslet, *The Lowells and Their Seven Worlds,* Boston, 1945, especially 345 ff.
10. Greenslet, *op. cit.,* 323.
11. A. L. Lowell, *op. cit.,* 5.
12. P. Lowell to Hector McPherson Jr., Nov. 16, 1903.
13. Greenslet, *op. cit.* 348.
14. Ibid., 348–49. The quotation is from a letter from W. Sturgis Bigelow to Augustus Lowell. See also A. L. Lowell, *op. cit.,* 12.
15. A. L. Lowell, *op. cit.,* 37–39.
16. Ibid., 38. The quotation is cited from Elizabeth Bisland, *Life and Letters of Lafcadio Hearn,* Vol. I, 459.
17. P. Lowell, *The Soul of the Far East,* New York, 1911 (first published in 1888), 51, 66, 111.
18. Ibid., 208.
19. Ibid., 210.
20. Greenslet, *op. cit.,* 354–55. The quotation is from a letter from Lowell to F. J. Stimson, one of his friends in the Dedham Polo Club.
21. *Funk and Wagnalls Standard College Dictionary,* New York, 1963. This is one of six definitions here and best seems to fit Lowell's usage of the word.
22. P. Lowell, "Great Discoveries and Their Reception," lecture text, *ca.* August 1916; *L.O.A.*
23. P. Lowell, "Commemoration Day Address," text dated May 30, 1903; *L.O.A.*
24. P. Lowell, "Great Discoveries."
25. Ibid.
26. Ibid.
27. Ibid.
28. P. Lowell, *The Soul of the Far East,* 7–9.
29. Ibid., 13–19.
30. Ibid., 194–95.

31. P. Lowell, *Mars,* London, 1895, 6–7.
32. Ibid., 8.
33. Ibid., 6.
34. P. Lowell, "Great Discoveries."
35. A. L. Lowell, *op. cit.,* 60; Greenslet, *op. cit.,* 358–59.

Notes to Chapter 3

1. W. L. Leonard to P. Lowell, Sept. 11, 1916. This letter, thanking Lowell for a pumpkin weighing nearly fifty pounds, is one of a number in the *L.O.A.* that refer to the products of Mars Hill's garden. Miss Wrexie Louise Leonard was Lowell's personal secretary from 1896 until his death in 1916 and frequently accompanied him on his travels.
2. P. Lowell to Montgomery Ward & Co., Jan. 29, 1903, and to A. C. McClurg Co., Aug. 19, 1907, and Oct. 19, 1909.
3. A. L. Lowell, *Biography of Percival Lowell,* New York, 1935, 102–06. The passage summarizes Percival's interest in botany.
4. P. Lowell to V. M. Slipher, Oct. 14, 1907.
5. A. L. Lowell, *op. cit.,* 153–56. The *L.O.A.* contains numerous letters referring to these trips and to Lowell's close friendship with Doe.
6. Most notably George R. Agassiz and Edward S. Morse, fellow members of Boston's intelligentsia. See A. L. Lowell, *op. cit.,* 146–47.
7. L. F. Ward to P. Lowell, March 31, 1907.
8. W. W. Campbell to V. M. Slipher, Nov. 14, 1905.
9. H. F. Ashurst to P. Lowell, May 22, 1907.
10. A. L. Lowell, *op. cit.,* 156.
11. P. Lowell to V. M. Slipher, June 4, 1904.
12. P. Lowell to V. M. Slipher, June 6, 1904. Hussey was the observatory's porter at the time.
13. Ibid. There are many examples of Lowell's concern for the comfort of his staff in the *L.O.A.*
14. Boston *Herald,* Feb. 12, 1894; *L.O.A.*
15. E. C. Pickering to P. Lowell, Feb. 25, 1894.
16. P. Lowell to E. C. Pickering, Nov. 16, 1894.
17. Nine years later, for example, the Lowell Observatory was still being referred to in the press as the "Harvard Observatory at Flagstaff." See New York *World,* May 31, 1903; *L.O.A.*
18. Douglass' pocket notebooks in which he logged his activities and observations on his Arizona trip are in the *L.O.A.*
19. See Appendix A.
20. A. L. Lowell, *op. cit.,* 71.
21. A. E. Douglass, "The Lowell Observatory and Its Work," draft dated Jan. 30, 1895, of article for *Popular Astronomy.*
22. A. E. Douglass to P. Lowell, telegram, March 9, 1894; see also Tombstone *Prospector,* March 10, 1894; *L.O.A.*
23. A. E. Douglass to P. Lowell, telegrams, March 19, 23, April 1, 2, 5 and 6, 1894.
24. P. Lowell to A. E. Douglass, telegrams, March 23, April 6, 7, 9, and 10, 1894.
25. P. Lowell to A. E. Douglass, April 10, 1894.
26. *Coconino Sun* (Flagstaff), April 12, 1894.
27. P. Lowell to A. E. Douglass, May 3, 1894.
28. *Coconino Sun,* May 10, 1894.
29. P. Lowell to A. E. Douglass, May 10, 1894.

30. P. Lowell to A. E. Douglass, April 16, 1894.
31. Now the Atchison, Topeka and Santa Fe Railway.
32. "Declaration to Professor A. E. Douglas [sic]," April 17, 1894; *L.O.A.*
33. P. Lowell to A. E. Douglass, telegram, April 21, 1894.
34. A. E. Douglass to P. Lowell, telegram, April 23, 1894.
35. P. Lowell to A. E. Douglass, telegram, April 24, 1894; see also *Coconino Sun,* May 3, 1894.
36. Ibid.
37. P. Lowell to A. E. Douglass, May 5, 1894.
38. A. L. Lowell, *op. cit.,* 72.
39. P. Lowell, "Two Stars," text of lecture given at Kingman, A.T., Oct. 20, 1911; *L.O.A.*
40. *Coconino Sun,* March 15, 1894.
41. Among numerous examples, see the *Arizona Silver Belt* (Gila County), March 10, 1894, the *Arizona Daily Gazette* (Phoenix), March 18, 1894, the Prescott *Courier,* March 30, 1894; *L.O.A.*
42. P. Lowell to A. E. Douglass, telegram, March 21, 1894.
43. A. E. Douglass to *Arizona Daily Gazette,* March 21, 1894.
44. G. W. Hull to A. E. Douglass, Feb. 18, 1894.
45. W. L. Van Horne to A. E. Douglass, Feb. 20, 1894.
46. C. G. Bell to A. E. Douglass, March 9, 1894.
47. W. A. Langhorne to A. E. Douglass, April 14, 1894.
48. F. D. Sugalls to A. E. Douglass, March 24, 1894.
49. Anton Proto to A. E. Douglass, April 6, 10, 1894.
50. W. R. Burrow to A. E. Douglass, March 24, 1894.
51. C. E. Hayes *et al* to A. E. Douglass, telegram, April 18, 1894.
52. *Coconino Sun,* April 26, 1894.
53. Ibid., citing the Williams *News.*
54. Ibid., citing the Prescott *Journal-Miner.*
55. Ibid., citing the Tucson *Star.*
56. Ibid.

Notes to Chapter 4

1. P. Lowell, *Mars,* London, 1895, v.
2. Otto Struve and Velta Zebergs, *Astronomy of the 20th Century,* New York, 1962, 143.
3. Lowell sent an expedition to Alianza, Chile, for the 1907 opposition of Mars; see Chapter 11 *infra.* He also contemplated a second expedition in 1911; see P. Lowell to O. Stone (Leander McCormick Observatory), Oct., 1910.
4. A. L. Lowell, *Biography of Percival Lowell,* New York, 1935, 70.
5. Ibid., 63, citing an unpublished introduction which P. Lowell apparently intended to use in his first volume of the *Annals of the Lowell Observatory.*
6. P. Lowell, "Atmosphere: In Its Effect on Astronomical Research," lecture text, *ca.* spring 1897; *L.O.A.*
7. W. H. Pickering, "Climate as Related to Astronomical Observations," *Popular Astronomy,* 3:465 (1896), and A. E. Douglass, "Atmosphere, Telescope and Observer," ibid., 5:64 (1897).
8. P. Lowell, "Atmosphere."
9. Ibid.
10. Ibid.
11. Struve and Zebergs, *op. cit.,* 123.
12. P. Lowell, "Atmosphere."

13. Ibid.
14. Ibid.
15. See Appendix A.
16. P. Lowell, "Atmosphere."
17. Ibid.
18. Ibid.
19. A. L. Lowell, *op. cit.*, 98–100.
20. P. Lowell to A. E. Douglass, April 21, 1897.
21. P. Lowell to V. M. Slipher, June 28, 1902.
22. See Appendix B.
23. P. Lowell to S. Newcomb, Dec. 29, 1902.
24. It is interesting to note that as late as 1971, an astronomer at the California Institute of Technology's Jet Propulsion Laboratory declared that "seeing" and its related phenomena "have been so intensively studied in the last few years that they are now fairly well understood—but not by astronomers!" See Andrew T. Young, "Seeing and Scintillation," *Sky and Telescope,* 42:139 (1971). Young's article, incidentally, includes this passage which, except for its tentative tone, seems startlingly Lowellian:

    "Evidently, there is an optimum aperture to use if maximum resolution is wanted. With too small a telescope, the seeing is good (apart from image motion), but the telescope size limits the resolution. With too big a telescope, larger and stronger turbulent areas are included, and seeing limits the resolution. The aperture must be matched to the seeing conditions."

    To which Lowell himself might have added: *Quod erat demonstrandum!*
25. A. L. Lowell, *op. cit.*, 61. Lowell claimed to be able to distinguish sixteen stars in the Pleiades with the naked eye; see E. P. Martz, Jr., "William Henry Pickering 1858–1938—An Appreciation," *Pop. Astron.*, 46:299 (1938).
26. P. Lowell to W. A. Cogshall, Sept. 23, 1903.
27. P. Lowell, "Means, Methods and Mistakes in the Study of Planetary Evolution," text dated April 13, 1905, of paper communicated to the Royal Astronomical Society Nov. 15, 1905; *L.O.A.*
28. Ibid. Note 24 is also pertinent here.
29. Ibid.
30. Ibid.
31. Ibid.
32. Ibid.
33. P. Lowell to A. E. Douglass, Dec. 13, 1894.
34. P. Lowell to A. E. Douglass, Jan. 2, 1895.
35. A. E. Douglass to P. Lowell, Jan. 14, 1895. The Sykes brothers, Godfrey and Stanley, were Flagstaff residents employed by the observatory to perform various nonobservational tasks on Mars Hill. Stanley became a master instrument maker, serving the observatory for more than fifty years and contributing significantly to its work. In 1899, Godfrey, the older brother, successfully floated the moveable wooden dome of the Lowell 24-inch telescope in tanks of water, although this system proved impractical and was used only briefly; see G. Sykes to A. E. Douglass, May 11, 1899.
36. P. Lowell to A. E. Douglass, March 5, 1895.
37. P. Lowell to A. E. Douglass, March 18, 1895.
38. A. E. Douglass to P. Lowell, March 25, 1895.
39. A. E. Douglass, "Notice to the Citizens of Flagstaff," text dated March 29, 1895; *L.O.A.*
40. A. E. Douglass to W. H. Pickering, Jan. 9, 1901.
41. A. E. Douglass to P. Lowell, March 18, 1895. The *Flagstaff Democrat* was a short-lived weekly rival to the *Coconino Sun*.
42. P. Lowell, *Annals of the Lowell Observatory,* I:4 (1898).
43. P. Lowell to A. E. Douglass, cablegram, March 4, 1896.

44. A. L. Lowell, *op. cit.*, 99–102.
45. See correspondence between V. M. Slipher and C. O. Lampland, October-November 1903.
46. P. Lowell to C. O. Lampland, Feb. 4, 1907.
47. P. Lowell to C. O. Lampland, June 10, 1908, to Ford Harvey, June 29, 1908, and to E. P. Ripley, June 29, 1908.
48. P. Lowell to W. A. Cogshall, Sept. 23, 1903.
49. P. Lowell to V. M. Slipher, Aug. 24, 1903.
50. P. Lowell, *Mars and Its Canals*, New York, 1906, 6–7.
51. Ibid., 12–15.
52. Ibid.
53. P. Lowell, "Is Mars Inhabited?" *Outlook*, April 13, 1907; *L.O.A.*
54. P. Lowell, "The Lowell Observatory and Its Work," text of lecture given at Prescott, A. T., Nov. 24, 1909; *L.O.A.* Ironically, the growth of Flagstaff's population to more than 30,000 in recent years, and the resulting increase in atmospheric pollution and more particularly night light, has forced Lowell Observatory to move its major telescopes to an isolated mesa twelve miles southeast of the city.

Notes to Chapter 5

1. P. Lowell, *Mars*, London, 1895, 153.
2. W. Whewell, *Of the Plurality of Worlds*, London, 1867. This fifth edition includes the prefaces of all preceding editions as well as Whewell's *A Dialogue on the Plurality of Worlds*, a supplement in which he replied point by point to criticisms resulting from the earlier editions of his work.
3. C. A. Stedtfeldt, "Can Organic Life Exist in the Planetary System Outside the Earth?" *Publications of the Astronomical Society of the Pacific*, 35:91 (1894).
4. C. Flammarion, "Can Organic Life Exist in the Solar System Anywhere But on the Planet Mars?" ibid., 37:214 (1894).
5. E. S. Holden, "The Lowell Observatory in Arizona," ibid., 36:160 (1894).
6. Ibid., quoting P. Lowell, "The Lowell Observatory and Its Work," Boston *Commonwealth*, May 26, 1894.
7. Ibid. Recent studies by several astronomers indicate that Jupiter radiates about 2.5 times as much thermal energy as it receives from the sun.
8. Ibid.
9. Ibid. Venus did not long remain so modest for Lowell, however; see Chapter 8 *infra*.
10. Lowell Observatory has made systematic visual and/or photographic studies of Mars at every opposition since 1894. Since early 1969, its Planetary Research Center has directed an International Planetary Patrol involving five other observatories in various parts of the world. The project, funded by the U.S. National Aeronautics and Space Administration, is designed to keep Mars and the other planets when available under 24-hour, day-to-day photographic surveillance.
11. P. Lowell, *Annals of the Lowell Observatory*, I:3 (1898).
12. P. Lowell, "Means, Methods and Mistakes in the Study of Planetary Evolution," text dated April 13, 1905; *L.O.A.*
13. P. Lowell, *Mars and Its Canals*, New York, 1906, 175.
14. P. Lowell, *Annals*, III:xiii (1905).
15. P. Lowell, "Means, Methods and Mistakes."
16. P. Lowell, *Mars as the Abode of Life*, New York, 1908, 146–47.
17. *The New York Times*, Dec. 8, 1907; *L.O.A.*
18. P. Lowell, "Is Mars Inhabited?" *Outlook*, April 13, 1907; *L.O.A.*

19. P. Lowell, *Annals*, I:79 (1898).

20. P. Lowell, *Mars*, 110–11.

21. P. Lowell, "Opposition of 1898–99," Supplement I, *Annals*, III:1 (1905).

22. P. Lowell, *Annals*, I:4 (1898).

23. Ibid., 406.

24. Ibid., 77.

25. P. Lowell, *Mars*, 85 ff.

26. W. H. Pickering, "The Seas of Mars," in *Annals*, I:77 (1898).

27. P. Lowell, *Mars*, 76–78.

28. P. Lowell to A. E. Douglass, Sept. 18, 1895. See also P. Lowell, *Mars*, 21 ff.

29. P. Lowell, *Annals*, I:376 (1898).

30. H. Masursky *et al*, "Mariner 9 Television Reconnaissance of Mars and Its Satellites: Preliminary Results," *Science*, 175:294 (1972).

31. A. E. Douglass, "Search for New Satellites of Mars," in *Annals*, I:377 (1898).

32. P. Lowell, *Annals*, I:79 (1898).

33. P. Lowell to W. H. Pickering, Nov. 25, 1913.

34. P. Lowell, *Mars*, 148–49.

35. P. Lowell, *Annals*, III:47 (1905).

36. P. Lowell, "Mars: Forecasts and Fulfillments," text *ca*. August 1916; *L.O.A.*

37. A. E. Douglass, "The Lowell Observatory and Its Work," draft dated Jan. 30, 1895, of an article for *Popular Astronomy; L.O.A.*

38. P. Lowell, *Mars and Its Canals*, 209.

39. A. E. Douglass, *Annals*, II:441 (1900). Antoniadi was a well-known English astronomer and a frequent critic of Lowell's martian ideas.

40. A. E. Douglass, "Canals in the Dark Areas," in *Annals*, I:253 (1898).

41. P. Lowell, *Mars*, 187.

42. Ibid., 199–200.

43. Lowell, *Annals*, I:44 (1898).

44. P. Lowell, *Mars*, 90–91, and E. E. Barnard, "Measures of the Polar Caps of Mars," *Pop. Astron.*, 2:433 (1895). Also see P. Lowell to H. H. Turner, March 18, 1903. Astronomers now accept Barnard's interpretation, noting the dust storms that rage over the planet during its perihelial oppositions.

45. A. E. Douglass, "Description of Terminator Observations," in *Annals*, I:338 (1898).

46. Ibid., 337. See also A. E. Douglass to P. Lowell, Dec. 5, 1894.

47. P. Lowell to A. E. Douglass, Dec. 13, 1894.

48. W. H. Pickering to P. Lowell, Sept. 16, 1894.

49. W. H. Pickering to P. Lowell, Sept. 22, 1894, and A. E. Douglass to P. Lowell, March 7, 1895.

50. W. H. Pickering, "Observations of the Planet Jupiter and Its Satellites—1894 and 1895," in *Annals*, II:1 (1900).

51. A. E. Douglass, "Eigenschein and Zodiacal Light Observations," draft dated April 8, 1895; *L.O.A.*

52. A. E. Douglass, "Atmospheric Currents," draft dated Jan. 1, 1895; *L.O.A.*

53. "Witnesses to Invention," document dated Oct. 15, 1894 and signed by P. Lowell and W. H. Pickering; *L.O.A.*

54. P. Lowell to A. E. Douglass, Jan. 5, 1895.

55. A. E. Douglass to P. Lowell, July 2, 1894.

56. P. Lowell to A. E. Douglass, July 17, 1894.

57. P. Lowell to A. E. Douglass, Aug. 4, 1894.

58. P. Lowell to A. E. Douglass, July 28, 1894.

Notes to Chapter 6

1. P. Lowell, *Annals of the Lowell Observatory,* I:201 (1898).
2. Ibid., 215.
3. P. Lowell, "Mars: Forecasts and Fulfillments," text *ca.* August 1916; *L.O.A.*
4. P. Lowell, *Mars,* London, 1895, 7-8.
5. P. Lowell to A. E. Douglass, March 25, 1895. (Punctuation added for clarity)
6. Ibid.
7. P. Lowell, "Mars," *Astronomy and Astrophysics,* 13:538 (1894).
8. P. Lowell, "Mars," *Popular Astronomy,* 2:1 (1894); "The Polar Snows," ibid., 2:52 (1894); "Spring Phenomena," ibid., 2:97 (1894); "Atmosphere," ibid., 2:153 (1894); "The Canals. I." ibid., 2:255 (1895); and "Oases," ibid., 2:343 (1895).
9. A. E. Douglass, "The Lowell Observatory and Its Work," *Pop. Astron.,* 2:395 (1895).
10. P. Lowell, *Mars and Its Canals,* New York, 1906, 4-5.
11. P. Lowell, *Mars as the Abode of Life,* New York, 1908, 3-12, and *The Evolution of Worlds,* New York, 1909, 1-30.
12. P. Lowell, *Mars,* 2.
13. Ibid., 6.
14. Unidentified newspaper, *ca.* August 1896; *L.O.A.*
15. A good example in the *L.O.A.* is a full-page Sunday supplement article, unidentified newspaper, *ca.* April 1904, in which martians are pictured as an artist imagined they were conceived by Lowell, astronomer Sir Robert Ball, scientist-theologian Emanuel Swedenborg, fiction writer H. G. Wells, and others.
16. P. Lowell, "Is Mars Inhabited?" *Outlook,* April 13, 1907, *L.O.A.*
17. P. Lowell to K. W. Ontbrank, Nov. 1, 1916.
18. P. Lowell to Austin Publishing Co., Nov. 11, 1905.
19. P. Lowell, *Mars,* 7-21.
20. Ibid., 29-30.
21. H. Masursky *et al,* "Mariner 9 Television Reconnaissance of Mars and Its Satellites: Preliminary Results," *Science,* 175:294 (1972). See also R. N. Watts Jr., "Some Mariner 9 Observations of Mars," *Sky and Telescope,* 43:208 (1972).
22. P. Lowell, *Mars,* 43-49, 64-65, 170.
23. A. J. Kliore *et al,* "Mariner 9 S-Band Occultation Experiment: Initial Results on the Atmosphere and Topography of Mars," *Science,* 175:313 (1972).
24. P. Lowell, *Mars,* 43-45.
25. P. Lowell, *Annals,* I:45 (1898).
26. P. Lowell, *Mars,* 43.
27. Ibid., 58 ff.
28. Ibid.
29. T. D. Parkinson and D. M. Hunten, "Martian Dust Storm: Its Depth on 25 November 1971," *Science,* 175:323 (1972), and A. L. Hammond, "The New Mars: Volcanism, Water, and a Debate Over Its History," ibid., 179:463 (1973).
30. Samuel Glasstone, *Book of Mars,* Washington, D.C., 1968, 84.
31. P. Lowell, *Mars,* 49 ff. See also *Annals,* I:247 (1898).
32. Kliore *et al, op. cit.*
33. P. Lowell, *Mars and Its Canals,* 18. The quotation is virtually identical in his earlier *Mars,* 74.
34. P. Lowell, *Mars,* 53-58. Earth's critical velocity is about 6.9 mps.
35. Ibid., 75.

36. R. H. Steinbacher *et al*, "Mariner 9 Science Experiments: Preliminary Results," *Science*, 175:293 (1972), and M. B. McElroy, "Mars: an evolving atmosphere," ibid., 175:443 (1972).
37. P. Lowell, *Mars*, 76.
38. Ibid., 76–78.
39. Ibid.
40. P. Lowell to W. Kaempffert, Nov. 24, 1906.
41. C. Sagan *et al*, "Variable Features on Mars: Mariner 9 Global Results," report to the annual meeting of the American Astronomical Society's Division of Planetary Sciences, Tucson, Arizona, March 21, 1973.
42. H. H. Bates, "The Chemical Composition of Mars' Atmosphere," *Publications of the Astronomical Society of the Pacific*, 38:300 (1894).
43. A. E. Douglass to P. Lowell, Jan. 16, 1895.
44. P. Lowell, *Mars*, 80–84.
45. P. Lowell, *Annals*, I:45 (1898).
46. J. H. Poynting, "On Prof. Lowell's Method for Evaluating Surface-Temperatures of the Planets," *Collected Scientific Papers*, Cambridge, 1920. (Paper first published in *The Philosophical Magazine*, July 1907.)
47. P. Lowell to C. O. Lampland, Dec. 14, 1907.
48. S. C. Chase Jr. *et al*, "Infrared Radiometry Experiment on Mariner 9," *Science*, 175:308 (1972), and B. C. Murray *et al*, "Geological Framework of the South Polar Region on Mars," *Icarus*, 17:328 (1972).
49. R. A. Hanel *et al*, "Infrared Spectroscopy Experiment on the Mariner 9 Mission: Preliminary Results," *Science*, 175:305 (1972).
50. P. Lowell, *Mars*, 49–53.
51. P. Lowell, *Annals*, I:47 (1898).
52. P. Lowell, "The Surface Temperature of Mars From a New Determination of the Solar Heat Received By It," text of paper read to the American Academy of Arts and Sciences Dec. 12, 1906; *L.O.A.* See also Lowell, "Insolation," *Lowell Observatory Bulletin 30*, June 22, 1907.
53. *Scientific American*, July 20, 1907; *L.O.A.*
54. S. C. Chase Jr. *et al*, *op. cit.* For Mariner 6 and 7, see "First Findings From the Mariner Flybys," *Sky and Telescope*, 38:232 (1969).
55. P. Lowell, *Mars*, 97–111.
56. Glasstone, *op. cit.*, 126 ff., 219.
57. P. Lowell, *Mars*, 113.
58. Ibid., 114.
59. P. Lowell, *Mars and Its Canals*, 119.
60. Ibid., 84. Also P. Lowell, *Mars*, 119–20.
61. Glasstone, *op. cit.* 111 ff.
62. Lowell's usual greeting to Schiaparelli; see P. Lowell to G. V. Schiaparelli, July 26, 1907, and March 16, 1909, for examples. Lowell addressed Flammarion as *cher ami Martien;* the Italian was "master," the Frenchman "friend."
63. P. Lowell, *Mars and Its Canals*, 115.
64. P. Lowell, *Mars*, 128.
65. Ibid.
66. Ibid., 129.
67. P. Lowell, "Oases," *Pop. Astron.*, 2:348 (1895).
68. P. Lowell, *Annals*, I:95 (1898), and *Mars*, 130–34.
69. P. Lowell, *Annals*, I:97 (1898).
70. Ibid., 192.
71. P. Lowell, *Mars*, 140.
72. Ibid., 149–54.
73. Ibid., 154 ff., 191.

74. P. Lowell, *Annals,* I:193 (1898).
75. P. Lowell, "Areography," text dated April 4, 1902; *L.O.A.*
76. P. Lowell, *Annals,* I:194 (1898).
77. P. Lowell, *Mars,* 154–65.
78. P. Lowell, *Annals,* I:199 (1898).
79. P. Lowell, *Mars,* 188–91.
80. Ibid., 196.
81. Ibid., 176–83.
82. Ibid., 184–85.
83. Ibid., 186.
84. P. Lowell to S. Newcomb, March 15, 1903. (Punctuation added)
85. P. Lowell, *Mars,* 197–200.
86. P. Lowell, *Annals,* I:250 (1898).
87. P. Lowell, *Mars,* 201–02.
88. Ibid., 202–07.
89. Ibid., 207.
90. Ibid., 209.
91. Ibid., 210.
92. Ibid., 211–12.

Notes to Chapter 7

1. Unidentified newspaper, *ca.* October 1896; *L.O.A.* There are literally hundreds of such short "filler" clippings in seven large scrapbooks in the Archives.
2. Unidentified newspaper, *ca.* spring 1905; *L.O.A.*
3. Unidentified newspaper, *ca.* spring 1905; *L.O.A.*
4. Unidentified newspaper, *ca.* summer 1905; *L.O.A.*
5. San Francisco *Weekly Post,* March 13, 1895; *L.O.A.*
6. E. S. Holden, editor's note to E. E. Hale's "Latest News From Mars," *Publications of the Astronomical Society of the Pacific,* 41:116 (1895). Hale's article appeared previously in the Boston *Commonwealth* and the *Scientific American.*
7. W. W. Campbell, "Mars, by Percival Lowell," *Pubs. Astron. Soc. Pac.* 51:207 (1896).
8. Ibid. A typewritten, hand-annotated English text of Schiaparelli's early martian memoir, *Astronomical and Physical Observations on the Axis of Rotation and on the Topography of the Planet Mars,* Royal Academy of the Lincei, 1878, in the Lowell Observatory library may be one of Pickering's translations.
9. Campbell, *op. cit.*
10. Ibid. Lick astronomer J. M. Schaeberle, like Pickering, had seen vague streak-like features in the martian dark areas, but unlike Pickering had drawn no analogies from them.
11. Ibid.
12. Ibid.
13. Ibid.
14. A. E. Douglass, "The Lick Review of 'Mars,'" *Popular Astronomy,* 4:199 (1896). The draft of this article, in both Lowell's and Douglass' handwriting, is in the *L.O.A.*
15. Ibid.
16. Ibid.
17. Ibid.
18. Ibid.
19. W. W. Payne, "The Planet Mars," *Pop. Astron.,* 3:345 (1896).

20. Unidentified newspaper, *ca.* spring 1896; *L.O.A.*
21. "Astronomer Barnard Goes to Yerkes Observatory," General Notes, *Pop. Astron.,* 3:46 (1895). Holden, incidentally, resigned as Lick director in 1897, also apparently because of conflicts with members of the Lick staff; see "Lick Observatory Trouble," ibid., 5:335 (1897).
22. C. A. Young, "Mr. Lowell's Theory of Mars," *Cosmopolitan,* October 1895; *L.O.A.*
23. C. A. Young, "Is Mars Inhabited?" Boston *Herald,* Oct. 18, 1896; *L.O.A.* This article was reprinted in *Pubs. Astron. Soc. Pac.,* 53:306 (1896).
24. Ibid.
25. Ibid.
26. Unidentified newspaper, *ca.* spring 1896; *L.O.A.* The item notes the "almost journalistic timeliness" of du Maurier's novel as serialized in *Harper's,* January through June 1897.
27. *Science,* May 12, 1897; *L.O.A.*
28. H. N. Russell, "The Heavens in March, 1901," unidentified newspaper; *L.O.A.*
29. Unidentified newspaper, *ca.* summer 1895; *L.O.A.* A number of other unidentified clippings of about this same period refer to opposition to Lowell's ideas within the academy.
30. P. Lowell, *Mars and Its Canals,* New York, 1906, viii.
31. Ibid., ix.
32. P. Lowell, "The Atmosphere of Mars," text dated January 1916 of an article for *Scientia* (Italy); *L.O.A.*
33. P. Lowell, "The On-Coming of a Martian Winter," undated text of article for the *Burr-McIntosh Monthly; L.O.A.*
34. A. E. Douglass to *Associated Press,* telegram, Oct. 24, 1896.
35. C. V. Van Anda to P. Lowell, May 18, 1905.
36. P. Lowell to U.S. Treasury Dept., Sept. 25, 1915, and to W. G. McAdoo, Jan. 26, 1916.
37. E. S. Holden, "The Lowell Observatory in Arizona," *Pubs. Astron. Soc. Pac.,* 36:160 (1894).
38. P. Lowell to F. J. Stimson, quoted in Ferris Greenslet, *The Lowells and Their Seven Worlds,* Boston, 1945, 355.
39. St. Louis *Globe-Democrat,* Jan. 24, 1909; *L.O.A.*
40. *Japan Daily Mail* (Tokyo), Feb. 4, 1909; *L.O.A.*
41. L. F. Ward to P. Lowell, March 31, 1907.
42. *Science,* March 29, 1907; *L.O.A.*
43. *The Independent,* Dec. 27, 1906; *L.O.A.*
44. *Catholic World,* February, 1907; *L.O.A.*
45. *The New York Times,* for example, used 2,300 words in its Dec. 19, 1908, review of *Mars as the Abode of Life,* and the *Catholic World's* review of *Mars and Its Canals,* cited above, ran 5,900 words.
46. E. E. Hale, "Latest News From Mars," *Pubs. Astron. Soc. Pac.,* 41:116 (1895).
47. E. E. Hale, "Easter Sermon," Boston *Commonwealth,* April 20, 1895; *L.O.A.*
48. Boston *Globe,* Nov. 11, 1906; *L.O.A.*
49. Boston *Transcript,* Nov. 9, 1906; *L.O.A.*
50. Boston *Transcript,* Oct. 16, 1906; *L.O.A.*
51. M. L. Crowell to P. Lowell, Nov. 17, 1906.
52. Anonymous to P. Lowell, undated (found with other letters relating to the 1906 lectures); *L.O.A.*
53. Anonymous to P. Lowell, undated (found with other letters relating to the 1906 lectures); *L.O.A.*
54. Boston *Herald,* Oct. 21, 1906; *L.O.A.*

Notes to Chapter 8

1. Edith H. Leonard to A. E. Douglass, Dec. 30, 1895, and A. Lawrence Lowell to A. E. Douglass, Dec. 31, 1895, quoting Lowell.
2. A. L. Lowell, *Biography of Percival Lowell,* New York, 1935, 92.
3. "Agreement," April 25, 1895; *L.O.A.* See also unidentified newspapers, *ca.* July-August, 1896; *L.O.A.*
4. P. Lowell to A. E. Douglass, May 27, 1896.
5. Boston *Journal,* Aug. 10, 1896. See also "The Lowell Observatory," *Popular Astronomy,* 4:163 (1896), and T.J.J. See, "A Sketch of the New 24-Inch Refractor," ibid., 4:297 (1896).
6. A. E. Douglass, *Annals of the Lowell Observatory,* II:208 (1900).
7. A. E. Douglass to P. Lowell, Nov. 23, 24, and 25, 1896.
8. A. E. Douglass, "Drawings of Jupiter's Third Satellite," *Pop. Astron.,* 5:308 (1897).
9. A. E. Douglass, "An Ascent of Popocatepetl," ibid., 5:505 (1898).
10. P. Lowell, *Annals,* III:100 (1905).
11. New York *Herald,* Sept. 6, 1896, and unidentified newspapers, *ca.* September 1896; *L.O.A.* See also "Companion to Sirius Visible Again," *Pop. Astron.* 4:220 (1896).
12. San Francisco *Chronicle,* Oct. 31, 1896.
13. T.J.J. See to A. E. Douglass, Dec. 1, 1897.
14. Chicago *Tribune,* April 17, 1897, and unidentified newspapers, *ca.* April-May 1897; *L.O.A.*
15. Boston *Transcript,* October, 1896; *L.O.A.*
16. G. H. Pettengill and R. B. Dyce, "A Radar Determination of the Rotation of the Planet Mercury," *Nature,* 206:1240 (1965).
17. H.-S. Liu and J. A. O'Keefe, "Theory of Rotation for the Planet Mercury," *Science,* 150:1717 (1965).
18. A. M. Clerke, *A Popular History of Astronomy During the Nineteenth Century,* London, 1902, 250–51.
19. Boston *Transcript,* November, 1896; *L.O.A.*
20. Ibid.
21. P. Lowell, "Detection of Venus' Rotation Period and of the Fundamental Physical Features of the Planet," *Pop. Astron.,* 4:281 (1896); and "Determination of the Rotation and Surface Character of the Planet Venus," *Monthly Notices of the Royal Astronomical Society,* 57:148 (1897).
22. Clerke, *op. cit.,* 255.
23. "Venus, 1896–97," Lowell Observatory Observation Logbook, *L.O.A.*
24. E. S. Holden to A. E. Douglass, Aug. 19, 1897.
25. E. E. Barnard to A. E. Douglass, May 5, 1898.
26. Unidentified newspaper, *ca.* April 1897; *L.O.A.*
27. A. G. Clark, *Science,* 5:768 (1897).
28. P. Lowell, "The Markings on Venus," *Astronomische Nachrichten,* No. 3823. Lowell's paper is dated July 3, 1902.
29. P. Lowell, "New Observations of the Planet Mercury," *Memoirs of the American Academy of Arts and Sciences,* XII:433 (1898). The paper was presented May 12, 1897.
30. P. Lowell to A. E. Douglass, April 21, 1897.
31. P. Lowell to A. E. Douglass, May 5 and June 15, 1897.
32. P. Lowell to A. E. Douglass, July 19, 1897.
33. P. Lowell to A. E. Douglass, Sept. 20, 1897.

34. W. L. Putnam to T. J. J. See, and to W. A. Cogshall, Sept. 29, 1897. Putnam was married to Lowell's sister, Elizabeth.
35. T. J. J. See to A. E. Douglass, Oct. 7, 1897.
36. W. L. Putnam to T. J. J. See, Dec. 6, 1897.
37. W. L. Leonard to A. E. Douglass, June 6, 1899.
38. A. L. Lowell, *op. cit.*, 98–102.
39. H. H. Turner to P. Lowell, Jan. 19, 1901.
40. A. E. Douglass to D. A. Drew, Oct. 18, 1899.
41. A. A. Belopolsky, "Ein versuch die Rotationgeschwindigkeit des Venusaequators auf spectrographischen Wege zu bestimmen," *Astronomische Nachrichten,* No. 3641 (1900).
42. P. Lowell to V. M. Slipher, Nov. 5, 1902.
43. P. Lowell to A. E. Douglass, Dec. 11, 1900.
44. See Chapter 9 *infra.*
45. P. Lowell to H. S. Pritchett, Sept. 30, 1902.
46. P. Lowell to H. S. Pritchett, March 23, 1903.
47. V. M. Slipher, "Spectrographic Investigation of the Rotation of Venus," *Lowell Observatory Bulletin 3,* June, 1903.
48. P. Lowell, *The Evolution of Worlds,* New York, 1909, 88.
49. V. M. Slipher to P. Lowell, Sept. 6, 1903.
50. P. Lowell, "Venus, 1903," *Lowell Observatory Bulletin 6,* Jan. 1, 1904.
51. P. Lowell to E. W. Maunder, Nov. 28, 1903.
52. P. Lowell to Hasket Derby, March 11, 1907. Derby was Lowell's Boston opthalmologist.
53. P. Lowell to "the literary heir(s)" of Charles A. Young, Feb. 13, 1908. Lowell used the 1902 edition of Young's book.
54. P. Lowell to editor, *The Observatory,* May 12, 1913.
55. R. M. Goldstein, "Radar Observations of Venus," *Astronomical Journal,* 69:12 (1964); R. B. Dyce, G. H. Pettengill and I. I. Shapiro, "Radar Determinations of the Rotations of Venus and Mercury," ibid., 72:351 (1967); P. Goldreich and S. J. Peale, "Spin-Orbit Coupling in the Solar System II: The Resonant Rotation of Venus," ibid., 72:662 (1967); and R. L. Carpenter, "A Radar Determination of the Rotation of Venus," ibid., 75:61 (1970).
56. T. J. J. See to A. E. Douglass, Sept. 17, 1897 (two letters).
57. P. Lowell to A. E. Douglass, June 25, 1897.
58. T. J. J. See to A. E. Douglass, Oct. 17, 1897.
59. S. L. Boothroyd to W. L. Putnam, June 17, 1898.
60. Ibid.
61. W. A. Cogshall to W. L. Putnam, June 23, 1898.
62. W. A. Cogshall to A. E. Douglass, June 23, 1898.
63. A. E. Douglass to W. L. Putnam, June 28, 1898.
64. W. L. Putnam to A. E. Douglass, June 10 and 14, 1898, and telegram, July 8, 1898.
65. A. E. Douglass to H. S. Pritchett, Feb. 19, 1899, and to C. G. Comstock, *ca.* spring 1899.
66. P. Lowell to W. D. McPherson, June 12, 1907.
67. A. E. Douglass to J. Jastrow, Jan. 9, 1901.
68. A. E. Douglass to W. L. Putnam, March 12, 1901; Andrew Ellicott Douglass Papers, Special Collections, University of Arizona Library. I am much indebted to Daphne Overstreet of the University of Arizona's Tree Ring Laboratory for calling my attention to the existence of this letter in the extensive Douglass collection.

69. A. E. Douglass to Mrs. P. Lowell, Aug. 17, 1917; A. E. Douglass Papers.
70. W. A. Cogshall to P. Lowell, Feb. 2, 1903 (with enclosures).
71. P. Lowell to A. E. Douglass, Oct. 14, 1901.
72. El Paso *Herald*, Oct. 21, 1916; *L.O.A.*
73. Gilliam Press Syndicate (E. Leslie Gilliam) to A. E. Douglass, Jan. 9, 1901.
74. D. E. Parks to A. E. Douglass, Jan. 17, 1901. (Punctuation added)
75. G. P. Serviss, column, unidentified newspaper, January 1901; *L.O.A.*
76. Ibid.

Notes to Chapter 9

1. A. M. Clerke, *A Popular History of Astronomy During the Nineteenth Century,* London, 1902, 400. The date, incidentally, also marked the 1901 opposition of Mars.
2. P. Lowell to A. E. Douglass, Feb. 27, 1901. Danish astronomer Tycho Brahe made a classic study of the exceptionally bright nova that appeared in Cassiopeia in 1572, achieving such brilliance that it was visible in full daylight for several weeks.
3. P. Lowell, *The Evolution of Worlds,* New York, 1909, 12–15.
4. P. Lowell to A. E. Douglass, March 22, 1901.
5. P. Lowell, *Annals of the Lowell Observatory,* III:144 (1905).
6. W. A. Cogshall to P. Lowell, June 24, 1901.
7. P. Lowell to W. A. Cogshall, July 7, 1901.
8. J. S. Hall, "V. M. Slipher's Trail-blazing Career," *Sky and Telescope,* 39:84 (1970). See also *Arizona Daily Sun* (Flagstaff), Nov. 10, 1969.
9. P. Lowell to W. A. Cogshall, Sept. 27, 1902, and to C. O. Lampland, Oct. 13, 1902.
10. Otto Struve and Velta Zebergs, *Astronomy of the 20th Century,* New York, 1962, 147.
11. Ibid. See also P. Lowell to V. M. Slipher and to W. A. Cogshall, June 26, 1906, and John A. Miller to V. M. Slipher, Jan. 4, 1908. The other two Lawrence Fellows were John C. Duncan, 1905–06, later professor of astronomy at Harvard University and Wellesley College and author of a popular textbook; and Kenneth P. Williams, 1907–08, who returned to Indiana to teach celestial mechanics and eventually became an authority on Civil War history.
12. See Chapter 14 *infra.*
13. P. Lowell to V. M. Slipher, May 23, 1904.
14. Hall, *op. cit.*
15. P. Lowell to V. M. Slipher, Sept. 17 and Nov. 14, 1901.
16. P. Lowell to V. M. Slipher, Oct. 21, 1901.
17. P. Lowell to V. M. Slipher, Oct. 28, 1901.
18. P. Lowell to V. M. Slipher, March 10, 1902.
19. P. Lowell to V. M. Slipher, March 26, 1902.
20. P. Lowell to V. M. Slipher, Dec. 18, 1901.
21. P. Lowell to V. M. Slipher, Jan. 4, 1902.
22. P. Lowell to H. S. Pritchett, Sept. 30, 1902, and to C. A. Young, Oct. 16 and 20, 1902.
23. P. Lowell to J. A. Brashear, Oct. 6, 1902. Keeler was a well-known astronomer who had done early spectrographic work at Allegheny Observatory and later at Lick where he was director for the two years preceding his death in 1900.
24. P. Lowell to V. M. Slipher, Oct. 21, 1902.
25. P. Lowell to V. M. Slipher, Nov. 5, 1902.

26. P. Lowell, "Spectrographic Proof of the Rotations of Jupiter, Saturn and Mars," text dated December 1902 of paper for the American Association for the Advancement of Science; *L.O.A.*

27. P. Lowell to V. M. Slipher, Oct. 21, 1902, Nov. 17, 1902, and August 24, 1903.

28. P. Lowell to unidentified addressee, Oct. 27, 1913.

29. P. Lowell and V. M. Slipher, "Spectrographic Discovery of the Rotation Period of Uranus," *Lowell Observatory Bulletin 53* (1912). See also P. Lowell to V. M. Slipher, Nov. 19, 1910.

30. V. M. Slipher to P. Lowell, Aug. 17, 1903.

31. V. M. Slipher to P. Lowell, April 3, 1904.

32. V. M. Slipher, "On the Spectrum of Uranus and Neptune," *Lowell Observatory Bulletin 13,* (1904).

33. T. Dunham Jr., "Note on the Spectrum of Jupiter and Saturn," *Publications of the Astronomical Society of the Pacific,* 45:42 (1933).

34. Clerke, *op. cit.,* 277. See also W. W. Campbell, "The Spectrum of Mars," *Pubs. Astron. Soc. Pac.,* 37:228 (1894).

35. P. Lowell to V. M. Slipher, Sept. 26, 1902.

36. P. Lowell to V. M. Slipher, Oct. 4, 1902.

37. P. Lowell to V. M. Slipher, Oct. 11 and 13, 1902.

38. P. Lowell to V. M. Slipher, Oct. 28, 1902.

39. P. Lowell to V. M. Slipher, Oct. 11, 1902.

40. P. Lowell to V. M. Slipher, Oct. 24, 1902.

41. P. Lowell to G. Cramer Dry Plate Co., April 8, 1903, and to V. M. Slipher, Sept. 10 and Oct. 13, 1904.

42. V. M. Slipher to P. Lowell, Oct. 19, 1904.

43. P. Lowell, "A New Method of Testing Spectroscopically a Martian Atmosphere," *Lowell Observatory Bulletin 17* (1905).

44. Ibid.

45. V. M. Slipher, "An Attempt to Apply Velocity Shift to Detecting Atmospheric Lines in the Spectrum of Mars," ibid.

46. W. S. Adams and T. Dunham Jr., "The B-band of Oxygen in the Spectrum of Mars," *Publications of the American Astronomical Society,* 7:171 (1933).

47. H. Spinrad, G. Munch and L. Kaplan, "Spectrographic Determination of Water Vapor on Mars," *Astrophysical Journal,* 137:1319 (1963).

48. P. Lowell to V. M. Slipher, Jan. 3, 1905, and July 31 and August 26, 1916.

49. P. Lowell, "A New Method of Testing Spectroscopically a Martian Atmosphere," *op. cit.* See also P. Lowell to R. J. Wallace, Jan 29 and Feb. 5, 1907, and Feb. 25, 1908, and E. E. Barnard to V. M. Slipher, April 20 and May 26, 1906.

50. V. M. Slipher to P. Lowell, Feb. 4, 1908.

51. V. M. Slipher to P. Lowell, Feb. 20, 1908.

52. P. Lowell to V. M. Slipher, Feb. 26, 1908.

53. Buffalo (N.Y.) *Times,* Boston *Globe,* London *Daily Mail* and other publications, Feb. 26, 1908; *L.O.A.*

54. *Nature,* March 5 and 26, 1908.

55. P. Lowell to Capt. W. J. Barnette, to C. Flammarion, to H. Deslandres and to N. Lockyer, March 2-5, 1908. Barnette was superintendent of the U.S. Naval Observatory.

56. S. Newcomb to V. M. Slipher, Jan. 2 and Oct. 5, 1907.

57. P. Lowell to V. M. Slipher, March 16, 1908.

58. *The New York Times,* March 1, 1908; *L.O.A.*

59. W. W. Campbell to V. M. Slipher, May 1 and 11 and Sept. 4, 1908.

60. J. A. Miller to V. M. Slipher, Nov. 17, 1908.

61. V. M. Slipher to P. Lowell, Feb. 20, 1908.
62. V. M. Slipher to P. Lowell, April 26, 1908.
63. P. Lowell to V. M. Slipher, May 4 and 19, 1908.
64. P. Lowell, to *Encyclopedia Britannica,* Feb. 26, 1908.
65. P. Lowell, *Mars as the Abode of Life,* New York, 1908, 136–38.
66. V. M. Slipher to E. B. Frost, Oct. 13, 1908. Frost was director of Yerkes Observatory.
67. F. R. Moulton to V. M. Slipher, Oct. 22, 1908. Slipher at this time was corresponding with Moulton about obtaining the Ph.D. degree from the University of Chicago, a degree he received the following year from Indiana University.
68. S. Newcomb to V. M. Slipher, Dec. 9, 1908, and V. M. Slipher to S. Newcomb, Dec. 2 and 15, 1908.
69. P. Lowell to V. M. Slipher, March 5, 1908.
70. P. Lowell to V. M. Slipher, June 29, 1908.
71. P. Lowell to V. M. Slipher, Dec. 8, 1908.
72. Ibid. See also F. W. Very, "Measurements of the Intensification of Aqueous Bands in the Spectrum of Mars," *Lowell Observatory Bulletin 36* (1909).
73. F. W. Very, "Quantitative Measurements in the Intensification of Great B in the Spectrum of Mars," *Lowell Observatory Bulletin 41,* Sept. 16, 1909.
74. J. A. Miller to V. M. Slipher, Jan. 2, 1909.
75. W. W. Campbell, "Observations on the Spectrum of Mars," *Annual Report of the Smithsonian Institution,* 1910, 16–17.
76. W. W. Campbell to V. M. Slipher, Oct. 4, 1909.
77. P. Lowell, supplement to *Lowell Observatory Bulletin 43* (undated). Lowell here cites C. G. Abbott's article, "A Shelter for Observers on Mt. Whitney," *Smithsonian Miscellaneous Collection,* Vol. 5, Part 4, 506.
78. P. Lowell to Mark Wicks, Jan 25, 1910.
79. F. W. Very, "New Measures of Martian Absorption Bands on Plate Rm 3076," and "Intensification of Oxygen and Water Vapor Bands in the Martian Spectrum," *Lowell Observatory Bulletins 49* (Aug. 26, 1910) and *65* (April 16, 1914). See also F. W. Very to P. Lowell, Oct. 12 and Nov. 13, 1909, and to V. M. Slipher, March 19, 1914.
80. P. Lowell to R. W. Willson, May 5, 1914.
81. L. E. Jewell to P. Lowell, June 25, 1905, and D. T. MacDougal to P. Lowell, Sept. 2, 1907.
82. V. M. Slipher to W. A. Cogshall, Feb. 25, 1908.
83. P. Lowell to L. B. Buchanan, Aug. 4, 1915.
84. G. P. Kuiper, "Planetary Atmospheres and Their Origin," in G. P. Kuiper (ed.), *The Atmospheres of the Earth and the Planets,* Chicago, 1952, 399. Not necessarily actual lichens, however, as Kuiper pointed out in "On the Martian Surface Features," *Pubs. Astron. Soc. Pac.,* 67:271 (1955), in correcting an interpretation of his work. In "Visual Observations of Mars, 1956," *Astrophysical Journal,* 125:307 (1956), he refers simply to "some very hardy vegetation."
85. V. M. Slipher to J. A. Miller, Oct. 18, 1908.
86. J. A. Miller to V. M. Slipher, Nov. 17, 1908.
87. W. W. Campbell to V. M. Slipher, Oct. 4, 1909.
88. P. Lowell, "Motion of Molecules in the Tail of Halley's Comet," *Lowell Observatory Bulletin 48,* Sept. 5, 1910. See also P. Lowell to V. M. Slipher, June 20 and 22 and July 8, 1910, to David Gill, July 12, 1910, to James Dewar, July 8, 1910, and to a "Dr. Nutting" (U.S. Bureau of Standards), July 10 and 14, 1910. Arrhenius' book, *Worlds in the Making,* New York, 1908, was Slipher's birthday gift to Lowell in 1908.
89. P. Lowell to N. Lockyer, and to H. Deslandres, Nov. 4, 1910.
90. A. Pannekoek, *The History of Astronomy,* New York, 1961, 434–35.

91. V. M. Slipher to P. Lowell, July 2 and Aug. 17, 1903, and May 14, 1904, and P. Lowell to V. M. Slipher, Oct. 21 and 28, 1901 and May 23, 1904.

92. Struve and Zebergs, *op. cit.,* 372.

93. V. M. Slipher, "Peculiar Star Spectra Suggestive of Selective Absorption of Light in Space," *Lowell Observatory Bulletin 51,* December 1909.

94. J. Hartmann to V. M. Slipher, March 6, 1910; J. C. Kapteyn to V. M. Slipher, Oct. 20, 1909; P. Lowell to V. M. Slipher, April 11, 1913 (with enclosed letter from K. Schwartzschild), E. Hertzsprung to V. M. Slipher, April 16, 1911; and correspondence between E. B. Frost and V. M. Slipher between Oct. 18, 1908 and Jan. 25, 1910.

95. P. Lowell to J. C. Kapteyn, Oct. 1, 1909.

96. J. C. Duncan to V. M. Slipher, Dec. 23, 1912. See also Struve and Zebergs, *op. cit.,* 373-76. Campbell felt the lines originated in the atmospheres of the stars.

97. P. Lowell, "Epitome of Results at the Lowell Observatory—April, 1913–April, 1914," *Lowell Observatory Bulletin 59,* April 16, 1914.

98. V. M. Slipher, "On the Spectrum of the Nebula in the Pleiades," *Lowell Observatory Bulletin 55,* Dec. 20, 1912. See also V. M. Slipher to F. W. Very, Sept. 5, 1910; to P. Lowell, Dec. 16 and 17, 1912; to J. C. Duncan, Dec. 29, 1912; and to W. W. Campbell, March 14, 1913.

99. J. A. Miller to V. M. Slipher, April 24, 1913.

100. P. Lowell to V. M. Slipher, Dec. 24, 1912, and May 2, 1913; E. A. Fath to V. M. Slipher, Feb. 10, 1913; and J. C. Duncan to V. M. Slipher, Feb. 17, 1913.

101. P. Lowell to V. M. Slipher, Jan. 29 and Feb. 8, 1909.

102. V. M. Slipher to G. R. Agassiz, Jan. 15, 1917.

103. V. M. Slipher to P. Lowell, Feb. 26, 1909.

104. V. M. Slipher to P. Lowell, Dec. 3, 1910.

105. V. M. Slipher to W. W. Campbell, undated (*ca.* spring 1913).

106. V. M. Slipher to E. A. Fath, March 2, 1911, and to E. Hertzsprung, May, 1914.

107. V. M. Slipher to P. Lowell, Dec. 28, 1912, and Jan 2, 1913. See also V. M. Slipher to E. A. Fath, Jan. 18, 1913, and to J. M. Schaeberle, Aug. 3 and 23, 1911.

108. V. M. Slipher to P. Lowell, Feb. 3, 1913. See also V. M. Slipher, "The Radial Velocity of the Andromeda Nebula," *Lowell Observatory Bulletin 58* (undated).

109. P. Lowell to V. M. Slipher, Feb. 8, 1913.

110. V. M. Slipher to P. Lowell, April 12, May 14 and 16, 1913; and to J. C. Duncan, June 11, 1913.

111. W. W. Campbell to V. M. Slipher, Dec. 16, 1916, and V. M. Slipher to W. W. Campbell, undated reply.

112. Hall, *op. cit.,* 86.

113. William Bonner, *The Mystery of the Expanding Universe,* New York, 1964, 1.

114. V. M. Slipher to P. Lowell, telegram, May 24, 1914.

115. V. M. Slipher, "The Detection of Nebular Rotation," *Lowell Observatory Bulletin 62* (1914). See also telegrams between Slipher and P. Lowell, Nov. 20, 1915.

116. V. M. Slipher to J. A. Miller, July 2, 1913.

117. V. M. Slipher to E. Hertzsprung, May 8, 1914; to P. Lowell, Dec. 28, 1912; and to J. C. Duncan, Dec. 29, 1912.

118. V. M. Slipher, "Structure and Motions of Spiral Nebulae," *Proceedings of the American Philosophical Society,* 56:403 (1917). See also Hall, *op. cit.,* 86. Lowell may have influenced Slipher's thinking here, for as early as 1915 he had cited Slipher's spectra as evidence that "the spiral nebulae are not the prototype of our system, but of something larger and quite different, other galaxies of stars." See P. Lowell, "Nebular Motion," text dated Nov. 23, 1915 of lecture to Boston's Melrose Club; *L.O.A.*

119. V. M. Slipher to P. Lowell, May 16, 1913; to E. B. Frost, Nov. 4, 1913; and to A. S. Eddington, Feb. 5, 1922.
120. C. O. Lampland, "Observations of Nebulae for Position and Proper Motion," *Lowell Observatory Bulletin 73* (1915).
121. V. M. Slipher to J. C. Duncan, June 11, 1913; and to E. B. Frost, Nov. 4, 1913.
122. J. A. Miller to V. M. Slipher, March 2, 1916.
123. V. M. Slipher to P. Lowell, May 16, 1913.
124. V. M. Slipher to E. B. Frost, Nov. 4, 1913.
125. V. M. Slipher to F. W. Very, Jan. 12, 1917.
126. Hall, *op. cit.*
127. V. M. Slipher to E. B. Frost, Oct. 22, 1914.

Notes to Chapter 10

1. See Chapter 13 *infra.*
2. See Chapter 14 *infra.*
3. P. Lowell to Houghton, Mifflin and Co., March 19, 1903.
4. E. Ledger, *The Nineteenth Century,* 315:773 (1903), and Garrett P. Serviss, column, unidentified newspaper, July 18, 1903.
5. P. Lowell to The Macmillan Co., Dec. 10, 1906, and Feb. 11, 1908; and to E. C. March, March 12 and 16, 1909.
6. Herman Zeiger to P. Lowell, April 13, 1909.
7. Houghton, Mifflin and Co. to P. Lowell, Sept. 27, 1907.
8. A. L. Lowell, *Biography of Percival Lowell,* New York, 1935, 97–99.
9. I. T. Headland to P. Lowell, Aug. 12, 1907; see also Boston *Transcript,* Oct. 5, 1907.
10. P. Lowell to I. T. Headland, Sept. 17, 1907.
11. S. L. Miller, "A Production of Amino-Acids under Possible Primitive Earth Conditions," *Science,* 117:528 (1953). The significance of Miller's experiment to the question of extraterrestrial life is discussed in I. S. Shklovskii and Carl Sagan, *Intelligent Life in the Universe,* New York, 1966, 229–31.
12. P. Lowell, *Mars and Its Canals,* New York, 1906, 355–58.
13. E. Haeckel, *The Wonders of Life,* New York, 1905, 193 ff. Some of Lowell's evolutionary thinking also came from Haeckel, *Last Words on Evolution,* London, 1906, a copy of which, autographed to Lowell by Haeckel, is in the Lowell Observatory library.
14. Lowell, *Mars and Its Canals,* 352.
15. Ibid., 348.
16. Ibid., 358.
17. Ibid., 355, 358–59.
18. P. Lowell, *Mars as the Abode of Life,* New York, 1908, 35–37.
19. P. Lowell, *Mars and Its Canals,* 18.
20. P. Lowell, *Mars as the Abode of Life,* 98, 56–59.
21. P. Lowell, *Mars and Its Canals,* 87–88.
22. P. Lowell, *Mars as the Abode of Life,* 84–86. See also note, "Temperature Deduced From Heat Retained," 251–55, where he calculates a mean martian temperature of 47.7 degrees Fahrenheit. His result is 46.3 degrees F. with Arrhenius' formula.
23. P. Lowell, *Mars and Its Canals,* 19.
24. P. Lowell, *Mars as the Abode of Life,* 90 ff.
25. Ibid., 96.
26. Ibid., 98–103.
27. P. Lowell, *Mars and Its Canals,* 16.
28. Ibid., 153.

29. Ibid., 131 ff. Lowell used the fourth edition (1894) of Dana's book.

30. P. Lowell, *Mars as the Abode of Life*, 121.

31. Ibid., 122. See also W. L. Leonard to E. Huntington, Nov. 11, 1913.

32. Ibid., 125–26. See also P. Lowell to L. F. Ward, April 10, 1907. Ward's report, "The Petrified Forest of Arizona," is in the Smithsonian Institution's *Annual Report for 1899*, 289–301, (1901).

33. P. Lowell, *Mars and Its Canals*, 156 ff.

34. P. Lowell, *Mars as the Abode of Life*, 285–86.

35. Ibid., 208–11.

36. Ibid., 211. See also P. Lowell to Langdon Smith, June 26, 1906.

37. Ibid., 216.

38. P. Lowell, *Mars and Its Canals*, 165.

39. P. Lowell, *Annals of the Lowell Observatory*, III:262 (1905).

40. P. Lowell, "Is Mars Inhabited?" *Outlook*, April 13, 1907; *L.O.A.*

41. New York *World*, May 31, 1903; *L.O.A.* See also P. Lowell, "A Projection on Mars," *Lowell Observatory Bulletin 1*, (June 1903). Of the seventy-five *Bulletins* published in Lowell's lifetime, Lowell himself contributed in whole or in part to forty-six of them. See also N.Y. *American* to P. Lowell, May 28, 1903.

42. London *Daily Mail*, August 1903; *L.O.A.*

43. P. Lowell, *Mars and Its Canals*, 290.

44. Ibid., 286–87.

45. Ibid., 303.

46. Ibid., 299.

47. Ibid., 375. See also *Mars as the Abode of Life*, 181. Later measurements by astronomer J. H. Focus at Athens Observatory indicated a rate for this wave of darkening of about 21.7 miles per day; see Shklovskii and Sagan *op. cit.*, 279.

48. P. Lowell, "The Cartouches of the Canals of Mars," *Lowell Observatory Bulletin 12* (undated).

49. P. Lowell, *Mars and Its Canals*, 374.

50. Ibid., 375.

51. P. Lowell to V. M. Slipher, Nov. 10, 1903, and C. O. Lampland, Nov. 10 and Dec. 9, 1903; and V. M. Slipher to P. Lowell, March 4, 1904.

52. P. Lowell to S. Newcomb, March 15, 1903, Oct. 18, 1905, and May 15, 1907. While Pickering had observed linear martian markings, he did not accept Lowell's interpretation of them and advanced a number of explanations of his own for them, some hardly less imaginative than Lowell's. See "Pickering vs. Lowell," *The New York Times*, Dec. 22, 1907, a review of Pickering's "Different Explanations for the Canals of Mars" in the January 1908 *Harper's*.

53. P. Lowell, *Mars and Its Canals*, 77–78.

54. The relatively vivid coloration of Mars reported by Lowell and other early observers is now thought to be the result of a combination of chromatic aberration produced by refracting telescopes, then in wide use, and of psychophysiological phenomena related to human vision. The problem is discussed in Shklovskii and Sagan, *op. cit.*, 273.

55. P. Lowell, *Annals*, III:265 (1905).

56. P. Lowell, *Mars and Its Canals*, 126.

57. Ibid., 119–27. See also P. Lowell, "Mare Erythraeum," *Lowell Observatory Bulletin 7* (undated).

58. J. E. Evans and E. Walter Maunder, "Experiments as to the Actuality of the 'Canals' observed on Mars," *Monthly Notices of the Royal Astronomical Society*, 53:488 (1903).

59. G. P. Serviss, column, unidentified newspaper, October 1903; *L.O.A.*

60. P. Lowell, *Mars and Its Canals*, 202. See also *Bulletin de la Societe Astronomique de France*, June 1905, 283.

61. P. Lowell to S. Newcomb, Oct. 27, 1905. Lowell here is arguing against Newcomb's choice of Maunder to write an *Encyclopedia Britannica* article on Mars.

62. P. Lowell, *Mars and Its Canals,* 202.

63. P. Lowell, "Experiment on the Visibility of Fine Lines," *Lowell Observatory Bulletin 2* (1903), and V. M. Slipher and C. O. Lampland, "Notes on Visual Experiment," *Bulletin 10* (1903). See also P. Lowell, *Mars as the Abode of Life,* 271–79.

64. P. Lowell, *Mars as the Abode of Life,* 274–75.

65. P. Lowell to E. W. Maunder, Nov. 28, 1903.

66. P. Lowell, *Mars as the Abode of Life,* 275.

67. P. Lowell to S. Newcomb, March 15, 1903. See also P. Lowell, *Mars and Its Canals,* 209.

68. P. Lowell, *Mars and Its Canals,* 292–93.

69. P. Lowell, "Means, Methods and Mistakes in the Study of Planetary Evolution," text dated April 15, 1905; *L.O.A.*

70. Ibid.

71. P. Lowell, *Mars and Its Canals,* 199–200.

72. Ibid., 200-01. See also P. Lowell, "Width of the Double Canals of Mars With Different Apertures," *Lowell Observatory Bulletin 5* (undated).

73. P. Lowell, *Mars and Its Canals,* 197–98.

74. G. P. Serviss, "The World's Battle for Life," unidentified newspaper, May 1904; *L.O.A.*

75. E. M. Antoniadi to P. Lowell, Sept. 9, 1909.

76. E. M. Antoniadi to P. Lowell, Sept. 26, 1909.

77. E. M. Antoniadi to P. Lowell, Oct. 9, 1909.

78. P. Lowell to E. M. Antoniadi, Nov. 2, 1909.

79. E. M. Antoniadi to P. Lowell, Nov. 15, 1909.

80. E. M. Antoniadi, report of director, Mars Section, *Journal of the British Astronomical Association,* 20:189 (1910).

81. G. E. Hale, *addendum,* ibid., 20:191 (1910).

82. See for example, "Canals of Mars Optical Illusions," *Daily Chronicle* (Victoria, B.C.), Feb. 9, 1916, and "No Canals or Life on Mars—Prof. Arrhenius Attacks Popular Idea of Prof. Lowell," Boston *Sunday Advertiser* and *American,* Oct. 6, 1918.

83. R. J. Trumpler, "Visual and Photographic Observations of Mars," *Publications of the Astronomical Society of the Pacific,* 36:263 (1924).

84. J. H. Jeans, *The Universe Around Us,* New York, 1929, 334.

85. V. M. Slipher to J. H. Jeans, Aug. 21, 1930.

86. E. M. Antoniadi, *La Planète Mars,* Paris, 1930. See Shklovskii and Sagan, *op. cit.,* 278, 282, for reproductions of some of Antoniadi's martian drawings.

87. R. B. Leighton, "Mars Pictures From Mariners 6 and 7," *Sky and Telescope,* 38:212 (1969). Leighton's reference to the "rough form" of the pictures relates to the fact that such photographs, received on earth by telemetry from the spacecraft in the form of "bits" of electronic data, can often be enhanced by computer techniques, somewhat as retouching enhances conventional photographs.

88. H. Masursky *et al,* "Mariner 9 Television Reconnaissance of Mars and Its Satellites: Preliminary Results," *Science,* 175:294 (1972); W. K. Hartmann, "The New Mariner Map of Mars," *Sky and Telescope,* 44:77 (1972); and C. W. Hord, "Mariner 9 Ultraviolet Topographic Measurements of Mars," report to the annual meeting of the American Astronomical Society's Division of Planetary Sciences, Tucson, Arizona, March 21, 1973.

89. P. Lowell, *Annals,* III:262 (1905). See also *Lowell Observatory Bulletins 8, 9, 14, 17* and *19* (1903–04).

90. A. L. Lowell, *op. cit.,* 148.

91. P. Lowell to V. M. Slipher, Nov. 10, 1903.

Notes to Chapter 11

1. P. Lowell, *Mars as the Abode of Life,* New York, 1908, 155; see also P. Lowell to John T. Trowbridge, Nov. 12, 1909.

2. Otto Struve and Velta Zebergs, *Astronomy of the 20th Century,* New York, 1962, 50–51.

3. Agnes M. Clerke, *A Popular History of Astronomy During the Nineteenth Century,* London, 1902, 281–82.

4. W. H. Pickering to P. Lowell, Sept. 24, 1894; see also P. Lowell, "The Canals of Mars—Photographed," *Lowell Observatory Bulletin 21* (1905).

5. P. Lowell to C. O. Lampland, Jan. 8, 1903.

6. P. Lowell to C. O. Lampland, Nov. 10, 1903.

7. P. Lowell, "The Canals of Mars—Photographed," ibid.; *Mars and Its Canals,* New York, 1906, 275–76; and "On a New Method of Sharpening Celestial Photographs and Applied With Success To Mars," *Lowell Observatory Bulletin 31* (1907). Also C. O. Lampland, "On Photographing the Canals of Mars," in *Bulletin 21* (1905).

8. P. Lowell to C. O. Lampland, Jan. 12, 1904.

9. P. Lowell to Charles R. Cross, Oct. 21, 1913.

10. Struve and Zebergs, *op. cit.,* 150.

11. P. Lowell to W. H. Bullock, July 29, 1907.

12. P. Lowell to W. H. Pickering, Jan 14., 1916.

13. P. Lowell to V. M. Slipher, April 2, 1904.

14. P. Lowell to C. O. Lampland, May 16, 1904.

15. W. L. Leonard to V. M. Slipher, June 29, 1904.

16. P. Lowell to V. M. Slipher, Aug. 4 and Sept. 25, 1904.

17. P. Lowell, "North Polar Cap of Mars—November 1904 to May 1905," *Lowell Observatory Bulletin 20* (undated); "The South Polar Cap of Mars in 1905," *Bulletin 26* (1906); "Position of the Axis of Mars," *Bulletin 24* (1905); and *Mars and Its Canals,* 124ff.

18. E. S. Morse, *Mars and Its Mystery,* Boston, 1906; and "My 34 Nights on Mars," New York *World,* Oct. 7, 1906; *L.O.A.*

19. Photographic Album #3, *L.O.A.,* contains snapshots of Michelson at the eyepiece of the 24-inch telescope.

20. *Coconino Sun* (Flagstaff), May 27, 1905.

21. P. Lowell, "The Beginning of the New North Polar Cap of Mars," *Lowell Observatory Bulletin 22* (1905). See also *The New York Times,* May 22, 1905; *L.O.A.*

22. B. J. Jenkins to P. Lowell, June 5, 1905.

23. P. Lowell, "The Canals of Mars—Photographed."

24. P. Lowell, *Mars and Its Canals,* 277.

25. Ibid., 276–77.

26. P. Lowell, "The Canals of Mars Photographed," *Astronomische Nachrichten,* No. 4035 (1905).

27. There are more than fifty clippings in the *L.O.A.* relating to the 1905 photography of the martian canals. Unfortunately few are identifiable by exact date or title of publication.

28. A. L. Lowell, *Biography of Percival Lowell,* New York, 1935, 118. Also *Pall Mall Gazette* (London), July 24, 1907; *L.O.A.*

29. London *Daily Graphic,* June 1905; *L.O.A.*

30. G. P. Serviss, column, unidentified newspaper, June, 1905; *L.O.A.*

31. G. P. Serviss, "Photographs of Mars," column, unidentified newspaper, June 1905; *L.O.A.*

32. Ibid.

33. P. Lowell to C. V. Van Anda, undated, *ca.* November 1905; to E. F. Strother (*The World's Work*), Nov. 10, 1905; and to the *Illustrated London News,* and *Knowledge and Scientific News,* Nov. 1, 1905.

34. "The Canals of Mars Photographed," *Scientific American,* June 1905; *L.O.A.*

35. P. Lowell, "The Canals of Mars—Photographed," *Popular Astronomy,* 8:479 (1905).

36. Hector McPherson Jr., "Mars and Its Canals," *Scottish Review,* Nov. 23, 1905; *L.O.A.*

37. "Photographs of the Canals of Mars," *Knowledge and Scientific News,* March 1906; *L.O.A.*

38. P. Lowell to S. Newcomb, Oct. 18, 1905.

39. W. H. Wesley, "Photographs of Mars," *Observatory,* 28:314 (1905).

40. Ibid.

41. Ibid.

42. H. H. Turner, "From an Oxford Note-Book," ibid., 28: 336 (1905).

43. A. C. D. Crommelin, address, *Journal of the British Astronomical Association,* 38:452 (1905).

44. *Pall Mall Gazette* (London), Nov. 17, 1905; *L.O.A.*

45. R. J. Strutt, "Celestial Photography and the 'Canals' of Mars," London *Tribune,* Feb. 13, 1906; *L.O.A.*

46. George R. Downing, "Photographing the Canals of Mars," unidentified newspaper, *ca.* October 1905; *L.O.A.* Downing was a professor at Columbia University in New York City.

47. Boston *Herald* and San Francisco *Call,* Feb. 1, 1906; *L.O.A.*

48. Boston *Post* and Boston *Transcript,* Feb. 2, 1906, and *Transcript,* March 3, 1906; *L.O.A.*

49. Lowell in later years exhibited the photographic work of his observatory at public libraries, museums and college campuses, and even sent displays to Canada and Europe to be shown at astronomical meetings. Some of these displays have been placed in the observatory's library at Flagstaff. Slides of photographs and spectrograms were sometimes a part of his lectures on astronomical subjects.

50. P. Lowell to V. M. Slipher, June 18, 1909.

51. P. Lowell, *Mars and Its Canals,* 277.

52. P. Lowell to V. M. Slipher, Sept. 11 1906.

53. *Coconino Sun,* Nov. 3, 1906.

54. Boston *Herald,* Nov. 28, 1906; *L.O.A.* Peary reached 87° N in 1906, and the pole itself on April 4, 1909.

55. P. Lowell, *Mars and Its Canals,* 54.

56. Boston *Herald,* Nov. 28, 1906; *L.O.A.*

57. P. Lowell to V. M. Slipher, Oct. 18, 1905.

58. Clerke, *op. cit.,* 464. Lowell's proposed telescope would have been the world's largest for a period. The largest up to that time had been Lord Rosse's 72-inch speculum metal reflector at Parsonstown, Eire. The 100-inch Hooker reflector would not be operational until 1919.

59. P. Lowell to G. W. Ritchey, undated, *ca.* March 1906.

60. P. Lowell to G. W. Ritchey, April 13, 1906.

61. "Agreement" by Alvan Clark and Sons (per H. Nolte), April 1908; *L.O.A.* See also P. Lowell to C. O. Lampland, April 9, 1908.

62. P. Lowell to A. Clark and Sons, Dec. 2, 1909.

63. P. Lowell to C. O. Lampland, Feb. 2, 1907.

64. "Agreement" by David P. Todd, Feb. 28, 1907; *L.O.A.*

65. E. C. Slipher, *A Photographic Study of the Brighter Planets,* Flagstaff, 1964, 22.

66. There are many clippings of news items and articles relating to the expedition in the *L.O.A.*, most of them contained in what I have tentatively labeled "Clipping Book No. 4." For a listing of identifiable publications in these clipping books, see Appendix C *infra*.

67. P. Lowell to E. H. Clement, May 18, 1907.

68. E. C. Slipher to V. M. Slipher, May 22, 1907.

69. E. C. Slipher, *The Photographic Story of Mars*, Cambridge, Mass., 1962, 61–63.

70. D. Todd to P. Lowell, cablegram, June 28, 1907. Lowell had devised a code in "cablese" by which seeing, weather conditions and reports of observations could be described in a few compound words.

71. D. Todd to P. Lowell, cablegram, July 18, 1907.

72. P. Lowell to D. Todd, July 26, 1907.

73. E. C. Slipher to V. M. Slipher, July 28, 1907.

74. Ibid.

75. D. Todd to P. Lowell, Aug. 8, 1907.

76. M. L. Todd, "Photographing the Canals on Mars," *Nation*, Sept. 19, 1907; *L.O.A.* See also Boston *Herald*, Oct. 11, 1907, and "Astronomer Tells What He Knows of the War Star and Its Inhabitants," St. Louis *Times*, Jan. 7, 1908; *L.O.A.*

77. New York *Evening Post*, Aug. 29, 1907.

78. E. C. Slipher, *The Photographic Story of Mars*, 61. See also *Associated Press* dispatch, July 22, 1907, as publshed in a number of newspapers; *L.O.A.*

79. P. Lowell to G. V. Schiaparelli, July 2 and Aug. 28, 1907, and to C. Flammarion, Aug. 28, 1907.

80. P. Lowell to L. F. Ward, Aug. 1, 1907; also text dated Aug. 6, 1907 of article for *Nature*, Aug. 23, 1907, and London *Daily Telegraph* and *The New York Times*, Aug. 30, 1907; *L.O.A.*

81. P. Lowell, "The South Polar Cap of Mars—1907," *Lowell Observatory Bulletin 29* (1907); "The North Polar Cap of Mars—March-June 1907," *Bulletin 30* (1907); "The Size of the South Polar Cap of Mars—1907," *Bulletin 35* (1908); "Position of the Axis of Mars," *Bulletin 33* (1908); and *Bulletin 28*, untitled, (1907).

82. P. Lowell to D. Todd, April 11, 1907.

83. S. Bolton to P. Lowell, May 12, 1907.

84. P. Lowell to Max Wolf, Jan. 3, 1907, and A. Berberich, Feb. 7 and 20, 1907.

85. "Saturn's Atmosphere," New York *Herald*, May 1907; *L.O.A.*

86. P. Lowell, "Insolation," *Lowell Observatory Bulletin 30* (1907).

87. *Coconino Sun* (Flagstaff), March 2, April 2 and 11, and Aug. 23, 1907; see also correspondence between W. L. Leonard, who was in Flagstaff that year, and Kenneth MacDonald, particularly for Oct. 30, 1907.

88. New York *Herald* and Boston *Globe*, June 27, 1907; *L.O.A.* The *Globe* noted that "unfortunately Nicola Tesla hasn't quite completed his invention so that the news cannot be telegraphed to Mars." Lowell's second LL.D. was conferred by Clark University in 1909.

89. S. S. McClure to P. Lowell, Sept. 6, 1907.

90. P. Lowell to S. S. McClure, Sept. 7, 1907.

91. P. Lowell to *McClure's*, Nov. 13, 1906, and to *Harper's*, Dec. 3, 1906.

92. P. Lowell to editor, *Century*, Sept. 7, 1907.

93. P. Lowell to G. R. Agassiz, Sept. 16, 1907.

94. P. Lowell to V. M. Slipher, Sept. 23, 1907. The offer was for $500.

95. P. Lowell to G. R. Agassiz, Sept. 16, 1907. In 1905, the Royal Society's assistant secretary Robert Harrison suggested "emphasizing" the lines in the Mars photographs for publication, and Lowell suggested that Wesley oversee

the work, stipulating, however, that it be stated in the captions that the lines had been emphasized. See P. Lowell to W. Wesley, Oct. 16, 1905, and to R. Harrison, Oct. 17, 1905.

96. G. R. Agassiz to P. Lowell, Sept. 27, 1907.

97. R. U. Johnson to P. Lowell, Oct. 7, 1907.

98. P. Lowell to R. U. Johnson, Oct. 8, 1907.

99. P. Lowell to editor, *Century*, Sept. 28, 1907. The technique has been used on occasion for astronomical photographs in books and other publications; see Clerke, *op. cit.*, frontispiece and title page, for an example. In 1962, E. C. Slipher, in a limited edition of his *The Photographic Story of Mars, op. cit.*, used direct prints from photographic plates in place of the usual halftone engravings to assure the best possible definition of detail.

100. P. Lowell to E. C. Slipher, Oct. 1, 1907.

101. P. Lowell, *Mars and Its Canals*, 271–77, and *Mars as the Abode of Life*, 153–55.

102. P. Lowell to editor, *Century*, Sept. 14, 1907.

103. D. Todd to P. Lowell, Sept. 14, 1907.

104. P. Lowell to editor, *Cosmopolitan*, Aug. 25, 1907.

105. P. Lowell to S. S. Chamberlain, Sept. 28, 1907.

106. P. Lowell to S. S. Chamberlain, Oct. 2, 1907.

107. S. S. Chamberlain to P. Lowell, Oct. 11, 1907.

108. D. Todd to P. Lowell, Oct. 15, 1907.

109. P. Lowell to C. O. Lampland, Nov. 20, 1907.

110. P. Lowell to A. L. Everett, Dec. 19, 1907.

111. P. Lowell to A. L. Everett, Dec. 21, 1907.

112. P. Lowell to D. Todd, Dec. 31, 1907.

113. P. Lowell to E. C. Slipher, Jan. 6, 1908.

114. P. Lowell to A. L. Everett, Jan 10, 1908.

115. P. Lowell to A. L. Everett, Feb. 5, 1908.

116. P. Lowell to E. C. Slipher, Feb. 8, 1908.

Notes to Chapter 12

1. G. V. Schiaparelli, quoted by W. H. Wright in "Mars," *Encyclopedia Britannica*, New York, 1929, Vol. 14, 957–58. Wright attributed the quotation to a translation by W. H. Pickering of Schiaparelli's *Mars*, 92–94.

2. G. V. Schiaparelli, interview datelined "Rome, Aug. 3, 1907" with A. W. Pearce, unidentified newspaper; *L.O.A.*

3. A clipping of a newspaper cartoon in the *L.O.A.* shows Flammarion gazing out of a window at an obviously populated Mars. The caption reads: "July First: Camel Flimflammarion announces that he has distinctly seen people swimming in the canals of Mars." Whoever sent the clipping to Lowell scrawled across the cartoon: "Goes one better than you!"

4. W. E. Garrett Fisher, "Science in 1906," London *Tribune*, December 1906; *L.O.A.*

5. W. Kaempffert, "What We Know About Mars," *McClure's*, March 1907; *L.O.A.*

6. E. H. Clement, "Other Views of Mars," Boston *Transcript*, Sept. 25, 1907; *L.O.A.*

7. "Pickering vs. Lowell," *The New York Times*, Dec. 22, 1907; *L.O.A.*

8. A. R. Wallace, *Man's Place in the Universe*, London, 1903, and *Is Mars Habitable?*, London, 1907.

9. See, for example, E. M. Antoniadi to P. Lowell, Oct. 9, 1909.

10. O. Struve and V. Zebergs, *Astronomy of the 20th Century*, New York, 1962, 143.

11. Lowell, for some years had signed himself facetiously as "Mars" in correspondence with his close friend, Prof. William E. Storey of Clark University. See P. Lowell to W. E. Storey ("Quadric"), April 4 and Nov. 11, 1903; and W. E. Storey to P. Lowell, March 13, 1907.

12. H. McPherson, *Astronomers of Today*, London, 1905, 206–16.

13. Philadelphia *Press*, December 1906; *L.O.A.*

14. Knoxville (Tenn.) *Sentinel*, July 1907; *L.O.A.*

15. New York *Herald*, Dec. 30, 1906; *L.O.A.*

16. *The New York Times*, Nov. 13, 1907.

17. London *Daily News*, July 4, 1907.

18. The first attempt to cut a canal across the Isthmus of Panama was begun in 1878, the year after Schiaparelli's discovery of the martian canals, by a French company headed by engineer Ferdinand de Lessups, builder of the Suez Canal. It failed, but subsequently the United States, after lengthy diplomatic negotiations and a brief and strange revolution in Panama itself, acquired the rights to build the canal. Construction began in May 1904 and after many obstacles were overcome, the first ship passed through the canal Aug. 3, 1914.

19. Lillian Whiting, "Looking for Lights on Mars," St. Louis *Post-Dispatch*, June 16, 1907; *L.O.A.*

20. Providence (R.I.) *Journal*, June 1907; *L.O.A.* The Cape Cod ship canal was begun in 1909 and completed in 1914.

21. P. Lowell, *Mars and Its Canals*, New York, 1906, 180.

22. R. J. Wallace to P. Lowell, Aug. 11, 1907. Wallace, a member of the Yerkes Observatory staff, here reports to Lowell that E. E. Barnard's attempts to photograph Mars with the Yerkes 40 inch in 1907 were "an absolute failure. No definition or anything else." Barnard, however, obtained some notable photographs in 1909; see G. P. Kuiper and M. R. Calvert, "Barnard's Photographs of Mars," *Astrophysical Journal*, 105:215 (1947).

23. Unidentified newspaper, *ca.* March 1907; *L.O.A.*

24. Ray Long to P. Lowell, Aug. 3, 1910.

25. There are, for example, in the *L.O.A.* appeals from the Flat Earth Society ("The Earth is Flat and Stands Flat. Prove It!"), and a communication dated July 14, 1907, and signed only "Steward 10 and three," which declares that "those on Mars are Ghosts—who were Murdered Here for Religious Reasons—and food—by the cannable [sic] missionary murderers."

26. W. E. Garrett Fisher, London *Tribune*, June 1906; *L.O.A.*

27. "Lowell's Observations of the Planet Mars," *Nature*, 74:587 (1906); *L.O.A.*

28. Boston *Herald*, Dec. 6, 1906; *L.O.A.*

29. C. N. Chadburne to P. Lowell, April 27, 1907.

30. H. Menke to P. Lowell, May 1, 1907.

31. E. H. Forbes to P. Lowell, Aug. 15, 1907.

32. W. H. Orr to P. Lowell, Aug. 19, 1907.

33. Marie Gould to P. Lowell, June 5, 1910.

34. Leroy Hughbanbank to P. Lowell, Aug. 12, 1910.

35. R. A. Willons to P. Lowell, July 31, 1905.

36. Leo Holcomb to P. Lowell, Sept. 27, 1907.

37. J. B. Hastings to P. Lowell, March 17, 1907.

38. H. G. Tompkins to P. Lowell, June 22, 1907, and P. Lowell to H. G. Tompkins, July 2, 1907.

39. S. M. Thompson to P. Lowell, March 11, 1907.

40. C. G. Rupert to P. Lowell, Aug. 23, 1910.

41. Alice H. Sill to P. Lowell, Aug. 21, 1907.
42. See Chapter 14 *infra*.
43. E. G. McLean to P. Lowell, Oct. 15, 1909.
44. Ellen G. Hunter to P. Lowell, Jan. 3, 1910. V. M. Slipher's answer, dated Jan. 9, 1910, was that "no satisfactory explanation has yet been offered" for the biblical phenomenon.
45. T. Nance to P. Lowell, Sept. 23, 1907.
46. C. P. Taft to "Mr. W. S. Leonard" [sic], Sept. 19, 1909.
47. C. J. Kleman to P. Lowell, Dec. 19, 1912.
48. J. M. Early to P. Lowell, undated, 1913.
49. G. W. Cooper to P. Lowell, March 27, 1915.
50. R. Connors to P. Lowell, March 27, 1915.
51. J. H. Henderson to P. Lowell, Jan. 2, 1916.
52. H. Jacob to P. Lowell, Feb. 21, 1916.
53. W. H. Brackett to P. Lowell, Oct. 10, 1916.
54. See Appendix C, and Bibliographical Essay *infra*. The majority of identifiable clippings in the *L.O.A.* date between 1905 and 1909, and most of these, in turn, are dated in 1907.
55. P. Lowell, "Is Mars Inhabited?" *Outlook*, April 13, 1907; *L.O.A.*
56. Ibid. (editor's introduction).
57. "Are There Martians?" London *Daily Chronicle*, December 1907; *L.O.A.*
58. Ibid.
59. Ibid.
60. P. Lowell to Mark Wicks, April 1, 1909 and Oct. 28, 1910. Wick's book was "an astronomical story" entitled *To Mars via the Moon*.
61. P. Lowell to Mark Wicks, Feb. 27, 1908.
62. P. Lowell, letter to editor, *Nature*, March 19, 1908; *L.O.A.*
63. Ibid.
64. P. Lowell, *Mars as the Abode of Life*, New York, 1908, 209.
65. Forest R. Moulton, *Consider the Heavens*, Garden City, N.Y., 1935. 104–05.
66. E. Noble, "Does the Universe Exist for Man Alone?" Boston *Sunday Herald*, June 28, 1908; *L.O.A.*
67. Ibid.
68. Ibid.
69. Lillian Whiting, *op. cit.*
70. C. Flammarion, interview with A. W. Pearce, *op. cit.*
71. Philadelphia *Record*, May 5, 1907; *L.O.A.*
72. L. F. Ward to P. Lowell, March 31, 1907.
73. Ibid. Ward fixed the date of his visit as June 4, 1901.
74. Ibid.
75. P. Lowell, *Mars and Its Canals*, 377–78.
76. P. Lowell, "Commemoration Day Address," text dated May 30, 1903: *L.O.A.*
77. London *Tribune*, June 1906; *L.O.A.*
78. *Saturday Evening Mail*, Feb. 9, 1907; *L.O.A.* Sherman, of course, noted that "war is hell!"
79. *Japan Daily Mail*, May 29, 1906; *L.O.A.*
80. E. H. Clement to P. Lowell, April 22, 1907. Lowell acknowledged the poem as "His Martian Executiveship" and as "minister plenipotentiary for Mars on Earth" in his reply to Clement, April 25, 1907.
81. "Alumni Day at Tufts," New York *Evening Post*, June 28, 1907; *L.O.A.*
82. Chicago *Record*, July 24, 1907; *L.O.A.*

83. C. Ferguson, "Raising Crops on Mars," Kansas City *Times*, Sept. 16, 1907; *L.O.A.*
84. Chicago *Examiner*, Sept. 6, 1907; *L.O.A.*
85. New York *Christian Advocate*, September 1907; *L.O.A.*
86. P. Lowell to W. N. Niles, Oct. 9, 1906.
87. G. R. Wieland to P. Lowell, Feb. 19, 1907.
88. Boston *Transcript* and New York *Staats-Zeitung*, July 3, 1907; *L.O.A.*
89. *Associated Press*, July 22, 1907; *L.O.A.*
90. P. Lowell, text dated Aug. 8, 1907, of an article for *Nature*, Aug. 23, 1907; *L.O.A.*
91. Ibid.
92. London *Daily Telegraph*, Aug. 31, 1907; *L.O.A.*
93. *The New York Times*, Aug. 31, 1907; *L.O.A.*
94. Ibid.
95. "Scouts Mars Theory," *The New York Times*, Aug. 31, 1907; *L.O.A.*
96. *The New York Times*, Sept. 5, 1907; *L.O.A.*
97. Hector McPherson Jr., London *Chronicle*, July 5, 1907; *L.O.A.*
98. E. V. Heward, "Mars: Is It a Habitable World?" *Fortnightly Review*, August 1907; *L.O.A.*
99. M. L. Todd, "Photographing the Canals of Mars," *Nation*, Sept. 16, 1907; *L.O.A.*
100. Los Angeles *Times*, Sept. 22, 1907; *L.O.A.*
101. *The Sphere* (London) September 1907; *L.O.A.*
102. Indianapolis *Star*, Sept. 15, 1907; *L.O.A.*
103. W. L. Leonard to editor, New York *Tribune*, Sept. 14, 1907.
104. *Scientific American*, Sept. 28, 1907; *L.O.A.*
105. *The New York Times*, Sept. 30, 1907; *L.O.A.*
106. Ibid.
107. "Doubts Martian Canals," *The New York Times*, Nov. 8, 1907; *L.O.A.*
108. *The New York Times*, Nov. 8, 1907; *L.O.A.*
109. P. Lowell ("Mathematician"), "The Evidence for Life on Mars," *Scientific American*, Oct. 26, 1907; *L.O.A.*
110. P. Lowell to American Philosophical Society, May 10, 1907, and to S. Newcomb, May 15, 1907.
111. P. Lowell to Abbe Th. Moreaux, June 11, 1907, and to W. D. McPherson, June 12, 1907.
112. P. Lowell to H. H. Turner, May 8, 1907.
113. Ibid.
114. P. Lowell to R. U. Johnson, Sept. 14, 1907.
115. P. Lowell to R. U. Johnson, Oct. 10, 1907.
116. P. Lowell to R. U. Johnson, Oct. 29, 1907.
117. Ibid.
118. P. Lowell to R. W. Gilder, Nov. 1, 1907.
119. Ibid.
120. P. Lowell to W. R. Hearst, Nov. 4, 1907.
121. Boston *Transcript*, Oct. 5, 1907; *L.O.A.*
122. Boston *Herald*, Oct. 11, 1907, and New York *Tribune*, Oct. 24, 1907; *L.O.A.*
123. *The New York Times*, Nov. 4 and 5, 1907; *L.O.A.*
124. Boston *Herald*, Nov. 29, 1907; *L.O.A.*
125. *The New York Times*, Dec. 8, 1907; *L.O.A.*

126. Boston *Herald*, Dec. 22, 1907; *L.O.A.*
127. Washington (D.C.) *Star*, Jan. 4, 1908; *L.O.A.* See also P. Lowell to V. M. Slipher, Jan. 8, 1908, and to W. J. Barnette, Jan. 21 and March 2, 1908. The text of this lecture, dated Jan. 3, 1908, is in the *L.O.A.*
128. C. Flammarion, "Photographies de Mars," *Societe Astronomique de France*, November 1907, and "Der Planet Mars und die Photographie," unidentified journal, May 16, 1908; *L.O.A.*
129. P. Lowell to L. F. Ward, April 14, 1908.
130. "Calls Mars Theory Rash and Fantastic," Pittsburgh (Pa.) *Press*, March 1908; *L.O.A.*
131. Ibid.
132. Prescott (Ariz.) *Journal-Miner*, Jan. 18, and *Coconino Sun* (Flagstaff), Jan. 19 and Feb. 13, 1908; *L.O.A.*
133. Boston *Transcript*, April 28, 1908; *L.O.A.* See also A. L. Lowell, *Biography of Percival Lowell*, New York, 1935, 147. The priest, Dr. Tomaya Suga, A. Lawrence notes, cut himself slightly.
134. See correspondence among P. Lowell, V. M. Slipher, W. L. Bryant and W. A. Cogshall between February 1908 and June 1909.
135. A. Lawrence Lowell, *op. cit.*, 148–49. See also unidentified newspaper, Sept. 20, 1908; *L.O.A.* There are few references to Lowell's marriage and honeymoon in the *L.O.A.*
136. Advertising brochure, The Macmillan Co., *ca.* December 1908; *L.O.A.*
137. Boston *Herald*, Dec. 14, 1908; *L.O.A.*
138. Boston *Transcript*, Dec. 29, 1908; *L.O.A.*
139. P. Lowell, "The Canali Novae of Mars," *Lowell Observatory Bulletin 45* (1910). See also P. Lowell to E. M. Antoniadi, Jan. 18, 1910.
140. Lowell Observatory to editor, *Scientific American*, May 19, 1910.
141. *The New York Times* to P. Lowell, telegram, Oct. 17, 1909.
142. G. A. Briggs, "Periodic Variations in Martian Dust Storm Activity and the Layered Deposits of the Polar Region," report to the annual meeting of the American Astronomical Society's Division of Planetary Sciences, Tucson, Arizona, March 21, 1973.
143. P. Lowell, draft of telegram to *The New York Times, ca.* October 1909.

Notes to Chapter 13

1. *The New York Times*, Nov. 13, 1907; *L.O.A.* The *Times* also noted the transit of Mercury that occurred Nov. 14, 1907.
2. Saturn's rings are presented edge-on to the earth twice during its 29½-year revolution around the sun.
3. *The New York Times*, Nov. 9, 1907; *L.O.A.*
4. "Astronomers Split on Saturn's Rings," *The New York Times*, Nov. 11, 1907; *L.O.A.* The item also quotes John A. Brashear of Allegheny Observatory as opposing the "falling-in" idea.
5. P. Lowell to C. O. Lampland, Nov. 20, 1907.
6. *The New York Times*, Nov. 21, 1907; *L.O.A.*
7. P. Lowell to C. O. Lampland, Nov. 29, 1907. See also P. Lowell, "Tores of Saturn," *Lowell Observatory Bulletin 32* (1907).
8. P. Lowell, "Saturn's Tores," *Scientific American*, December 1907; *L.O.A.*
9. Boston *Transcript*, Nov. 13 and 27, 1907; *L.O.A.*
10. A. L. Lowell, *Biography of Percival Lowell*, New York, 1935, 134.
11. See *Lowell Observatory Bulletins 1–75* (1903–1916); *Memoirs of the Lowell Observatory*, Vols. I and II (1915); and *Memoirs of the American Academy of Arts and Sciences*, Vol. XIV, No. 1 (1913).
12. P. Lowell, *The Solar System*, New York, 1903, 115.

13. P. Lowell, *Mars and Its Canals,* New York, 1906, 15.
14. P. Lowell, *Mars as the Abode of Life,* New York, 1908, 1–2.
15. Ibid.
16. Lowell claimed to have found mathematical errors in the Chamberlin-Moulton formulation; see P. Lowell to F. R. Moulton, June 15, 1906.
17. A. M. Clerke, *A Popular History of Astronomy During The Nineteenth Century,* London, 1902, 313. Accounting for the sun's radiation was a major problem for astronomers and physicists before the discovery of radioactivity and the advent of nuclear physics.
18. O. Struve and V. Zebergs, *Astronomy of the 20th Century,* New York, 1962, 169–70. For the Jean-Jeffreys version, see J. H. Jeans, *The Universe Around Us,* New York, 1929.
19. Indianapolis *Star,* Feb. 28, 1909; Muncie (Ind.) *Star,* March 4, 1909; and Salt Lake City *News,* March 27, 1909, among others in the *L.O.A.*
20. Omaha (Neb.) *World-Herald,* March 1, 1909, and the Minneapolis *Journal,* Feb. 28, 1909; *L.O.A.*
21. Denver (Colo.) *Republican,* Feb. 28, 1909; *L.O.A.*
22. Salt Lake City (Utah) *News,* March 27, 1909; *L.O.A.*
23. Boston *Transcript,* March 5, 1909; *L.O.A.*
24. P. Lowell to A. C. D. Crommelin, March 27, 1909.
25. W. H. Pickering, letter to editor, Boston *Transcript,* March 15, 1909; *L.O.A.*
26. P. Lowell, *Mars as the Abode of Life,* 6.
27. P. Lowell, *The Evolution of Worlds,* 12.
28. P. Lowell, *Mars as the Abode of Life,* 4–6.
29. P. Lowell, *The Evolution of Worlds,* 33 ff.
30. Ibid., 42 ff.
31. Ibid., 55–58.
32. Ibid., 39, 49–50.
33. Clerke, *op. cit.,* 376.
34. Struve and Zebergs, *op. cit.,* 168–70.
35. Clerke, *op. cit.,* 401 ff. The early classification of "white" and "green" nebulae resulted from spectroscopic observations reported in 1868 by Sir William Huggins. Huggins attributed the hue of "green" nebulae to an unknown substance which he called "nebulum."
36. P. Lowell, *Mars as the Abode of Life,* 11–12.
37. Ibid., 12–14. Of course the earth's ocean basins and continents are no longer considered permanent features even "in fundamentals" since the development of the theories of plate techtonics and continental drift in the late 1960s. Nor is the contraction theory of mountain-building held today.
38. Ibid., 12.
39. Ibid., 7–9.
40. Ibid., 15–16.
41. Ibid., 16–19.
42. Ibid., 23.
43. There are, of course, serious obstacles to this aspect of the tidal friction theory and indeed, despite the success of the Apollo manned landings on the moon in the late 1960s and the early 1970s, the problem of the origin of the earth-moon system is still very much at issue.
44. Ibid., 26–27.
45. Ibid., 27.
46. Ibid., 29–30. Lowell may have had in mind something like the familiar Coriolis effect which relates to rotating spheroids and results from the differential velocities of points on their surfaces between the equator and the poles.
47. Ibid., 31–32.
48. Ibid., 35–36, 39.
49. Ibid., 14.

50. Ibid., 42–43.

51. Ibid., 45–46.

52. Ibid., 47–48.

53. Ibid., 48.

54. Ibid., 49.

55. Ibid., 51. In a note to this passage (237) entitled "Effect of Increased Carbon Dioxide Upon Plants," Lowell cites experiments reported by H. T. Brown and F. Escombe reported in *Proceedings of the Royal Society*, LXX (1902).

56. P. Lowell, *The Evolution of Worlds*, 177.

57. P. Lowell to F. A. Lucas, June 25, 1909.

58. P. Lowell, *The Evolution of Worlds*, 155.

59. S. Arrhenius, *Worlds in the Making*, New York, 1908, 212 ff. Arrhenius suggested that simple living organisms such as spores or bacteria might have seeded the earth with life from other planets in the universe, arguing that terrestrial winds sometimes wafted microorganisms into the earth's stratosphere where they might be launched into space by electrical forces.

60. P. Lowell, *Mars as the Abode of Life*, 66.

61. Ibid., 67, 69.

62. Ibid., 70.

63. Ibid., 71.

64. Ibid., 71–72.

65. Ibid., 89, 112–15, and Note 18, 265–71. Lowell advanced this view earlier in "Mars on Glacial Epochs," *Proceedings of the American Philosophical Society*, XXXIV, No. 164 (1897). The 1971 Mariner 9 results suggest to some scientists that mass anomalies have periodically shifted Mars' axial tilt, causing climatic changes. See B. C. Murray, "Mars From Mariner 9," *Scientific American*, 228:61 (1973).

66. P. Lowell, *Mars as the Abode of Life*, 108–09.

67. Ibid., 109–10.

68. Ibid., 110.

69. Ibid., 111–112.

70. Ibid., 111, 115. ff.

71. Ibid., 135.

72. P. Lowell, *The Evolution of Worlds*, 203, 213–226.

73. Ibid., 58.

74. Ibid., 51 ff.

75. Ibid., 127. Pluto would prove an exception.

76. The rotation of Neptune, of course, is now known to be direct.

77. Lowell here seems to give a new twist to an old problem. An original retrograde rotation for all the planets is implicit in Keplerian-Newtonian mechanics which require that velocities of orbiting particles decrease as distance from the primary increases. Laplace sought to reconcile this with the observed motions in the solar system by postulating that the rings of nebular material from which the planets formed revolved all of a piece like a giant wheel, outer particles moving faster than inner ones and thus imparting a direct, or "forward" rotation to an accreting planet. In the late nineteenth century, France's H. Faye tried to resolve the problem by suggesting that all the planets except Uranus and Neptune were formed before the sun reached its full development, while Max Wolf proposed that tidal friction had reversed the original retrograde rotations, although he did not suggest that the planets thereby were turned topsy-turvy on their axes. See Clerke, *op. cit.*, pp. 315, 321–22.

78. Ibid., 138–153.

79. Ibid., 154. Clerke, *op. cit.*, 308, notes the "diffidence" with which Laplace presented his nebular hypothesis in 1796.

80. C. L. Poor, *Science*, 31:506 (1910).

81. E. Blackwelder, "Mars as the Abode of Life," *Science*, April 23, 1909, 659.

82. Ibid., 660.
83. Ibid.
84. Ibid., 661.
85. Ibid.
86. P. Lowell, "Mars as the Abode of Life," *Science*, Sept. 10, 1909, 338.
87. T. J. J. See, "Fair Play and Toleration in Science," *Science*, May 28, 1909, 858–60.
88. Ibid.
89. J. Barrell, "Fair Play and Toleration in Criticism," *Science*, July 2, 1909, 23.
90. F. R. Moulton, "Remarks on Recent Contributions to Cosmogony," *Science*, July 23, 1909, 113-17.
91. T. J. J. See, "Geology and Cosmology," *Science*, Oct. 8, 1909, 479–80.
92. P. Lowell, *Mars as the Abode of Life*, 338–40.
93. P. Lowell to A. A. Michelson, June 15, 1909; to L. F. Ward, Oct. 26, 1910; and to J. E. Gore, July 1, 1909.
94. F. R. Moulton, "A Reply to Dr. Lowell," *Science*, Nov. 5, 1909, 639.
95. Ibid., 640.

Notes to Chapter 14

1. A. M. Clerke, *A Popular History of Astronomy During the Nineteenth Century*, London, 1902, 286.
2. P. Lowell, "Memoir on Saturn's Rings," *Memoirs of the Lowell Observatory*, Vol. II (1915). See also *Lowell Observatory Bulletins* 44 (1910), 64 (1915), and 68 (1915).
3. P. Lowell, "The Origin of the Planets," *Memoirs of the American Academy of Arts and Sciences*, Vol. XIV, No. 1 (1913). It is interesting to note that as recently as 1968, England's S. F. Dermott suggested that "a marked preference for near commensurability found among pairs of mean motions in the solar system reflects a condition of formation, secondary bodies forming preferentially in near-commensurate orbits." To test this he and colleague A. P. Lenham, assuming the terrestrial planets and asteroids were formed before the Jovian outer planets, analyzed the distribution of the orbital periods of the asteroids and found "a preference for these periods to be near-commensurate with that of Mars." See S. F. Dermott and A P. Lenham, "Stability of the Solar System: Evidence From the Asteroids," *The Moon*, 5:294 (1972).
4. P. Lowell, "The Origin of the Planets."
5. Ibid., plates II and IV. In Plate IV, incidentally, Lowell indicates a "Planet X_2" at 75 a.u.
6. Nearly every basic textbook on astronomy describes Bode's law and its role in planetary discovery. Good discussions are found in Gunter G. Roth, *The System of Minor Planets*, New York, 1962, 19-20, and in Morton Grosser, *The Discovery of Neptune*, Cambridge, Mass., 1962, 30 ff.
7. B. Peirce, "Investigations into the action of Neptune to Uranus," *Proceedings of the American Academy of Arts and Sciences*, I:65 (1848). Galle discovered Neptune with the Berlin Observatory's 9-inch refractor on Sept. 23, 1846 only 55 minutes of arc from the position given him by Leverrier.
8. A good discussion of Peirce's position is presented in B. A. Gould, *Report on the History of the Discovery of Neptune*, Washington, D.C., 1850. See also J. F. W. Herschel, *Introduction to Astronomy*, London, 1858, 533 ff., and W. M. Smart, "John Couch Adams and the Discovery of Neptune," *Occasional Notes of the Royal Astronomical Society*, 2:33 (1947). Most textbooks and histories of astronomy review the controversy in a more general way, using as the prime source Astronomer Royal Sir George Airy's "Account of some circumstances historically connected with the discovery of the planet exterior to Uranus," *Monthly Notices of the Royal Astronomical Society*, 7:121 (1846). Airy's "Account" also appears in *Astronomische Nachrichten*, Nos. 585 and 586.

9. P. Lowell, "Memoir on a Trans-Neptunian Planet," *Memoirs of the Lowell Observatory*, Vol. I (1915), 8, 69, 100.

10. Grosser, *op. cit.*, 123, citing a letter from Leverrier to Galle, Oct. 1, 1846.

11. Clerke, *op. cit.*, 306.

12. P. Lowell, *The Solar System*, New York, 1903, 115.

13. H.-E. Lau, "La Planete Transneptunienne," *L'Astronomie*, 28:276 (1914); and M. A. Gaillot, "Tables nouvelles des mouvements d'Uranus et de Neptune," *Annales de L'Observatoire de Paris*, 28 (1910). Pickering, incidentally, predicted at least seven planets between 1908 and 1932, designating them, *O, P, Q, R, S, T,* and *U,* and publishing more or less complete sets of orbital elements for all except planet *T.*

14. P. Lowell, "Memoir on a Trans-Neptunian Planet," 7.

15. U. J. J. Leverrier, "Variations séculaires de éléments des orbites des quatre planètes Jupiter, Saturn, Uranus et Neptune," *Annales de L'Observatoire de Paris*, Vol. 11 (1876).

16. A. Gaillot, "Tables Nouvelles des mouvements d'Uranus et de Neptune," *op. cit.* Vol. 28 (1910).

17. P. Lowell, "Memoir on a Trans-Neptunian Planet," 32.

18. Ibid., 7.

19. P. Lowell to C. Flammarion, Feb. 14, 1905, and E. C. Pickering to P. Lowell, Feb. 8 and July 26, 1905.

20. P. Lowell to V. M. Slipher, June 26, 1906.

21. W. S. Harshman to P. Lowell, March 21, 1905.

22. W. T. Carrigan to P. Lowell, April 11, 1905.

23. It should be remembered that the often laborious computations involved in astronomical work in Lowell's day were performed by human, rather than electronic computers.

24. P. Lowell to W. T. Carrigan, Jan. 23, 1907.

25. P. Lowell to W. T. Carrigan, Feb. 1, 1907.

26. P. Lowell to W. T. Carrigan, Feb. 8, 1907.

27. P. Lowell to W. T. Carrigan, Feb. 14, 1907.

28. P. Lowell to W. T. Carrigan, March 7 and 13, and May 18, 1908.

29. P. Lowell to W. H. Pickering, Nov. 16, 1908.

30. P. Lowell to W. T. Carrigan, Nov. 19, 1908.

31. Ibid.

32. P. Lowell to C. O. Lampland, Dec. 28, 1908.

33. P. Lowell to W. T. Carrigan, Jan. 19, 1909.

34. P. Lowell to W. T. Carrigan, May 3, 1909.

35. P. Lowell to W. T. Carrigan, May 11, 1909.

36. W. H. Pickering, "A Search for a Planet Beyond Neptune," *Annals of the Astronomical Observatory of Harvard College*, 61:113 (1909). Lowell's annotated copy is in the Lowell Observatory library.

37. P. Lowell to A. C. D. Crommelin, July 28, 1909.

38. P. Lowell to C. O. Lampland, Dec. 1, 1910.

39. P. Lowell to C. O. Lampland, telegram, March 13, 1911.

40. W. L. Leonard to C. O. Lampland, March 22, 1911.

41. P. Lowell to C. O. Lampland, April 27, 1911.

42. P. Lowell to C. O. Lampland, telegram, July 8, 1911.

43. P. Lowell to V. M. Slipher, telegram, July 13, 1911.

44. P. Lowell to C. O. Lampland, July 27, 1911.

45. P. Lowell to C. O. Lampland, telegram, June 5 and 6, 1912.

46. P. Lowell to V. M. Slipher, telegram, June 12, 1912

47. P. Lowell to C. O. Lampland, June 12, 1912.

48. P. Lowell to C. O. Lampland, telegrams, June 14 and 20 and July 10, 1912; and to J. C. Duncan, May 28, 1909.
49. P. Lowell to C. O. Lampland, Aug. 9, 1912.
50. P. Lowell to C. O. Lampland, Aug. 24, 1912.
51. P. Lowell to C. O. Lampland, telegram, Sept. 12, 1912.
52. P. Lowell to C. O. Lampland, Sept. 21, 1912.
53. W. L. Leonard to V. M. Slipher, Nov. 8, 1912.
54. Constance Lowell to V. M. Slipher, Nov. 8, 1912.
55. W. L. Leonard to V. M. Slipher, Dec. 12, 1912.
56. Constance Lowell to V. M. Slipher, Sept. 16, 1913.
57. P. Lowell to C. O. Lampland, telegrams (2), Jan. 22, 1913.
58. P. Lowell to C. O. Lampland, telegram, Jan. 29, 1913.
59. P. Lowell to C. O. Lampland, telegrams, Feb. 3 and 5, 1913.
60. P. Lowell to C. O. Lampland, Feb. 8, 1913. For epoch Jan. 1, 1850, the elements were: Mean longitude at epoch, 11°40′; longitude at perihelion, 186°; eccentricity, .228; major axis, 47.5 a.u.; longitude, descending node, 110°59′; inclination of orbit on orbit of Uranus, 7°18′; and mass, 2.4 (earth = 1).
61. P. Lowell to C. O. Lampland, Feb. 21, 1913.
62. P. Lowell to E. M. Doe, telegram, Feb. 25, 1913.
63. P. Lowell to A. Gaillot, May 22, 1913, and to C. O. Lampland, July 10, 1913.
64. P. Lowell to C. O. Lampland, July 24 and Aug. 21, 1913; and E. L. Williams to P. Lowell and C. O. Lampland, Aug. 4, 1913.
65. P. Lowell to C. O. Lampland, Aug. 21, 1913.
66. E. L. Williams to P. Lowell, telegrams, Jan. 3, 7, and 29, and Feb. 7, 1914.
67. Constance Lowell to V. M. Slipher, April 26 and May 14, 1914.
68. P. Lowell to C. O. Lampland, May 5, 1914.
69. A. L. Lowell, *Biography of Percival Lowell,* New York, 1935, 175.
70. P. Lowell to C. O. Lampland, Aug. 15, 1914.
71. P. Lowell to V. M. Slipher, Aug. 14, 1914.
72. P. Lowell to C. O. Lampland, Dec. 21, 1914.
73. P. Lowell to C. O. Lampland, telegram, undated, *ca.* Jan. 14, 1915.
74. P. Lowell to J. Trowbridge, Dec. 9, 1914, and Jan. 11, 1915, and to V. M. Slipher, Dec. 21, 1914. Lowell had sought to stack the committee, proposing his old friend W. E.Storey and the amiable and neutral S. W. Burnham of Yerkes Observatory as members. When Trowbridge suggested Harvard's E. C. Pickering, however, Lowell threatened to withdraw the medal unless he was named chairman of the committee and allowed to name the third member.
75. P. Lowell to E. B. Wilson, Feb. 4, 1915.
76. Oddly enough, Miss Williams did not want her name to appear in the Planet X memoir; see W. L. Leonard to P. Lowell, April 21, 1915.
77. P. Lowell, "Memoir on a Trans-Neptunian Planet," 100.
78. Ibid.
79. Ibid., 69.
80. Ibid., 103.
81. Ibid.
82. Ibid., 105.
83. Ibid.
84. P. Lowell to C. O. Lampland, July 15, 1915.
85. "The Discovery of a Solar System Body Apparently Trans-Neptunian," *Lowell Observatory Observation Circular,* March 13, 1930.
86. *The New York Times,* March 15, 1930.
87. Ibid.

88. E. W. Brown, "On the Predictions of Trans-Neptunian Planets from the Perturbations of Uranus," *Proceedings of the National Academy of Science,* 16:364 (1930).
89. Ibid.
90. Ibid.
91. H. N. Russell, communication to A. L. Lowell published in A. L. Lowell, *op. cit.,* 203–05. See also Russell's article on Pluto in the December 1930 *Scientific American.*
92. E. C. Bower, "On the Orbit and Mass of Pluto With an Ephemeris for 1931–32," *Lick Observatory Bulletin 437* (1931).
93. Ibid. See also P. K. Seidelmann *et al,* "The Mass of Pluto," *Astronomical Journal,* 76:488 (1971).
94. C. J. Cohen *et al,* "New Orbit for Pluto and Analysis of Differential Corrections," ibid., 72:973 (1967).
95. W. H. Pickering, "The Next Planet Beyond Neptune," *Popular Astronomy,* 36:143 (1928).
96. C. W. Tombaugh, "The Discovery of Pluto," in Harlow Shapley (ed.), *Source Book in Astronomy 1900–1950,* Cambridge, Mass., 1960, 69–74.
97. Bower, *op. cit.*
98. Cohen *et al, op. cit.*
99. Tombaugh, *op. cit.,* 70.
100. Bower, *op. cit.*
101. Tombaugh, *op. cit.,* 74.
102. Russell, *op. cit.*
103. A. L. Lowell, *op. cit.,* 192.

Notes to Chapter 15

1. P. Lowell to E. M. Doe, Jan. 6, 1911.
2. P. Lowell to E. M. Doe, Nov. 9, 1912.
3. P. Lowell, "Immigration Versus the United States," text of address delivered at Phoenix, Arizona, Feb. 17, 1916; *L.O.A.*
4. P. Lowell to E. M. Doe, Dec. 18, 1914. In 1906 Lowell supported a bill in the Massachusetts legislature to permit hotels to serve liquor until midnight on weekdays, declaring its defeat would "foister a system of hypocrisy upon the community...." (P. Lowell to G. Cushing, March 27, 1906). He used alcohol, of course, keeping liquor and wine both at his observatory and his Boston home. His taste, however, is probably not reflected in an order in which he specified wine of the same vintage, "but not from the same cask in which the cockroach was preserved in." (P. Lowell to Moquin Restaurant and Wine Co., Dec. 9, 1912).
5. P. Lowell to R. B. Sims, May 23, 1913.
6. P. Lowell to E. M. Doe, Jan. 2 and 17, 1916. In the second letter, Lowell wryly commended Doe for not using the check "to send Democratic politicians to Washington."
7. P. Lowell to V. M. Slipher, Nov. 3, 1905, and to E. M. Doe, Nov. 23, 1905, and Jan. 2, 17, and 30, 1906. Funston had obtained one loan from Lowell and was seeking another. As early as 1904, incidentally, Lowell suggested starting up a "straight Republican" newspaper in Flagstaff with Mayor Leo Vercamp and merchant-banker David Babbitt in opposition to the *Sun.* Such a paper, he noted, could also assure accuracy and economy in printing the observatory *Bulletins.* (See P. Lowell to C. O. Lampland, June 16, 1904).
8. P. Lowell to E. M. Doe, and to A. T. Cornish, Oct. 6, 1905, and to E. P. Ripley, Dec. 28, 1905, and June 26, 1906. Cornish proposed the project and Lowell sought to interest Ripley, president of the Santa Fe Railway, in it. Lowell was not always so flippant about environmental concerns. "Do not cut the white oak," he instructed V. M. Slipher June 12, 1912, in regard to cutting fence posts on observatory land, and "Please see that wild flowers on hill are never picked," he wired him June 30, 1916.

9. P. Lowell to Harvard Union, and to Victorian Club, Nov. 17, 1910; also lecture texts "Aspects of Socialism," dated Dec. 8, 1910, and "Astronomy Today," dated Dec. 13, 1910; *L.O.A.*

10. P. Lowell, "Two Stars," text of lecture delivered at Kingman, A.T., Oct. 20, 1911; *L.O.A.*

11. Ibid.

12. Ibid.

13. P. Lowell to W. L. Leonard, undated, quoted in W. L. Leonard, *Percival Lowell: An Afterglow,* Boston, 1921, 94. Unfortunately none of the letters Miss Leonard published in her fond memoir bears a date, although some can be roughly dated from internal evidence.

14. P. Lowell to H. Kobold, Feb. 1, 1909.

15. P. Lowell to A. L. Lowell, Sept. 20, 1909.

16. P. Lowell to H. Kobold, Nov. 15, 1909.

17. E. Stover to P. Lowell, May 20, 1910. Breen, a pioneer Arizona newspaper-man, was then editor of Flagstaff's *Coconino Sun.*

18. P. Lowell to E. Stover, May 24, 1910.

19. E. Stover to P. Lowell, May 27, 1910.

20. I. S. Shklovskii and C. Sagan, *Intelligent Life in the Universe,* New York, 1966, 332–33.

21. D. A. Fraser to Lowell Observatory, May 20, 1910.

22. W. L. Bennett to Lowell Observatory, May 9, 1910.

23. C. Matson to Lowell Observatory, May 3, 1910.

24. P. Lowell to W. T. Sedgewick, and to H. C. Wilson, July 22, 1910. See also P. Lowell, "Schiaparelli," *Pop. Astron.,* 18:456 (1910).

25. The 40-inch reflector Lowell had ordered from Alvan Clark and Sons in April 1908 was still not working satisfactorily in July 1910. See P. Lowell to V. M. Slipher, July 18, 1910.

26. See *Lowell Observatory Bulletins 46, 50, 56, 59, 60, 69, 71,* and *72,* all relating to observations of Mars from 1910 through 1915.

27. P. Lowell to E. A. McCullough, June 11, 1912.

28. P. Lowell to C. E. Housden, May 23, 1913.

29. P. Lowell to editor, *Knowledge,* May 23, 1913.

30. P. Lowell to V. Emanuel, telegram, Sept. 15, 1913.

31. P. Lowell to V. Emanuel, Sept. 15, 1913.

32. W. L. Leonard to V. Emanuel, Oct. 8, 1913.

33. J. R. Worthington to P. Lowell, Oct. 7, 1907, and Oct. 30, 1909.

34. P. Lowell to J. R. Worthington, Nov. 11, 1913. For Jarry-Desloges' charts, see his *Observations des Surfaces Planetaires,* Paris, 1908, Vol. 1, 46–47.

35. P. Lowell to W. H. Pickering, Nov. 25, 1913.

36. P. Lowell to W. Kaempffert, June 2, 1915.

37. P. Lowell to W. Kaempffert, Nov. 10, 1915.

38. P. Lowell to W. Kaempffert, Dec. 15, 1915.

39. P. Lowell to W. Kaempffert, Jan. 31, 1916.

40. P. Lowell to W. Kaempffert, Feb. 2, 1916.

41. P. Lowell to editor, *Scientific American,* March 1, 1916.

42. P. Lowell, "Means, Methods and Mistakes in the Study of Planetary Evolution," text dated April 13, 1905, of paper communicated to the Royal Astronomical Society, Nov. 15, 1905; *L.O.A.*

43. P. Lowell, "The Lowell Observatory and Its Work," text of lecture delivered at Prescott, A.T., Nov. 24, 1909; *L.O.A.*

44. P. Lowell, *The Evolution of Worlds,* New York, 1909, 179–80.

45. P. Lowell, "Recent Discoveries About Mars and the Martians," lecture text, *ca.* February 1910; *L.O.A.*

46. P. Lowell, "Two Stars."

47. P. Lowell, "Great Discoveries and Their Reception," lecture text *ca.* August 1916; *L.O.A.*

48. Ibid.

49. Ibid.

50. Ibid.

51. P. Lowell, "The Far Horizons of Science," lecture text *ca.* August, 1916; *L.O.A.*

52. P. Lowell, "Mars and the Earth," lecture text *ca.* August 1916; *L.O.A.*

53. P. Lowell, "Mars: Forecasts and Fulfillments," lecture text *ca.* August 1916; *L.O.A.*

54. *The New York Times,* Oct. 18, 1916; *L.O.A.*

55. Ibid.

56. *The New York Times,* Oct. 29, 1916; *L.O.A.*

57. Ibid.

58. Dr. M. G. Fronske, attending physician at Lowell's death, recalled some 56 years later in a conversation with the author that Lowell had become angry with a servant shortly before his fatal stroke. Probably he was physically exhausted as well, for he had spent the previous night at the telescope with E. C. Slipher observing Jovian satellites. A. L. Lowell, *Biography of Percival Lowell,* New York, 1935, 194, reported that "Before he became unconscious he said that he always knew it would come thus, but not so soon."

59. P. Lowell to H. Cupbridge, Jan. 5, 1910, in which Lowell acknowledges the dedication of Cupbridge's song, "Marcia, the Martian Maid."

60. R. A. Lupoff, introduction to E. R. Burroughs, *John Carter of Mars,* New York, 1970, 7-18. Burroughs, of course, is better known for his later series on *Tarzan of the Apes,* a circumstance which may be of some interest in intellectual history.

61. Shklovskii and Sagan, *op. cit.,* 276.

62. F. R. Moulton, *Consider the Heavens,* Garden City, N.Y., 1935, 23.

63. Ibid., 87–88.

64. The American Institute of Aeronautics and Astronautics devoted its 12th annual meeting, May 23–25, 1966, to the theme, "The Search for Extraterrestrial Life." The climax was a panel discussion among such well-known humanists as Dr. Max Lerner of Brandeis University, Dr. Harold D. Lasswell of Yale, and the Reverend Joseph J. Lynch, S.J., of Fordham University on the implications of extraterrestrial life. Father Lynch, incidentally, suggested that an extraterrestrial civilization would have a class of "spiritual" beings analogous to terrestrially conceived "angels." The symposium, which the author attended, was held at Anaheim, Calif., at the Disneyland Hotel.

65. J. Barzun, *Science: The Glorious Entertainment,* New York, 1964, 65.

66. W. Sullivan, *We Are Not Alone,* New York, 1966, 197 ff. See also Shklovskii and Sagan, *op. cit.,* 379 ff.

67. F. R. Moulton, *op. cit.,* 104.

68. Ibid., 105.

69. Struve and V. Zebergs, *Astronomy of the 20th Century,* New York, 1962, 156–57.

70. G. P. Kuiper, "Planetary Atmospheres and Their Origin," in G. P. Kuiper (ed.), *The Atmospheres of the Earth and the Planets,* Chicago, 1952, 363.

71. S. C. Chase *et al,* "Infrared Radiometry Experiment on Mariner 9," *Science,* 175:308 (1972). See also B. C. Murray *et al,* "Geological Framework of the South Polar Region of Mars," *Icarus,* 17:328 (1972).

72. Kuiper, "Planetary Atmspoheres," *op. cit.,* 399, 403–04. See also Kuiper, "On the Martian Surface Features," *Publications of the Astronomical Society of the Pacific,* 67:271 (1955), and "Visual Observations of Mars, 1956," *Astrophysical Journal,* 125:307 (1956).

73. P. Lowell to V. M. Slipher, June 2, 1916. See also G. P. Kuiper, *Harvard Card 831* (1947); M. B. McElroy, "Mars: an evolving atmosphere," *Science*, 175:443 (1972); R. Hanel *et al*, "Investigations of the Martian Environment by Infrared Spectroscopy on Mariner 9," *Icarus*, 17:423 (1972).

74. G. P. Kuiper, "Planetary Atmospheres," *op. cit.*, 358.

75. R. Hanel, *op. cit.* See also R. Hanel *et al*, "Infrared Spectroscopy Experiment on the Mariner 9 Mission: Preliminary Results," *Science*, 175:305 (1972).

76. Shklovskii and Sagan, *op. cit.*, 279.

77. "Red Plastic Snow," *Scientific American*, February 1970, 46.

78. D. B. McLaughlin, "Interpretation of Some Martian Surface Features," *Astronomical Journal*, 59:328 (1954); "Changes on Mars as evidence of wind deposition and volcanism," *ibid.*, 60:261 (1955); and "The volcanic and aeolian hypothesis of Martian features," *Pubs. Astron. Soc. Pac.*, 68:211 (1956).

79. C. Sagan, "Variable Features on Mars: Preliminary Mariner 9 Television Results," *Icarus*, 17:346 (1972); and "Variable Features on Mars: Mariner 9 Global Results," report to the annual meeting of the American Astronomical Society's Division of Planetary Sciences, Tucson, Arizona, March 21, 1973.

80. E. C. Slipher, *The Photographic Story of Mars*, Cambridge, Mass., 1962, 163.

81. Ibid. 70.

82. G. de Vaucouleurs, *The Planet Mars*, London, 1951, 73 ff.

83. G. de Vaucouleurs, *Physics of the Planet Mars*, New York, 1954, 41.

84. Many Mariner 9 photographs have been published in both the popular and professional media. See notably *Science*, Vol. 175, No. 4019, Jan. 21, 1972; *Sky and Telescope*, Vol. 43, No. 4, April 1972, and Vol. 44, No. 2, August 1972; *Scientific American*, Vol. 228, No. 1, January 1973; and the *National Geographic*, Vol. 143, No. 2, February 1973.

85. H. Masursky, press conference, June 14, 1972, quoted in W. K. Hartmann, "The New Mariner 9 Map of Mars," *Sky and Telescope*, 44:77 (1972).

86. R. L. Waterfield, *A Hundred Years of Astronomy*, New York, 1940, 50. Struve and Zebergs, *op. cit.*, cite the passage on p. 147.

87. Struve and Zebergs, *op. cit.*, 141.

88. More than 600 newspaper clippings relating to Lowell's death are in the *L.O.A.*

89. "Memorial to Percival Lowell," Council, Royal Astronomical Society of Canada, January 4, 1917; *L.O.A.*

90. G. R. Agassiz, "Percival Lowell (1855–1916)," *Proceedings of the American Academy of Arts and Sciences*, 52:847 (1917).

Bibliography

The Lowell Observatory Archives

This study of *Lowell and Mars* is based primarily on correspondence and other documents and materials in the Lowell Observatory Archives dating from 1894, when Percival Lowell established the observatory at Flagstaff, through 1916, the year of his death. These data consist of approximately 15,000 pages of letters, copies of letters, notes, records, texts and manuscripts; some photographs, and more than 3,000 clippings from newspapers and other periodicals.

In February 1970, when research for this study began, most of the material was contained in thirty "letter-file" boxes, fourteen "press-letter" books, eight scrapbooks, five copy pads, three file drawers, three photograph albums, and a number of small pocket notebooks. These, in turn, were stored randomly in a large vault in the basement of the observatory's main library building along with a larger body of similar material accumulated over subsequent years. A necessary preliminary to the work was to locate and separate the pertinent items from the mass of later and unrelated material in the vault and in other storage areas in the library, and then arrange them in rough chronological order.

During the course of the study, the documentary material was organized, indexed, and microfilmed in a cooperative project undertaken in June 1970 by Lowell Observatory and the American Institute of Physics. This project, for which I served as supervisor and consultant, resulted in the May 1973 publication of a microfilm edition of the "Early Correspondence of the Lowell Observatory 1894–1916." Nearly all the correspondence cited in this study is contained in that edition, copies of which have been made available at the observatory, at the A.I.P.'s Niels Bohr Library in New York City, at the American Philosophical Society Library in Philadelphia, and at several university libraries, including those at Boston University and Northern Arizona University in Flagstaff.

Letters and direct copies of letters make up the bulk of the collection. Generally, the approximately 6,500 letters contained in the letter-file boxes represent incoming correspondence received either at Flagstaff or at Lowell's Boston office, while the 7,000 or so direct copies in the press-letter books are of letters written by Lowell and various members of his staff. Because much of the correspondence was between Boston and Flagstaff, some of the original letters are duplicated in the press-letter books.

The letter-file boxes, which also contained bills, invoices, and other "housekeeping" records of the observatory as well as miscellaneous documents, were labeled by year, although not infrequently their contents were filed (or refiled) with some indifference to the labeling. Filing was more-or-less alphabetical, but no one criterion had been used over the years. These materials, along with about 200 letters hand-written largely by Lowell and Andrew E. Douglass in 1894–95, which had been pasted into a large scrapbook, have now been assembled, filed alphabetically by surname of writer or correspondent, and organized into the four following categories or collections:

The Percival Lowell Papers 1894–1916, including all documents written by or to Lowell.

The Andrew Ellicott Douglass Papers 1894–1901, including all documents written by or to Douglass, except those to or from Lowell.

The Vesto Melvin Slipher Papers 1901–1916, including all documents written by or to Slipher, except those to or from Lowell (there is no correspondence in the archives between Douglass and Slipher).

The Lowell Observatory Miscellaneous Papers 1894–1916, including documents originating from or received by the observatory as an institution as well as small collections of the correspondence of William L. Putnam, Carl O. Lampland, Earl C. Slipher, W. Louise Leonard, and other Lowell associates with persons other than Lowell, Douglass, or V. M. Slipher.

This is the form in which the materials are now—in 1973—filed in the Lowell Observatory Archives and in which they appear, although in a somewhat different sequence, in the microfilm edition.

The press-letter books contain direct copies, pressed onto the thin, translucent, silk-fiber pages, of letters and other writings by Lowell and members of his staff, again both in Flagstaff and Boston. The majority of the copied material is typewritten and usually bears holographic signatures or notations. The number of pages in the books containing copied material ranges from 19 to 996 and averages about 500 per book, often two letters per page. The contents of the books are in the main, in chronological order; together the fourteen books span the period from 1896 through November 8, 1916, four days before Lowell's death. The copied material in the books has been extensively indexed and included in the microfilm edition; the books themselves remain on file with other materials in the collection in the observatory's archives.

Seven scrapbooks in the observatory vault contain clippings from more than one hundred identifiable newspapers, magazines, or journals published during the period in many parts of the world, as well as many clippings that cannot be immediately identified. One 200-page scrapbook, prepared by a New York City clipping service, contains some 600 press notices of Lowell's death, the clippings neatly arranged on each page and carefully identified as to source. Another smaller scrapbook contains press reports and about thirty letters relating to Lowell's lecture series in Boston in 1906 and 1909 along with reviews of the books that emerged from them— *Mars as the Abode of Life* (1908) and *The Evolution of Worlds* (1909).

The remaining scrapbooks are filled with clippings ranging in length and scope from multi-page magazine articles to single-paragraph newspaper "fillers" and spanning the period from 1894 through 1909, with the majority dating from 1905 through 1907. Clippings from the earlier years are seldom identified or identifiable as to publication or exact date, although usually the month and year has been noted on the scrapbook page and in some cases the date at least can be approximated from the content of the clipping itself. After 1905, however, most of the clippings either are identifiable from their face, or have been identified holographically, probably by Lowell's secretary W. Louise Leonard. In addition, several hundred newspaper clippings, some dating as late as 1916, were found in the letter-file boxes, filed either individually or with letters in which they were enclosures. As published material, the clippings are not included in the microfilm edition of early Lowell Observatory correspondence, but they remain on file in the observatory's archives.

Three points must be made concerning these clippings. First, except in a few cases where the files of the particular publications were immediately available to me in Flagstaff or where copyrighted material was involved, I have assumed that these clippings are genuine and that they actually represent what they purport to represent to the extent to which this is determinable. It would be incredible, indeed, if they were spurious. Thus, in most instances I have accepted the clippings in the archives for what they appear to be. In any case, the task of verifying even the identified clippings in files of more than a hundred publications throughout the world, some of them long defunct, was precluded by practical considerations and, moreover, would be out of all proportion to its value to this study.

Second, the clippings in the archives are largely those which somehow came to the attention of Lowell, his family, friends, acquaintances, or members of his staff. Thus they are perhaps somewhat biased in favor of Lowell and his ideas, although a large if minor fraction of them are critical or skeptical in tone.

Finally, the clippings in the archives certainly represent only some portion of the published material appearing in the popular media concerning Lowell, his work, and his theories during his lifetime.

The file drawers contained notes, drafts, and texts of articles, papers and lectures by Lowell. Much of this material is in typewritten form, with corrections and notations on many of the pages in Lowell's distinctive handwriting. Many of the articles and papers have been published in popular or professional journals and are referred to by such publication when cited in this study. But a number of his longer papers and most of his lectures are not available in print, although excerpts may have been used in press reports from time to time. Texts of such lectures and papers cited in this study include the following, listed chronologically:

"Atmosphere: In Its Effect on Astronomical Research," *ca.* spring 1897.
"Areography," April 4, 1902.
"Commemoration Day Address," May 30, 1903.
"Means, Methods and Mistakes in the Study of Planetary Evolution," April 13, 1905.
"The Surface Temperature of Mars From a New Determination of the Solar Heat Received by It," Dec. 12, 1906.
"The Lesson of Mars," *ca.* January 1907.
"The Geography of Mars," Jan. 3, 1908.
"The Lowell Observatory and Its Work," Nov. 24, 1909.
"Recent Discoveries About Mars and the Martians," *ca.* February 1910.
"Comets," March 10, 1910.
"Aspects of Socialism," Dec. 8, 1910.
"Astronomy Today," Dec. 13, 1910.
"Two Stars," Oct. 20, 1911.
"The Atmosphere of Mars," *ca.* January 1916.
"Immigration Versus the United States," Feb. 17, 1916.

"Far Horizons of Science," *ca*. August 1916.
"Great Discoveries and Their Reception," *ca*. August 1916.
"Mars and the Earth," *ca*. August 1916.
"Mars: Forecasts and Fulfillments," *ca*. August 1916.

The last four of these texts comprise the series of lectures that Lowell gave at colleges and universities in the Pacific Northwest and in California in September and October 1916.

The remainder of the material in the archives is of lesser significance to this study. The copy pads and notebooks were used by Douglass between 1894 and 1897 for random notes and routine records relating largely to his responsibilities regarding the observatory's day-to-day operations. The photograph albums contain about a hundred small prints, some quite faded, that were taken mainly in the years 1894, 1896, 1905, and 1907. Some of these are reproduced in this present book. In addition to photographs of scenes of the observatory, its facilities and environs, there are also photographs of Lowell and various members of his staff, and of some of the visitors who sojourned briefly on Mars Hill in these years. Some record presumably festive excursions by Lowell and others into the countryside around Flagstaff.

The material, overall, was in excellent condition, although some pages in the copy books and a few copies in the press-letter books are very faint and difficult to read even with optical aids.

The collection, of course, does not constitute a complete record in any way, for the internal evidence of the material itself—the one-sidedness of some of the correspondence, for example, or the occurrence of references to letters and other papers that I have been unable to find—point to gaps in the documentation. These gaps seem to be particularly serious biographically in regard to Lowell's personal and social life. What evidence there is on this is consistent in indicating that Lowell maintained an affable but impersonal attitude toward everyone, even the women who might be presumed to have had an influence in his life, such as Miss Leonard, his personal secretary and sometime traveling companion for twenty years, or Constance Savage Keith, his long-time Boston neighbor whom he married in 1908 when he was 53 years old. There is almost nothing in the archives that refers to his marriage and only a few letters by Lowell that refer to his wife. Even Miss Leonard's extensive correspondence over the years is properly businesslike and secretarial, although sometimes somewhat gossipy. Nor do the archives reveal very much about Lowell's personal or social relationships with his family, his friends and associates in Boston and elsewhere, or even members of the observatory staff. On the other hand, there are some personal trivia—his taste in food and wine is well enough documented, for instance, and one can find that he wore a size 7 hat and silk underdrawers.

There are other serious gaps in the archival record as well, such as the notable absence of documentation of the circumstances surrounding Douglass' abrupt departure from the observatory in 1901. Yet, despite these deficiencies, the collection nonetheless provides more information on Lowell's life and work, and the development of his ideas, than has been heretofore available.

In the short bibliography that follows, I have listed only books cited in this present volume. Most general texts or surveys of astronomy published since 1894 that touch on planetary studies, as well as more recent books on extraterrestrial life, contain references to Lowell and his Mars theory. Usually these books can be found in any astronomical library and in most large general libraries.

Nor have I listed the other printed sources cited in this study, such as articles, papers, and reports in either popular or professional journals, as such a listing would be lengthy and would merely duplicate the already adequate identification of this material in the chapter notes.

Finally, the publications of the Lowell Observatory have been important sources for this study, and require a general citation here. These publications are:

Annals of the Lowell Observatory. 3 vols. 1898, 1900, 1905.
Lowell Observatory Bulletins. Nos. 1-75. 1903–1916.
Memoirs of the Lowell Observatory. 2 vols. 1915.

Cited Bibliography

ARRHENIUS, SVANTE. Worlds in the Making. New York: Harper & Brothers, 1908.

ANTONIADI, E. M. La Planète Mars. Paris: Hermann et Cie., 1930.

BARZUN, JACQUES. Science: The Glorious Entertainment. New York: Harper & Row, 1964.

BERRILL, N. J. Worlds Without End. New York: The Macmillan Company, 1964.

BONNER, WILLIAM. The Mystery of the Expanding Universe. New York: The Macmillan Company, 1964.

BURROUGHS, EDGAR RICE. John Carter of Mars (with introduction by Richard A. Lupoff). New York: Ballantine Books, Inc., 1970.

CLERKE, AGNES M. *A Popular History of Astronomy During the Nineteenth Century.* Fourth edition. London: Adam and Charles Black, 1902.

DANA, JAMES DWIGHT. *Manual of Geology.* Fourth edition. New York: American Book Co., 1894.

DE VAUCOULEURS, GERARD. *Physics of the Planet Mars.* New York: The Macmillan Company, 1954.

_____. *The Planet Mars.* Trans. Patrick Moore. Second edition. London: Faber and Faber Ltd., 1951.

FLAMMARION, CAMILLE. *La Planète Mars et ses conditions d'habitabilité.* Paris: Gauthier-Villars et Fils, 1892.

GLASSTONE, SAMUEL. *Book of Mars.* Washington, D.C.: National Aeronautics and Space Administration, 1968.

GOULD, BENJAMIN A. *Report on the History of the Discovery of Neptune.* Washington, D.C.: Smithsonian Institution, 1850.

GREENSLET, FERRIS. *The Lowells and Their Seven Worlds.* Boston: Houghton, Mifflin & Co., 1946.

GROSSER, MORTON. *The Discovery of Neptune.* Cambridge, Mass.: Harvard University Press, 1962.

HAECKEL, ERNST. *Last Words on Evolution.* Trans. Joseph McCabe. London: A. Owen & Co., 1906.

_____. *The Wonders of Life.* New York: Harper & Brothers, 1905.

HERSCHEL, JOHN F. W. *Outlines of Astronomy.* Eighth edition. London: Longmans, Green and Co., 1858.

HERSCHEL, WILLIAM. *The Scientific Papers of Sir William Herschel.* 2 vols. London: The Royal Society and the Royal Astronomical Society, 1912.

JARRY-DESLOGES, M. *Observations des Surfaces Planètaires.* 9 vols. Paris: Gauthier-Villars et Fils, 1908–1946.

JEANS, JAMES H. *The Universe Around Us.* New York: The Macmillan Company, 1929.

KOCH, HOWARD. *The Panic Broadcast.* New York: Avon Books, 1970.

KUIPER, GERARD P. (ed.). *The Atmospheres of the Earth and the Planets.* Revised edition. Chicago: University of Chicago Press, 1952.

LEONARD, W. LOUISE. *Percival Lowell: An Afterglow.* Boston: Richard G. Badger, 1921.

LOWELL, A. LAWRENCE. *Biography of Percival Lowell.* New York: The Macmillan Company, 1935.

LOWELL, PERCIVAL. *Chosön —The Land of Morning Calm.* Boston: Ticknor & Co., 1886.

_____. *Mars.* London: Longmans, Green and Co., 1895.

_____. *Mars and Its Canals.* New York: The Macmillan Company, 1906.

_____. *Mars as the Abode of Life.* New York: The Macmillan Company, 1908.

_____. *Noto.* Boston: Houghton, Mifflin and Company, 1891.

_____. *Occult Japan.* Boston: Houghton, Mifflin and Company, 1894.

_____. *The Evolution of Worlds.* New York: The Macmillan Company, 1909.

_____. *The Solar System.* Boston: Houghton, Mifflin and Company, 1903.

_____. *The Soul of the Far East.* New York: The Macmillan Company, 1911 (First published in 1888).

MCPHERSON, HECTOR JR. *Astronomers of Today.* London: Gall & Inglis, 1905.

MOORE, PATRICK. *Guide to Mars.* Second edition. London: Frederick Muller, Ltd., 1958.

MORSE, EDWARD S. *Mars and Its Mystery.* Boston: Little, Brown & Co., 1906.

MOULTON, FOREST R. *Consider the Heavens.* Garden City, N.Y.: Doubleday, Doran & Co., 1935.

PANNEKOEK, ANTONIE. *A History of Astronomy.* New York: Interscience Publishers, Inc., 1961 (first published in 1951).

POYNTING, JOHN H. *Collected Scientific Papers.* Cambridge: University Press, 1920.

ROTH, GÜNTER D. *The System of Minor Planets.* Trans. Alex Helm. New York: D. Van Nostrand Co., Inc., 1962.

SERVISS, GARRETT P. *Other Worlds.* New York: D. Appleton & Co., 1901.

SHAPLEY, HARLOW (ed.). *A Source Book in Astronomy 1900–1950.* Cambridge, Mass.: Harvard University Press, 1960.

SHKLOVSKII, I. S. and CARL SAGAN. *Intelligent Life in the Universe.* Trans. Paula Fern. New York: Dell Publishing Co., 1966.

SLIPHER, EARL C. *A Photographic Study of the Brighter Planets.* Flagstaff: Lowell Observatory and the National Geographic Society, 1964.

_____. *The Photographic Story of Mars.* Cambridge, Mass.; Sky Publishing Co., Inc., 1962.

STRUVE, OTTO and VELTA ZEBERGS. *Astronomy of the 20th Century.* New York: The Macmillan Company, 1962.

SULLIVAN, WALTER. *We Are Not Alone.* Signet edition. New York: McGraw-Hill Book Co., 1964.

THIEL, RUDOLPH. *And Then There Was Light.* Trans. Richard and Clara Wilson. Mentor edition. New York: The New American Library of World Literature, Inc., 1960.

WALLACE, A. R. *Is Mars Habitable?* London: The Macmillan Company, 1907.

_____. *Man's Place in the Universe.* London: McClure, Phillips and Company, 1903.

WATERFIELD, R. L. *A Hundred Years of Astronomy.* New York: The Macmillan Company, 1940.

WHEWELL, WILLIAM. *Of the Plurality of Worlds.* Fifth edition. London: Longmans, Green, Reader and Dyer, 1867 (first published in 1854).

YOUNG, CHARLES A. *Manual of Astronomy.* Boston: Ginn and Co., 1902.

Index